Computational Solid Mechanics for Oil Well Perforator Design

Other World Scientific Titles by Lee Wen Ho

Computer Simulation of Shaped Charge Problems
ISBN: 978-981-256-623-2

Computational Methods for Two-Phase Flow and Particle Transport
ISBN: 978-981-4460-27-9 (pbk)

Computational Solid Mechanics for Oil Well Perforator Design
ISBN: 978-981-3239-32-6

Computational Solid Mechanics for Oil Well Perforator Design

Wen Ho Lee

National Cheng Kung University, Taiwan
Feng Chia University, Taiwan

World Scientific

NEW JERSEY · LONDON · SINGAPORE · BEIJING · SHANGHAI · HONG KONG · TAIPEI · CHENNAI · TOKYO

Published by

World Scientific Publishing Co. Pte. Ltd.
5 Toh Tuck Link, Singapore 596224
USA office: 27 Warren Street, Suite 401-402, Hackensack, NJ 07601
UK office: 57 Shelton Street, Covent Garden, London WC2H 9HE

Library of Congress Cataloging-in-Publication Data
Names: Lee, Wen Ho, author.
Title: Computational solid mechanics for oil well perforator design / Wen Ho Lee,
 (National Cheng Kung University, Taiwan & Feng Chia University, Taiwan).
Description: New Jersey : World Scientific, 2018.
Identifiers: LCCN 2018010670 | ISBN 9789813239326 (hc : alk. paper)
Subjects: LCSH: Oil well casing--Design and construction--Data processing. |
 Drilling and boring machinery--Design and construction--Data processing. |
 FORTRAN (Computer program language) | MATLAB.
Classification: LCC TN871.22 .L44 2018 | DDC 622/.33820284--dc23
LC record available at https://lccn.loc.gov/2018010670

British Library Cataloguing-in-Publication Data
A catalogue record for this book is available from the British Library.

Copyright © 2018 by World Scientific Publishing Co. Pte. Ltd.

All rights reserved. This book, or parts thereof, may not be reproduced in any form or by any means, electronic or mechanical, including photocopying, recording or any information storage and retrieval system now known or to be invented, without written permission from the publisher.

For photocopying of material in this volume, please pay a copying fee through the Copyright Clearance Center, Inc., 222 Rosewood Drive, Danvers, MA 01923, USA. In this case permission to photocopy is not required from the publisher.

For any available supplementary material, please visit
http://www.worldscientific.com/worldscibooks/10.1142/10966#t=suppl

Desk Editors: Anthony Alexander/Yu Shan Tay

Typeset by Stallion Press
Email: enquiries@stallionpress.com

Dedicated to My Sister Lee, Yee Pee and Her Husband

Dr. Hwang, Yan Tung

Preface

Oil well perforator usually comprises high explosive, copper liner and steel case. During the perforation process, the liner jet has to penetrate through the steel pipe and the rock surrounding the steel pipe. Therefore, the computer code should have the capability to compute the multi-material problems. For calculating the liner deformation due to the compression from the high explosive burn, Lagrangian code is better than Eulerian code. Once the high explosive finishes the burn, it is better to use Eulerian code to follow the jetting of the liner and the penetration of the steel pipes.

In two-dimensional Lagrangian code, 'Forced Gradient' is a very good numerical method for solving the momentum equations. However, it cannot calculate the strong shock wave problem accurately even with the help from 'REZONE' which is described in Appendix A.

Eulerian method for solving the solid mechanic conservation equations is giving in Chapter 3. The momentum equations are derived from Newton's second law, while the energy equation is from the thermodynamic law. Computational method is provided with detailed step by step procedures.

In Chapter 4, a FORTRAN-77 program is provided for computing the shear modulus and yield strength for many materials. Also, another program for calculating two-dimensional and three-dimensional high explosive burn time and burn distance is described. The Hugoniot relations for gas, liquid and solid are derived in this chapter along with the theoretical spall strength.

The exact dimensions of seven perforators, four shaped charges and two explosive formed projectiles are given in Chapter 5 with MATLAB plotting programs. Two-dimensional Lagrangian method for radiation diffusion is discussed in Chapter 6.

K. B. Wallick, my colleague at Los Alamos National Laboratory, developed a 'REZONE' package for two-dimensional Lagrangian code which is described in Appendix A. T. R. Hill, also my colleague, developed a computer code to calculate the eigenvalue for nuclear system as discussed in Appendix B. I like to thank Prof. Y. C. Tai of National Cheng Kung University and Prof. C. M. Hsieh of National Kaohsiung Marine University for teaching me MATLAB. I also want to thank Prof. Y. C. Shiah of National Cheng Kung University and Prof. Y. M. Lee of Feng Chia University for their important suggestions in Chapter 3.

Special gratitude will go to Alberta Lee for her help in editing this book.

Wen Ho Lee
Rocklin, California, USA
fyl88@msn.com

November 2017

About the Author

I was born on December 21, 1939, in Nantow, Taiwan. I graduated from Texas A&M University with a Ph.D. in Mechanical Engineering in 1970. Till 1978, I worked for Idaho National Engineering Lab, Argonne National Lab, and private industries on the following assignments:

1. Thermal hydraulic problems for primary coolant and containment systems solved by numerical modeling of two-phase flow equations.
2. The computer magnetic disk head (read and write) and the foil bearing tape head (read and write) designed by solving the time-dependent Reynolds equation.
3. Atmospheric turbulent flow physics solved by using the spectral and the finite difference methods.

From 1978–1999, I worked for the University of California and at Los Alamos National Laboratory, New Mexico. My works are related to large codes development for solving engineering problems using two-dimensional Lagrangian and Eulerian methods and physics-related solutions for hydrodynamics, material strength, fracture and impact mechanics, radiation and neutron transports, and high explosive detonation wave problems.

Contents

Preface		vii
About the Author		ix
1.	Introduction	1
	Bibliography	6
2.	Lagrangian Method for Shock Wave and Stress Calculations	7
	2.1 Introduction	8
	2.2 The Governing Equations	9
	2.3 Calculation of Stress Deviators at t^{n+1}	26
	2.4 Correction of Stresses for Rigid Body Rotation During Δt	31
	2.5 Calculation of P^{n+1} and ϵ^{n+1}	32
	2.6 Calculation of Longitudinal Sound Speed	34
	2.7 Artificial Viscosity Used in the Two-Dimensional Lagrangian Code	36
	Bibliography	36
3.	Two-Dimensional Eulerian Method	37
	3.1 Introduction	38
	3.2 General Description of Physical Formulation	39
	3.2.1 The Conservation Equation for a Stress-Supporting Medium	39
	3.2.2 Equation of State	49
	3.2.3 Spall	59
	3.3 Computational Scheme	60

		3.3.1	General Discussion	60
		3.3.2	Summary of Calculation Procedure	61
		3.3.3	Lagrangian Phase	62
		3.3.4	Stress Calculation in the Plastic Regime of Flow	82
		3.3.5	Particle Transport and Remapping	90
		3.3.6	Computation for Spall	93
		3.3.7	Time Step Control	95
		3.3.8	The Logic for the Calculation Procedure	96
	3.4	Truncation Error Analysis		98
		3.4.1	Mass	99
		3.4.2	Radial Momentum	105
		3.4.3	Axial Momentum	108
		3.4.4	Internal Energy	111
		3.4.5	Deviatoric Stresses	115
	3.5	Equivalent Plastic Strain		117
	3.6	FCT Applied to Second-Order PIC		119
		3.6.1	Introduction	119
		3.6.2	Modified Mass Transport	119
		3.6.3	The Modified FCT Analysis	125
	Bibliography			128
4.	EOS, Constitutive Relationship and High Explosive			131
	4.1	Introduction to the Equation of State		133
	4.2	The $Mie - Grüneisen$ EOS and the Simple u_s, u_p Model		133
	4.3	The Osborne Model		135
	4.4	The Tillotson Equation of State		136
	4.5	Introduction to the Constitutive Relationship		136
	4.6	Quadratic Model		137
	4.7	Steinberg–Guinan Model		137
	4.8	Steinberg's New Model		140
		4.8.1	Program EOSGY	143
	4.9	High Explosive		144
		4.9.1	Introduction	144
		4.9.2	JWL Equation of State	145
		4.9.3	Small Variation of JWL EOS	146
		4.9.4	Computer Code for Two-Dimensional Programmed Burn	146

		4.9.5	Computer Code HEDET3 for Three-Dimensional Programmed Burn	150
	4.10		Derivation of the Hugoniot Relations	151
		4.10.1	Introduction	151
		4.10.2	Conservation of Mass	152
		4.10.3	Conservation of Momentum	155
		4.10.4	Conservation of Energy	156
	4.11		The Shock-Change Equations	158
		4.11.1	Introduction	158
		4.11.2	The Shock-Change Equation	159
		4.11.3	Summary of the Shock-Change Equation	168
	4.12		The Theoretical Spall Strength	168
	Bibliography			171
5.	The Exact Dimensions of the Perforators			173
	5.1	Introduction		174
	5.2	Perforator B		175
	5.3	Perforator E		176
	5.4	Perforator G		177
	5.5	Perforator L		178
	5.6	Perforator M		178
	5.7	Perforator N		180
	5.8	Perforator P		181
	5.9	The Penetrating Characteristic of the Penetrators		182
		5.9.1	Introduction	182
		5.9.2	Copper Liner of a Diameter 3.5 cm	183
		5.9.3	Perforators Described in Figs. 6 and 11 of the File EULE2D-Fig	186
		5.9.4	Perforators Described in Figs. 16, 26 and 31 of the File EULE2D-Fig	187
		5.9.5	Special Design of 4.3 cm Charge Diameter Shaped Charge	189
	5.10		Program Curve	190
	5.11		Plotting Programs Using MATLAB	195
	5.12		The Exact Dimensions of Explosive Formed Projectile and Shaped Charge	196
		5.12.1	Copper Explosive Formed Projectile	196
		5.12.2	Bi-conical Copper Liner Shaped Charge	197

	5.12.3	Small Charge Diameter Conical Shaped Charge	198
	5.12.4	Small Charge Diameter Shaped Charge with Wave Shaper	199
	5.12.5	Non-axisymmetric Tantalum EFP Warhead	200
	5.12.6	Copper Hemi-spherical Liner with Energetic Explosive	202
5.13	Computer Programs	205
Bibliography	206

6. Two-Dimensional Lagrangian Method for Radiation Diffusion   207

- 6.1 Introduction 209
- 6.2 Definition of Variable and Notation 210
- 6.3 The Governing Equation 215
- 6.4 Equation of State 216
- 6.5 Calculation Procedures and Finite Differences 216
- 6.6 Two-Dimensional Lagrangian Method for Radiation Diffusion Problems 225
  - 6.6.1 Introduction 225
  - 6.6.2 Finite Difference Approximation for the Radiation Diffusion Equation 225
  - 6.6.3 Monte Carlo Method 233
  - 6.6.4 Monte Carlo Procedure 237
- Bibliography 240

Appendix A   Rezone for Two-Dimensional Lagrangian Hydrodynamic Code   241

- A.1 Introduction 242
- A.2 The REZONE Model 243
  - A.2.1 Zone and Point Model 243
  - A.2.2 The Mass Model 243
  - A.2.3 Sub-zone Definition 245
- A.3 A Brief Description of the Rezone Method 246
  - A.3.1 The REZONE Code 246
  - A.3.2 The Displacement Pass 246
  - A.3.3 The Expansion Pass 247
  - A.3.4 The Vertex Pass 247
  - A.3.5 The Midpoint Pass 250

A.3.6	The Point 8 Pass	250
A.3.7	The Velocity Adjustment Pass	250
A.3.8	The Averaging Pass	250

A.4 Testing for a Rezone 252
    A.4.1 Philosophy 252
    A.4.2 Test Details — General Case 252
    A.4.3 Tests on Boundaries 256
    A.4.4 Additions to the Tests 257
    A.4.5 Limitations on the Testing 257
    A.4.6 Changes in Test Values 259
    A.4.7 General Remarks 259

A.5 The Displacement Pass 259
    A.5.1 Displacement Cases for the Interior Points 260
    A.5.2 Displacement on Boundary Points 261
    A.5.3 The Displacement Method 261
    A.5.4 Limitations on the Displacement Calculations 267
    A.5.5 Remarks 271

A.6 Expansion Pass 271
    A.6.1 Introduction 271
    A.6.2 Sub-mesh Storage 272

A.7 Rezone Method — General Case 274
    A.7.1 Preparation — Definition of the System 275
    A.7.2 Find Orientation of the System 275
    A.7.3 The Rezone Process 276
    A.7.4 The Rezone Cases 279
    A.7.5 Mapping Other Quantities 287
    A.7.6 Intersection Calculation 287

A.8 Rezone Method — Boundary Cases 288
    A.8.1 Constant R, Constant Z, and Slide Angle 288
    A.8.2 Free Surface Case 292
    A.8.3 Center of Mass Case 293

A.9 The Vertex, Midpoint, Point 8, and Velocity Passes 295
    A.9.1 Possible Cases 295
    A.9.2 The Vertex Pass 297
    A.9.3 The Midpoint Pass 297
    A.9.4 The Point 8 Pass 298
    A.9.5 The Velocity Adjustment Pass 298

A.10 The Averaging Pass 298
    A.10.1 Point Quantities 299

   A.10.2 Zone Quantities 300
 A.11 Completing the Rezone 300
   A.11.1 New Velocities 301
   A.11.2 New Zone Mass 301
   A.11.3 New Pressure 301
   A.11.4 New q Terms (Artificial Viscosity) 301
   A.11.5 Clear Flags 301
 A.12 Summary 301
 A.13 Final Remarks 302
 A.14 Examples 302
 A.15 Directed Kinetic Energy 305
 Bibliography 307

Appendix B Eigenvalue Calculations 309

 B.1 Introduction 310
 B.2 Standard Methods 313
 B.3 Group-Collapse Coarse Mesh Re-balance 314
 B.4 Whole System Group-wise Re-balance 318
 B.5 Variable Convergence Precision and Iteration Strategies . . 322
 B.6 Test Problem and Results 324
 B.7 Subcritical Searches 338
 B.8 Implementation of α Re-balance Acceleration 346
 B.9 Conclusion 356
 Bibliography 356

Appendix C Hugoniot Data and JWL EOS 359

Appendix D Supplementary Materials 367

Index 369

Chapter 1

Introduction

In the oil industry, a method of completing a well is to first run the casing pipe through the oil sand formation, set it below the producing horizon, and then cement it in place as shown in Fig. 1.1. Then the casing is perforated to allow the oil or gas to flow into the wellbore. Perforating is accomplished with a gun pipe equipped with shaped charges. Figure 1.1 shows the application of a typically one shaped charge per 2.54 cm in the pipe length direction. Any two-neighboring shaped charges are separated with an angle of 60, 90 or 120 degree in the azimuthal direction. The gun, a circular container that fits inside the casing, has electrically fired cartridges that discharge small high-velocity jets. These jets penetrate the casing wall, the outside cement, and the oil sand. By doing this, channels are opened up through which the oil or gas can flow into the production pipe.

The metal casing of an oil well bore is surrounded by cement which is in turn in contact with the hydrocarbon bearing rocks. Oil well perforators generally perforate oil well casings in one of two ways. Deep hole perforators are designed to produce a high level of perforation through the metal casing and cement into the hydrocarbon bearing rocks. Big hole perforators are designed to produce large holes in the casing only.

Both deep hole and big hole perforators use a form of shaped hollow charge. In its most common configuration a shaped charge consists of a cylindrical tubular casing containing a hollow metal liner, mounted so that its axis of symmetry is coincident with that of the casing. The liner shape is most commonly conical although other geometries such as hemispheres or trumpets can be used. The base of the liner is at the end of the cylinder facing the target and explosive is packed within the casing and around the outside of the liner. When the explosive is detonated at the end of the cylinder furthest from the target, a detonation front sweeps the liner

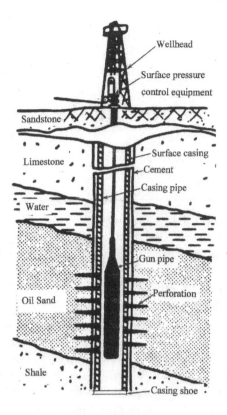

Fig. 1.1 The setup of the wellhead, pressure control, gun, casing, cement, and shaped charge arrangement.

causing it to collapse and produce a high velocity jet of liner material which is directed towards the target. A history of shaped charge warheads can be found in Ref. [1.1].

The hollow liner used in big hole perforators are generally parabolic in shape and are made of Cu(60% weight)/Zn(40% weight) brass. The apex of the liner has a hole in it which facilitates the formation of a large diameter jet (larger than if the liner surface continued all the way to the apex). For typical pipe diameters (on the order of 10.0 cm), big hole perforators have a diameter of approximately 4.2 cm with a hole of diameter 1.0 cm in the apex of the liner. This configuration is capable of producing a hole of approximately 2.0–2.5 cm in the oil well casing.

Perforating charge performance in producing both hole and perforation depth is related more to charge design than charge size. A good perforator

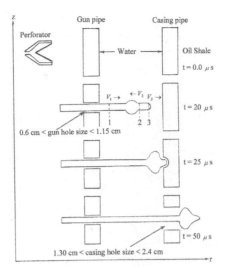

Fig. 1.2 The formation of the double velocity inverse gradients and the hole sizes in the gun and casing walls.

should have high-energy explosive with fine grain size and uniform density. When perforating charge explodes low order burn, large fragments of the charge cases will remain inside the gun tube. This is an indication that the perforating job quality is questionable.

The main concept of modern perforators is to create a small hole in the gun pipe and a large hole in the casing wall. As shown in Fig. 1.2, at time $t = 0.0$ μs, the perforator is located inside the gun with water between gun and casing walls. The outside of the casing is surrounded by oil shale or granite. At $t = 20$ μs, the shaped charge jet has already punched through the gun wall with a hole size between 0.6 cm and 1.15 cm. The jet inside the water region has three different velocity vectors, i.e., $v_1 > 0$, $v_2 < 0$ and $v_3 > 0$. The resulting bulge forms between points 1 and 2. The bulge reaches the casing wall without impact at 25 μs which is also the time when the maximum diameter of the bulge is produced. At $t = 50$ μs, the jet will penetrate through the casing wall and create a hole with a size between 1.30 cm and 2.40 cm.

The concept of using shaped charge for oil well casing perforation was introduced by McLemore [1.3] in 1946. For more than half a century, most of the shaped charge designs for the oil industry have been designed by trial and error through laboratory experiments. Recently, the two-dimensional

Eulerian code similar to the particle code [1.4] has been used to simulate the shaped charge jet formations as well as the jet penetration into the casing wall and the surrounding rocks. The computed results are within 8% error in comparison with the experimental data. Such successful simulations should be attributed to the development in numerical methods for solving a system of partial differential equations with a well-defined equation of state and constitutive law for real material in the last decade.

For an engineering problem which involves high explosive and a metal shell, e.g., shaped charge or an explosive formed projectile, one can use a good computer code to design a new device, provided this new device is of small deviation (say less than 5%) from an existing one which has been well tuned with pertinent parameters against the experimental data. All of the shaped charges described in this book are based on this concept. These oil well perforators include the charge diameters of 3.5 cm, 4.3 cm, 6.5 cm, and 8.2 cm using copper as the liner and RDX as the main charge with smaller charge diameter (e.g., 3.5 cm) used for deep wellbore perforation. Most of the liners are made from metallic powder and pressed into conical, hemispherical, trumpet, or bell shapes. For the small conical liner angle, penetrations into the rock outside the casing are deeper. However, the hole diameters are smaller. Hemispherical and bell-shaped liners tend to produce a larger hole that is shallow in penetration depth. A trumpet liner is somewhat in-between conical and hemispherical ones. In general, a larger hole in the casing wall and the outside rock is more desirable since this may allow more oil or gas to flow into the wellbore. The thickness and the shape of the outside steel case for the shaped charge also have some effect on the performance of the copper jet.

Since the clearance between the gun wall and the casing wall is small, the stand-off for the shaped charge is limited. As a result, the copper liner has limited time to form a jet which must posses a double inverse velocity gradient. The high explosive also has to sustain a harsh environment; that is, the temperature in the deep down-hole is usually above 260°C. These restrictions make the design of the oil well shaped charge much more difficult compared with the one used in a conventional weapon.

In this book, seven perforators are presented with their important characteristics pertinent to the optimum design for oil well perforation applications.

For multi-material problems, Lagrangian method is the most accurate tool for tracking the material interfaces, assuming that the material deformation is minimal. As soon as the deformation becomes large, the computational mesh starts to form a long thin zone or slender zone that will terminate the calculation due to an unaffordable small time step. For the last 40 years, the scientists working on Lagrangian codes spend most of their time trying to fix these mesh tangling problems. The Lagrangian method is presented in Chapter 2.

The shock wave and stress calculations in two-dimensional Lagrangian coordinate are discussed in Chapter 2. It uses Eulerian coordinate for the physical plane and Lagrangian coordinate for the logical plane. The forces due to the artificial viscosity are acting in the same direction as the moving shock wave.

Due to its fixed grid mesh, Eulerian method can handle large deformation. However, the accuracy of calculating the mass, the momentum, and the energy flux across the zone boundary can be challenging. The other difficulty associated with the Eulerian method is the tracking of the material interface that tends to be smeared and fuzzy. Therefore, most of the code developers using Eulerian equations are battling with the flux or the advection and the material interfacial problems. The Particle-in-Cell (PIC) method for tracking the material interfaces is described in Chapter 3.

For shaped charge problems, most of the wave codes use programmed burn for computing the chemical released energy from the high explosive. The programmed burn model, although not as accurate as the reactive burn model, is simple and robust. If one would modify the detonation speed with considering the burning front curvature's effect, then the programmed burn model will become more accurate for the insensitive explosive. In principle, the reactive burn model is more desirable due to its accuracy. However, in the large code calculation, it becomes impractical for the reactive burn approach because it requires an enormous large core memory and CPU. The programmed burn is presented in Chapter 4 along with a Fortran 77 computer code to calculate the burn times at each grid point for two-dimensional as well as three-dimensional problems. A computer code to calculate the equation of state and the constitutive relationship for many materials are also provided. The derivations for the Hugoniot conditions for gas, liquid and solid are also described in this chapter.

In Chapter 5, the detail dimensions of the case, the liner and the explosive of seven perforators are provided along with the computer codes which plot the perforators. The penetration of the jet into the gun tube, the casing wall, and the surrounding rock is discussed in this chapter. The exact dimensions of two explosive formed projectiles and four shaped charges are also described in the last part of this chapter.

In Chapter 6 we discuss the two-dimensional Lagrangian method for solving the radiation diffusion equations by using the finite difference and the Monte Carlo sampling.

In two-dimensional Lagrangian-code calculations, the mesh may distort and the calculation becomes very difficult. A rezone or some kind of combined Lagrangian/Eulerian calculations will smooth out the distortions of the mesh without stopping the problem. Only those points in the mesh that show distortion are rezoned. The rezone method is described in Appendix A. The time absorption alpha eigenvalue for critical size search in a nuclear system is presented in Appendix B.

Bibliography

1.1 Walters, WP and Zukas, JA. (1989). *Fundamentals of Shaped Charges*, Wiley-Interscience, New York.
1.2 Lee, WH. (2006). *Computer Simulation of Shaped Charge Problems*, World Scientific Publishing Co., Singapore.
1.3 McLemore, RL. (1946). *World Oil*, July 8, 1946.
1.4 Lee, WH and Painter, J. (1999) Material void-opening computation using particle method, *Int. J. Impact Eng.*, 22, 1-22.

Chapter 2

Lagrangian Method for Shock Wave and Stress Calculations

Notations

E specific internal energy per unit volume ($\frac{Mbar-cm^3}{cm^3}$)
e equivalent plastic strain (no unit)
G shear modulus of elasticity ($Mbar$)
I specific internal energy per unit mass ($\frac{Mbar-cm^3}{g}$)
j Jacobian of the transformation between (R, Z) and $(k, \ell)(cm^2)$
k Lagrangian coordinate, it is in logical plane and has no unit
ℓ Lagrangian coordinate, it is in logical plane and has no unit
M cell mass (g)
P pressure ($Mbar$)
Q artificial viscosity ($Mbar$)
R, r radial coordinate (cm)
S^{ij} stress deviator tensor ($Mbar$)
$S^{rr}, S^{zz}, S^{rz}, S^{\theta\theta}$ stress deviator components ($Mbar$)
s distance (cm)
t time (μsec)
Δt time step (μsec)
Δt_{max} maximum allowed time step (μsec)
U velocity in R direction ($\frac{cm}{\mu sec}$)
$\vec{u}$ velocity vector (U, V) ($cm/\mu sec$)
u_R displacement in the R direction (cm)
V velocity in Z direction ($\frac{cm}{\mu sec}$)
v volume (cm^3)
u_Z displacement in the Z direction (cm)
$\dot{W}$ rate of energy source due to work hardening ($\frac{Mbar-cm^3}{cm^3-\mu sec}$)
Y flow stress of elasticity ($Mbar$)
Z, z axial coordinate (cm)

Greek letters

δ_{ij} Kronecker delta (no unit)
ϵ specific internal energy per mass ($\frac{Mbar-cm^3}{g}$)
γ ratio of the specific heat, i.e., $\gamma = C_P/C_V$ (no unit)
η normalized density $= \frac{\rho}{\rho_0}$ (no unit)
θ angular coordinate
λ half of the grid size in R direction, i.e., $\lambda = \Delta R/2$ (cm)
μ normalized density minus one ($=\eta - 1 = \frac{\rho}{\rho_0} - 1$)(no unit)
ρ density ($\frac{g}{cm^3}$)
σ^{ij} stress tensor ($Mbar$)
$\sigma^{rr}, \sigma^{zz}, \sigma^{rz}, \sigma^{\theta\theta}$ stress components ($Mbar$)
τ specific volume ($\frac{cm^3}{g}$)
ω angle of clockwise rigid body rotation

Subscripts

0 initial value
e elastic regime
kk normal strains, i.e., $e_{rr}, e_{zz}, e_{\theta\theta}$
p plastic regime
R derivative with respect to R coordinate
t derivative with respect to time
Z derivative with respect to Z coordinate

Superscripts

n time at n time-step, i.e., $t^n = t_0 + n \cdot \Delta t$
kk normal stresses, i.e., $\sigma^{rr}, \sigma^{zz}, \sigma^{\theta\theta}$

2.1 Introduction

In this chapter, a simple finite difference approximation is used to solve the stress and strain problems in two-dimensional Lagrangian coordinate system. The fundamental governing equations are the same as that described in Chapter 3 which uses two-dimensional Eulerian coordinates. For small deformation problems, the Lagrangian approximation is more accurate than the Eulerian method. In Chapter 3, the finite difference grid is fixed but the advection through the grid boundary requires a lot of effort to do the calculation. Here, we do not have the advection problem, therefore, the Lagrangian scheme looks much clean and simple.

2.2 The Governing Equations

The governing equation in a two-dimensional cylindrical or plane Eulerian coordinate can be defined as follows:

Mass

$$\frac{D\rho}{Dt} + \rho\left(\frac{\partial U}{\partial R} + \alpha\frac{U}{R} + \frac{\partial V}{\partial Z}\right) = 0, \qquad (2.1)$$

Momentum

$$\rho\frac{DU}{Dt} = \frac{\partial \sigma^{RR}}{\partial R} + \frac{\partial \sigma^{RZ}}{\partial Z} + \frac{\alpha}{R}(\sigma^{RR} - \sigma^{\theta\theta}), \qquad (2.2)$$

$$\rho\frac{DV}{Dt} = \frac{\partial \sigma^{ZZ}}{\partial Z} + \frac{\partial \sigma^{RZ}}{\partial R} + \frac{\alpha}{R}\sigma^{RZ}, \qquad (2.3)$$

Energy

$$\rho\frac{D\epsilon}{Dt} = \sigma^{RR}\frac{\partial U}{\partial R} + \alpha\sigma^{\theta\theta}\frac{U}{R} + \sigma^{ZZ}\frac{\partial V}{\partial Z} + \sigma^{RZ}\left(\frac{\partial U}{\partial Z} + \frac{\partial V}{\partial R}\right), \qquad (2.4)$$

where $\alpha = 0$ for plane geometry and $\alpha = 1$ for cylindrical geometry with the following definitions

$$\frac{D}{Dt} = \frac{\partial}{\partial t} + U\frac{\partial}{\partial R} + V\frac{\partial}{\partial Z}, \qquad (2.5)$$

$$\sigma^{ij} = S^{ij} - P\delta_{ij}, \qquad (2.6)$$

where σ^{ij} is the stress tensor and δ_{ij} the Kronecker delta. The detail derivations of Eqs. (2.1)-(2.4) are given in Sections 3.2.1.1-3.2.1.3 of Chapter 3. For compression, P is negative of the normal stresses, therefore

$$P = -\frac{1}{3}\sigma^{kk} = -\frac{1}{3}(\sigma^{RR} + \sigma^{ZZ} + \sigma^{\theta\theta}), \qquad (2.7)$$

$$S^{ij} = \text{stress deviator}, \qquad (2.8)$$

and

$$\epsilon = \text{specific internal energy}. \qquad (2.9)$$

Now, let the transformation between the Eulerian and Lagrangian coordinates be

$$R = R(k, \ell, t), \tag{2.10}$$

$$Z = Z(k, \ell, t), \tag{2.11}$$

and

$$t = t, \tag{2.12}$$

where

$$R(k, \ell, 0) = k, \tag{2.13}$$

and

$$Z(k, \ell, 0) = \ell. \tag{2.14}$$

The Eulerian partial derivative operators transform as

$$\frac{\partial}{\partial R} = j^{-1}(Z_\ell \frac{\partial}{\partial k} - Z_k \frac{\partial}{\partial \ell}), \tag{2.15}$$

$$\frac{\partial}{\partial Z} = j^{-1}(Z_k \frac{\partial}{\partial \ell} - Z_\ell \frac{\partial}{\partial k}), \tag{2.16}$$

and

$$\frac{D}{Dt} = \frac{\partial}{\partial t} + j^{-1}[R_t(Z_\ell \frac{\partial}{\partial k} - Z_k \frac{\partial}{\partial \ell}) + Z_t(Z_k \frac{\partial}{\partial \ell} - Z_\ell \frac{\partial}{\partial k})], \tag{2.17}$$

where

$$j = R_k Z_\ell - R_\ell Z_k. \tag{2.18}$$

Equations (2.15) and (2.16) are derived in Eqs. (6.11) and (6.12) of Chapter 6. The velocities are defined as

$$U = R_t, \tag{2.19}$$

and

$$V = Z_t. \tag{2.20}$$

Another useful relation for carrying the transformation to Lagrangian coordinate is obtained by partially differentiating Eq. (2.18) with respect to t, it follows

$$j_t = R_k V_\ell - R_\ell V_k + Z_\ell U_k - Z_k U_\ell. \tag{2.21}$$

Then, from Eqs. (2.15) and (2.16),

$$U_R + V_Z = \frac{j_t}{j}. \tag{2.22}$$

For convenience, let

$$j' = (\frac{R}{k})^\alpha j, \tag{2.23}$$

then

$$j'_t = (\frac{R}{k})^\alpha j_t + \alpha(\frac{R}{k})^{\alpha-1}\frac{U}{k}j. \tag{2.24}$$

Dividing Eq. (2.24) by Eq. (2.23) we have

$$\frac{j'_t}{j'} = \frac{j_t}{j} + \alpha(\frac{R}{k})^{-1}\frac{U}{k} = \frac{j_t}{j} + \alpha\frac{U}{R}. \tag{2.25}$$

Finally, Eqs. (2.22) and (2.25) then imply

$$U_R + \alpha\frac{U}{R} + V_Z = \frac{j'_t}{j'}. \tag{2.26}$$

The left-hand side of Eq. (2.26) is readily recognized as $\vec{\nabla} \cdot \vec{u}$. Therefore, the divergence of the velocity vector in the Eulerian fixed-frame transforms in Lagrangian coordinates to the rate of growth of a differential volume element (for a specific mass element) divided by that differential volume element, i.e., the rate of growth of the dilatation. To return to the transformation of Eqs. (2.1)-(2.4), we first apply Eq. (2.26) to Eq. (2.1)

$$\frac{D\rho}{Dt} + \rho\frac{j'_t}{j'} = 0. \tag{2.27}$$

Integrating Eq. (2.27), we have

$$\frac{1}{\rho}\frac{D\rho}{Dt} + \frac{1}{j'}\frac{Dj'}{Dt} = 0, \qquad (2.28)$$

or

$$\rho j' = C, \qquad (2.29)$$

since at $t = 0$, $\rho = \rho_0$ and $j = R_k Z_\ell - R_\ell Z_k = (1)(1) - (0)(0) = 1$, therefore, we have

$$\rho j' = \rho_0 j'_0 = \rho_0 (\frac{R}{k})^\alpha j = \rho_0 (\frac{R}{R})^\alpha (1) = \rho_0(k, \ell). \qquad (2.30)$$

Now we apply Eqs. (2.15) and (2.16) to Eqs. (2.2) through (2.4) to obtain

$$\rho \frac{DU}{Dt} = j^{-1}[Z_\ell \sigma_k^{RR} - Z_k \sigma_\ell^{RR}] + j^{-1}[R_k \sigma_\ell^{RZ} - R_\ell \sigma_k^{RZ}] + \frac{\alpha}{R}(\sigma^{RR} - \sigma^{\theta\theta}), \qquad (2.31)$$

$$\rho \frac{DV}{Dt} = j^{-1}[R_k \sigma_\ell^{ZZ} - R_\ell \sigma_k^{ZZ}] + j^{-1}[Z_\ell \sigma_k^{RZ} - Z_k \sigma_\ell^{RZ}] + \frac{\alpha}{R}\sigma^{RZ}, \qquad (2.32)$$

and

$$\rho \frac{D\epsilon}{Dt} = \sigma^{RR} j^{-1}[Z_\ell U_k - Z_k U_\ell] + \alpha \sigma^{\theta\theta}\frac{U}{R} + \sigma^{ZZ} j^{-1}[R_k V_\ell - R_\ell V_k]$$

$$+ \sigma^{RZ} j^{-1}[R_k U_\ell - R_\ell U_k + Z_\ell V_k - Z_k V_\ell]. \qquad (2.33)$$

Finally, multiplying Eqs. (2.31) and (2.32) by j' and using Eq. (2.30) we have

$$\rho_0 \frac{DU}{Dt} = (\frac{R}{k})^\alpha [Z_\ell \sigma_k^{RR} - Z_k \sigma_\ell^{RR} + R_k \sigma_\ell^{RZ} - R_\ell \sigma_k^{RZ}] + \frac{\alpha}{R}\frac{\rho_0}{\rho}(\sigma^{RR} - \sigma^{\theta\theta}), \qquad (2.34)$$

and

$$\rho_0 \frac{DV}{Dt} = (\frac{R}{k})^\alpha [R_k \sigma_\ell^{ZZ} - R_\ell \sigma_k^{ZZ} + Z_\ell \sigma_k^{RZ} - Z_k \sigma_\ell^{RZ}] + \frac{\alpha}{R}\frac{\rho_0}{\rho}\sigma^{RZ}. \qquad (2.35)$$

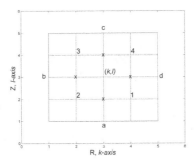

Fig. 2.1 The nomenclature for a set of four cells with the notations (k, ℓ), 1, 2, 3 and 4.

We will first difference the above two equations and then return to the conservation of energy, Eq. (2.33). The nomenclature for a set of four cells is illustrated in Fig. 2.1.

One method of differencing Eqs. (2.34) and (2.35) such that the results yield the so-called "Force Gradient" is described in Ref. [2.1] and will be repeated here.

To difference and apply Eqs. (2.34) and (2.35) to determine the accelerations of point (k, ℓ) in Fig. 2.1, consider the first term on the right-hand side of Eq. (2.34)

$$\left(\frac{R}{k}\right)^\alpha Z_\ell \sigma_k^{RR}. \tag{2.36}$$

We consider the stresses as known in some approximate sense at the center of each Lagrangian cell, i.e., at the points 1, 2, 3, and 4 in Fig. 2.1. We can then find their Lagrangian derivatives at the midpoint of the four straight lines that meet at the point (k, ℓ). These points are marked in Fig. 2.1 with an x. In particular, for the midpoint of the line (k, ℓ) to c, because σ^{RR} is differenciated with respect to k, the first term could be differenced as

$$\left[\frac{\frac{R_c+R}{2}}{\frac{k_c+k}{2}}\right]^\alpha \frac{Z_c - Z}{\ell_c - \ell} \frac{\sigma_4^{RR} - \sigma_3^{RR}}{k_4 - k_3}, \tag{2.37}$$

where all quantities without subscripts refer to the point (k, ℓ). Because we wish to apply Eqs. (2.34) and (2.35) to the (k, ℓ), we must average expression (2.37) with the gradient centered on the line (k, ℓ) to a; so finally we have for the difference term corresponding to expression (2.36) the term

$$\frac{1}{2}\left\{\left[\frac{\frac{R_c+R}{2}}{\frac{k_c+k}{2}}\right]^\alpha \frac{Z_c - Z}{\ell_c - \ell}\frac{\sigma_4^{RR} - \sigma_3^{RR}}{k_4 - k_3} + \left[\frac{\frac{R+R_a}{2}}{\frac{k+k_a}{2}}\right]^\alpha \frac{Z - Z_a}{\ell - \ell_a}\frac{\sigma_1^{RR} - \sigma_2^{RR}}{k_1 - k_2}\right\}. \tag{2.38}$$

In the second term on the right-hand side of Eq. (2.38), we notice that σ^{RR} is differentiated with respect to ℓ; therefore, the natural way to difference this term is to average the two gradients centered on the midpoints of lines (k, ℓ) to d, and (k, ℓ) to b. Hence the finite difference form of this term is

$$-\frac{1}{2}\left\{\left[\frac{\frac{R_d+R}{2}}{\frac{k_d+k}{2}}\right]^\alpha \frac{Z_d - Z}{k_d - k}\frac{\sigma_4^{RR} - \sigma_1^{RR}}{\ell_4 - \ell_1} + \left[\frac{\frac{R+R_b}{2}}{\frac{k+k_b}{2}}\right]^\alpha \frac{Z - Z_b}{k - k_b}\frac{\sigma_3^{RR} - \sigma_2^{RR}}{\ell_3 - \ell_2}\right\}. \tag{2.39}$$

If we now examine the denominator in Eq. (2.38), we see that if the Lagrangian mesh is initially rectangular and of uniform spacing the denominators will be, respectively, $\frac{1}{2}(v_4+v_3)$ and $\frac{1}{2}(v_1+v_2)$ where v_i is the volume of cell i. Then dividing Eq. (2.34) through by ρ_0, Eqs. (2.38) and (2.39) simplify to

$$\frac{(Z_c - Z)(\sigma_4^{RR} - \sigma_3^{RR})}{m_4 + m_3}\left[\frac{R_c+R}{2}\right]^\alpha + \frac{(Z - Z_a)(\sigma_1^{RR} - \sigma_2^{RR})}{m_1 + m_2}\left[\frac{R+R_a}{2}\right]^\alpha, \tag{2.40}$$

and

$$-\frac{(Z_d - Z)(\sigma_4^{RR} - \sigma_1^{RR})}{m_4 + m_1}\left[\frac{R_d+R}{2}\right]^\alpha - \frac{(Z - Z_b)(\sigma_3^{RR} - \sigma_2^{RR})}{m_3 + m_2}\left[\frac{R+R_b}{2}\right]^\alpha, \tag{2.41}$$

where m_i is the mass in cell i. In Cartesian coordinates, m_i will be the mass in a cell of unit thickness perpendicular to the R, Z plane; in cylindrical coordinates, the mass will be that in a torus of rectangular cross section having an "angular thickness" of 1 rad. The above method provides a scheme to difference all terms containing stress derivatives on the right-hand side of Eqs. (2.34) and (2.35); so it remains to consider those terms peculiar to cylindrical coordinates in these equations.

To get a "feel" for handling these terms in the difference equation, the origins in the differential equations should be reviewed. These terms arise in cylindrical coordinates because in the derivation of Eqs. (2.2) and (2.3) one first takes a small but finite mass element in cylindrical coordinates and analyzes the forces acting in the positive R and Z directions as contributed by all the stresses acting all the surfaces. More detail information is provided in Section 3.2.1.2. The last term on the right-hand side of Eq. (2.34) can be approximated as (remembering that we have divided Eqs. (2.34) and (2.35) through by ρ_0)

$$\frac{\alpha}{R\rho}(\sigma^{RR} - \sigma^{\theta\theta}) =$$

$$\frac{\alpha}{4}\left\{\frac{\sigma_1^{RR} - \sigma_1^{\theta\theta}}{\overline{R}_1\rho_1} + \frac{\sigma_2^{RR} - \sigma_2^{\theta\theta}}{\overline{R}_2\rho_2} + \frac{\sigma_3^{RR} - \sigma_3^{\theta\theta}}{\overline{R}_3\rho_3} + \frac{\sigma_4^{RR} - \sigma_4^{\theta\theta}}{\overline{R}_4\rho_4}\right\}, \qquad (2.42)$$

and the last term on the right-hand side of Eq. (2.35) is

$$\frac{\alpha}{R\rho}\sigma^{RZ} = \frac{\alpha}{4}\left\{\frac{\sigma_1^{RZ}}{\overline{R}_1\rho_1} + \frac{\sigma_2^{RZ}}{\overline{R}_2\rho_2} + \frac{\sigma_3^{RZ}}{\overline{R}_3\rho_3} + \frac{\sigma_4^{RZ}}{\overline{R}_4\rho_4}\right\}. \qquad (2.43)$$

$\overline{R}_i$ is the radial distance to the center of cell i. Because the mass is known for each cell, we can compute the above expressions in an easier way by

$$\frac{1}{\overline{R}_i\rho_i} = \frac{1}{\overline{R}_i\frac{m_i}{\overline{R}_iA_i}} = \frac{A_i}{m_i}, \qquad (2.44)$$

where A_i is the area of cell i. Thus, expressions (2.42) and (2.43) simplify to

$$\frac{\alpha}{R\rho}(\sigma^{RR} - \sigma^{\theta\theta}) = \frac{\alpha}{4}\{\frac{(\sigma_1^{RR} - \sigma_1^{\theta\theta})A_1}{m_1}$$

$$+\frac{(\sigma_2^{RR} - \sigma_2^{\theta\theta})A_2}{m_2} + \frac{(\sigma_3^{RR} - \sigma_3^{\theta\theta})A_3}{m_3} + \frac{(\sigma_4^{RR} - \sigma_4^{\theta\theta})A_4}{m_4}\}, \qquad (2.45)$$

and

$$\frac{\alpha}{R\rho}\sigma^{RZ} = \frac{\alpha}{4}\left\{\frac{\sigma_1^{RZ}A_1}{m_1} + \frac{\sigma_2^{RZ}A_2}{m_1} + \frac{\sigma_3^{RZ}A_3}{m_3} + \frac{\sigma_4^{RZ}A_4}{m_4}\right\}. \qquad (2.46)$$

Finally, we will introduce the notation of deviator stresses in the stress tensor before writing the difference equations in final form. By definition,

$$S^{ij} = \sigma^{ij} - \frac{1}{3}(\sigma^{RR} + \sigma^{\theta\theta} + \sigma^{ZZ})\delta_{ij}, \qquad (2.47)$$

or

$$S^{ij} = \sigma^{ij} + P\delta_{ij}. \tag{2.48}$$

The substitution of

$$\sigma^{ij} = S^{ij} - P\delta_{ij}, \tag{2.49}$$

in the difference equation permits convenient reversion to hydrodynamic calculation by setting all the deviators to zero. In final form, then, the complete difference equations for the acceleration in the R direction is

$$\frac{\Delta U}{\Delta t} = \left(\frac{\Delta U}{\Delta t}\right)_1 + \left(\frac{\Delta U}{\Delta t}\right)_2, \tag{2.50}$$

where

$$\left(\frac{\Delta U}{\Delta t}\right)_1 = \frac{(Z_c-Z)(S_4^{RR}-S_3^{RR}+P_3-P_4)}{m_4+m_3}\left[\frac{R_c+R}{2}\right]^\alpha$$

$$+\frac{(Z-Z_a)(S_1^{RR}-S_2^{RR}+P_2-P_1)}{m_1+m_2}\left[\frac{R_a+R}{2}\right]^\alpha$$

$$-\frac{(Z_d-Z)(S_4^{RR}-S_1^{RR}+P_1-P_4)}{m_4+m_1}\left[\frac{R_d+R}{2}\right]^\alpha$$

$$-\frac{(Z-Z_b)(S_3^{RR}-S_2^{RR}+P_2-P_3)}{m_3+m_2}\left[\frac{R+R_b}{2}\right]^\alpha$$

$$+\frac{(R_d-R)(S_4^{RR}-S_1^{RR})}{m_4+m_1}\left[\frac{R_d+R}{2}\right]^\alpha$$

$$+\frac{(R-R_b)(S_3^{RR}-S_2^{RR})}{m_3+m_2}\left[\frac{R+R_b}{2}\right]^\alpha$$

$$-\frac{(R_c-R)(S_4^{RR}-S_3^{RR})}{m_4+m_3}\left[\frac{R_c+R}{2}\right]^\alpha$$

$$-\frac{(R-R_a)(S_1^{RR}-S_2^{RR})}{m_1+m_2}\left[\frac{R+R_a}{2}\right]^\alpha$$

$$+\frac{\alpha}{4}[(S_1^{RR}-S_1^{\theta\theta})\frac{A_1}{m_1}+(S_2^{RR}-S_2^{\theta\theta})\frac{A_2}{m_2}$$

$$+(S_3^{RR}-S_3^{\theta\theta})\frac{A_3}{m_3}+(S_4^{RR}-S_4^{\theta\theta})\frac{A_4}{m_4}], \tag{2.51}$$

and

$$\left(\frac{\Delta U}{\Delta t}\right)_2 = \frac{(Z_c-Z)(Q_3-Q_4)}{m_4+m_3}\left[\frac{R_c+R}{2}\right]^\alpha$$

$$+\frac{(Z-Z_a)(Q_2-Q_1)}{m_1+m_2}\left[\frac{R_a+R}{2}\right]^\alpha$$

$$-\frac{(Z_d-Z)(Q_1-Q_4)}{m_4+m_1}\left[\frac{R_d+R}{2}\right]^\alpha$$

$$-\frac{(Z-Z_b)(Q_2-Q_3)}{m_3+m_2}\left[\frac{R+R_b}{2}\right]^\alpha. \tag{2.52}$$

In Eq. (2.52), the artificial viscosities Q_1, Q_2, Q_3 and Q_4 are non-zero when the cell is compressing or shock wave is running through the cell. Using the notations shown in Fig. 2.1, we will rewrite the Eq. (2.52) as

$$\left(\frac{\Delta U}{\Delta t}\right)_2 = D3UZ + D1UZ + D4UZ + D2UZ, \tag{2.53}$$

with

$$D3UZ = \frac{(Z_c-Z)(Q_3-Q_4)}{m_4+m_3}\left[\frac{R_c+R}{2}\right]^\alpha, \tag{2.54}$$

$$D1UZ = \frac{(Z-Z_a)(Q_2-Q_1)}{m_1+m_2}\left[\frac{R_a+R}{2}\right]^\alpha, \tag{2.55}$$

$$D4UZ = -\frac{(Z_d-Z)(Q_1-Q_4)}{m_4+m_1}\left[\frac{R_d+R}{2}\right]^\alpha, \tag{2.56}$$

and

$$D2UZ = -\frac{(Z-Z_b)(Q_2-Q_3)}{m_3+m_2}\left[\frac{R+R_b}{2}\right]^\alpha. \tag{2.57}$$

Modifications to the artificial viscosity are made so that the acceleration from the artificial viscosity will project onto the unit vector in the direction of local acceleration. Therefore, Eg.(2.54) is rewritten as

$$D3UZ = \frac{(R_c+R)^\alpha}{2(m_4+m_3)}[Q_3(Z_c-Z) - Q_4(Z_c-Z)], \tag{2.58}$$

or

$$D3UZ = \frac{(R_c+R)^\alpha}{2(m_4+m_3)}\left[\left(Q_3 \frac{Z_\ell U_k - R_\ell V_k}{U_k^2+V_k^2}U_k\right)_{at.point.3}\right.$$
$$\left.-\left(Q_4 \frac{Z_\ell U_k - R_\ell V_k}{U_k^2+V_k^2}U_k\right)_{at.point.4}\right], \tag{2.59}$$

or

$$D3UZ = \frac{(R_c+R)^\alpha}{2(m_4+m_3)}[(Z_{k,\ell+1} - Z_{k,\ell})(U_{k,\ell+1} - U_{k-1,\ell+1})$$

$$-(R_{k,\ell+1} - R_{k,\ell})(V_{k,\ell+1} - V_{k-1,\ell+1})]\cdot$$

$$\frac{Q_3(U_{k,\ell+1}-U_{k-1,\ell+1})}{(U_{k,\ell+1}-U_{k-1,\ell+1})^2+(V_{k,\ell+1}-V_{k-1,\ell+1})^2}$$

$$-\frac{(R_c+R)^\alpha}{2(m_4+m_3)}[(Z_{k,\ell+1} - Z_{k,\ell})(U_{k+1,\ell+1} - U_{k,\ell+1})$$

$$-(R_{k,\ell+1} - R_{k,\ell})(V_{k+1,\ell+1} - V_{k,\ell+1})]\cdot$$

$$\frac{Q_4(U_{k+1,\ell+1} - U_{k,\ell+1})}{(U_{k+1,\ell+1} - U_{k,\ell+1})^2 + (V_{k+1,\ell+1} - V_{k,\ell+1})^2}, \tag{2.60}$$

$$D1UZ = \frac{(R_a+R)^\alpha}{2(m_1+m_2)}\left[\left(Q_2 \frac{Z_\ell U_k - R_\ell V_k}{U_k^2+V_k^2}U_k\right)_{at.point.2}\right.$$
$$\left.-\left(Q_1 \frac{Z_\ell U_k - R_\ell V_k}{U_k^2+V_k^2}U_k\right)_{at.point.1}\right], \tag{2.61}$$

or

$$D1UZ = \frac{(R_a+R)^\alpha}{2(m_1+m_2)}[(Z_{k,\ell} - Z_{k,\ell-1})(U_{k,\ell} - U_{k-1,\ell})$$

$$-(R_{k,\ell} - R_{k,\ell-1})(V_{k,\ell} - V_{k-1,\ell})]\cdot$$

$$\frac{Q_2(U_{k,\ell}-U_{k-1,\ell})}{(U_{k,\ell}-U_{k-1,\ell})^2+(V_{k,\ell}-V_{k-1,\ell})^2}$$

$$-\frac{(R_a+R)^\alpha}{2(m_1+m_2)}[(Z_{k,\ell} - Z_{k,\ell-1})(U_{k+1,\ell} - U_{k,\ell})$$

$$-(R_{k,\ell} - R_{k,\ell-1})(V_{k+1,\ell} - V_{k,\ell})]\cdot$$

$$\frac{Q_1(U_{k+1,\ell} - U_{k,\ell})}{(U_{k+1,\ell} - U_{k,\ell})^2 + (V_{k+1,\ell} - V_{k,\ell})^2}, \tag{2.62}$$

$$D4UZ = \frac{(R_d+R)^\alpha}{2(m_4+m_1)}\left[\left(Q_4 \frac{Z_k U_\ell - R_k V_\ell}{U_\ell^2+V_\ell^2}U_\ell\right)_{at.point.4}\right.$$

$$-\left(Q_1 \frac{Z_k U_\ell - R_k V_\ell}{U_\ell^2 + V_\ell^2} U_\ell\right)_{at.point.1}\ ], \tag{2.63}$$

or

$$D4UZ = \frac{(R_d+R)^\alpha}{2(m_4+m_1)}[(Z_{k+1,\ell} - Z_{k,\ell})(U_{k,\ell+1} - U_{k,\ell})$$

$$-(R_{k+1,\ell} - R_{k,\ell})(V_{k,\ell+1} - V_{k,\ell})].$$

$$\frac{Q_4(U_{k,\ell+1}-U_{k,\ell})}{(U_{k,\ell+1}-U_{k,\ell})^2+(V_{k,\ell+1}-V_{k,\ell})^2}$$

$$-\frac{(R_d+R)^\alpha}{2(m_4+m_1)}[(Z_{k+1,\ell} - Z_{k,\ell})(U_{k,\ell} - U_{k,\ell-1})$$

$$-(R_{k+1,\ell} - R_{k,\ell})(V_{k,\ell} - V_{k,\ell-1})].$$

$$\frac{Q_1(U_{k,\ell} - U_{k,\ell-1})}{(U_{k,\ell} - U_{k,\ell-1})^2 + (V_{k,\ell} - V_{k,\ell-1})^2}, \tag{2.64}$$

$$D2UZ = \frac{(R_b+R)^\alpha}{2(m_3+m_2)}\left[\left(Q_3 \frac{Z_k U_\ell - R_k V_\ell}{U_\ell^2 + V_\ell^2} U_\ell\right)_{at.point.3}\right.$$

$$-\left(Q_2 \frac{Z_k U_\ell - R_k V_\ell}{U_\ell^2 + V_\ell^2} U_\ell\right)_{at.point.2}\ ], \tag{2.65}$$

or

$$D2UZ = \frac{(R_b+R)^\alpha}{2(m_3+m_2)}[(Z_{k,\ell} - Z_{k-1,\ell})(U_{k,\ell+1} - U_{k,\ell})$$

$$-(R_{k,\ell} - R_{k-1,\ell})(V_{k,\ell+1} - V_{k,\ell})].$$

$$\frac{Q_3(U_{k,\ell+1}-U_{k,\ell})}{(U_{k,\ell+1}-U_{k,\ell})^2+(V_{k,\ell+1}-V_{k,\ell})^2}$$

$$-\frac{(R_b+R)^\alpha}{2(m_3+m_2)}[(Z_{k,\ell} - Z_{k-1,\ell})(U_{k,\ell} - U_{k,\ell-1})$$

$$-(R_{k,\ell} - R_{k-1,\ell})(V_{k,\ell} - V_{k,\ell-1})].$$

$$\frac{Q_2(U_{k,\ell} - U_{k,\ell-1})}{(U_{k,\ell} - U_{k,\ell-1})^2 + (V_{k,\ell} - V_{k,\ell-1})^2}. \tag{2.66}$$

The acceleration in the Z direction is

$$\frac{\Delta V}{\Delta t} = \left(\frac{\Delta V}{\Delta t}\right)_1 + \left(\frac{\Delta V}{\Delta t}\right)_2, \tag{2.67}$$

where

$$\left(\frac{\Delta V}{\Delta t}\right)_1 = -\frac{(R_c-R)(S_4^{ZZ}-S_3^{ZZ}+P_3-P_4)}{m_4+m_3}\left[\frac{R_c+R}{2}\right]^\alpha$$

$$-\frac{(R-R_a)(S_1^{ZZ}-S_2^{ZZ}+P_2-P_1)}{m_1+m_2}\left[\frac{R_a+R}{2}\right]^\alpha$$

$$-\frac{(R_d-R)(S_4^{ZZ}-S_1^{ZZ}+P_1-P_4)}{m_4+m_1}\left[\frac{R_d+R}{2}\right]^\alpha$$

$$-\frac{(R-R_b)(S_3^{ZZ}-S_2^{ZZ}+P_2-P_3)}{m_3+m_2}\left[\frac{R+R_b}{2}\right]^\alpha$$

$$-\frac{(Z_d-Z)(S_4^{RZ}-S_1^{RZ})}{m_4+m_1}\left[\frac{R_d+R}{2}\right]^\alpha$$

$$+\frac{(Z-Z_b)(S_3^{RZ}-S_2^{RZ})}{m_3+m_2}\left[\frac{R+R_b}{2}\right]^\alpha$$

$$-\frac{(Z_c-Z)(S_4^{RZ}-S_3^{RZ})}{m_4+m_3}\left[\frac{R_c+R}{2}\right]^\alpha$$

$$-\frac{(Z-Z_a)(S_1^{RZ}-S_2^{RZ})}{m_1+m_2}\left[\frac{R+R_a}{2}\right]^\alpha$$

$$+\frac{\alpha}{4}\left[(S_1^{RZ})\frac{A_1}{m_1}+(S_2^{RZ})\frac{A_2}{m_2}+(S_3^{RZ})\frac{A_3}{m_3}+(S_4^{RZ})\frac{A_4}{m_4}\right], \qquad (2.68)$$

and

$$\left(\frac{\Delta V}{\Delta t}\right)_2 = -\frac{(R_c-R)(Q_3-Q_4)}{m_4+m_3}\left[\frac{R_c+R}{2}\right]^\alpha$$

$$-\frac{(R-R_a)(Q_2-Q_1)}{m_1+m_2}\left[\frac{R_a+R}{2}\right]^\alpha$$

$$-\frac{(R_d-R)(Q_1-Q_4)}{m_4+m_1}\left[\frac{R_d+R}{2}\right]^\alpha$$

$$-\frac{(R-R_b)(Q_2-Q_3)}{m_3+m_2}\left[\frac{R+R_b}{2}\right]^\alpha. \qquad (2.69)$$

Using the notations shown in Fig. 2.1, Eq. (2.69) becomes

$$\left(\frac{\Delta V}{\Delta t}\right)_2 = D3UR + D1UR + D4UR + D2UR, \qquad (2.70)$$

with

$$D3UR = \frac{(R_c - R)(Q_3 - Q_4)}{m_4 + m_3}\left[\frac{R_c + R}{2}\right]^\alpha, \qquad (2.71)$$

$$D1UR = \frac{(R - R_a)(Q_2 - Q_1)}{m_1 + m_2}\left[\frac{R_a + R}{2}\right]^\alpha, \qquad (2.72)$$

$$D4UR = -\frac{(R_d - R)(Q_1 - Q_4)}{m_4 + m_1}\left[\frac{R_d + R}{2}\right]^\alpha, \qquad (2.73)$$

and

$$D2UR = -\frac{(R - R_b)(Q_2 - Q_3)}{m_3 + m_2}\left[\frac{R + R_b}{2}\right]^\alpha. \qquad (2.74)$$

Similar to Eqs. (2.58)-(2.66), one gets

$$D3UR = -\frac{(R_c + R)^\alpha}{2(m_4 + m_3)}[Q_3(R_c - R) - Q_4(R_c - R)], \qquad (2.75)$$

or

$$D3UR = -\frac{(R_c+R)^\alpha}{2(m_4+m_3)}[\left(Q_3 \frac{Z_\ell U_k - R_\ell V_k}{U_k^2 + V_k^2} V_k\right)_{at.point.3}$$
$$- \left(Q_4 \frac{Z_\ell U_k - R_\ell V_k}{U_k^2 + V_k^2} V_k\right)_{at.point.4}], \qquad (2.76)$$

or

$$D3UR = -\frac{(R_c+R)^\alpha}{2(m_4+m_3)}[(Z_{k,\ell+1} - Z_{k,\ell})(U_{k,\ell+1} - U_{k-1,\ell+1})$$

$$-(R_{k,\ell+1} - R_{k,\ell})(V_{k,\ell+1} - V_{k-1,\ell+1})].$$

$$\frac{Q_3(V_{k,\ell+1} - V_{k-1,\ell+1})}{(U_{k,\ell+1} - U_{k-1,\ell+1})^2 + (V_{k,\ell+1} - V_{k-1,\ell+1})^2}$$

$$-\frac{(R_c+R)^\alpha}{2(m_4+m_3)}[(Z_{k,\ell+1} - Z_{k,\ell})(U_{k+1,\ell+1} - U_{k,\ell+1})$$

$$-(R_{k,\ell+1} - R_{k,\ell})(V_{k+1,\ell+1} - V_{k,\ell+1})].$$

$$\frac{Q_4(V_{k+1,\ell+1} - V_{k,\ell+1})}{(U_{k+1,\ell+1} - U_{k,\ell+1})^2 + (V_{k+1,\ell+1} - V_{k,\ell+1})^2}, \qquad (2.77)$$

$$D1UR = \frac{(R_a+R)^\alpha}{2(m_1+m_2)}[\left(Q_2 \frac{Z_\ell U_k - R_\ell V_k}{U_k^2 + V_k^2} V_k\right)_{at.point.2}$$

$$-\left(Q_1 \frac{Z_\ell U_k - R_\ell V_k}{U_k^2 + V_k^2} V_k\right)_{at.point.1}], \qquad (2.78)$$

or

$$D1UR = \frac{(R_a+R)^\alpha}{2(m_1+m_2)}[(Z_{k,\ell} - Z_{k,\ell-1})(U_{k,\ell} - U_{k-1,\ell})$$

$$-(R_{k,\ell} - R_{k,\ell-1})(V_{k,\ell} - V_{k-1,\ell})].$$

$$\frac{Q_2(V_{k,\ell}-V_{k-1,\ell})}{(U_{k,\ell}-U_{k-1,\ell})^2+(V_{k,\ell}-V_{k-1,\ell})^2}$$

$$-\frac{(R_a+R)^\alpha}{2(m_1+m_2)}[(Z_{k,\ell} - Z_{k,\ell-1})(U_{k+1,\ell} - U_{k,\ell})$$

$$-(R_{k,\ell} - R_{k,\ell-1})(V_{k+1,\ell} - V_{k,\ell})].$$

$$\frac{Q_1(V_{k+1,\ell} - V_{k,\ell})}{(U_{k+1,\ell} - U_{k,\ell})^2 + (V_{k+1,\ell} - V_{k,\ell})^2}, \qquad (2.79)$$

$$D4UR = \frac{(R_d+R)^\alpha}{2(m_4+m_1)}[\left(Q_4 \frac{Z_k U_\ell - R_k V_\ell}{U_\ell^2 + V_\ell^2} V_\ell\right)_{at.point.4}$$

$$-\left(Q_1 \frac{Z_k U_\ell - R_k V_\ell}{U_\ell^2 + V_\ell^2} V_\ell\right)_{at.point.1}], \qquad (2.80)$$

or

$$D4UR = \frac{(R_d+R)^\alpha}{2(m_4+m_1)}[(Z_{k+1,\ell} - Z_{k,\ell})(U_{k,\ell+1} - U_{k,\ell})$$

$$-(R_{k+1,\ell} - R_{k,\ell})(V_{k,\ell+1} - V_{k,\ell})].$$

$$\frac{Q_4(V_{k,\ell+1}-V_{k,\ell})}{(U_{k,\ell+1}-U_{k,\ell})^2+(V_{k,\ell+1}-V_{k,\ell})^2}$$

$$-\frac{(R_d+R)^\alpha}{2(m_4+m_1)}[(Z_{k+1,\ell} - Z_{k,\ell})(U_{k,\ell} - U_{k,\ell-1})$$

$$-(R_{k+1,\ell} - R_{k,\ell})(V_{k,\ell} - V_{k,\ell-1})].$$

$$\frac{Q_1(V_{k,\ell} - V_{k,\ell-1})}{(U_{k,\ell} - U_{k,\ell-1})^2 + (V_{k,\ell} - V_{k,\ell-1})^2}, \qquad (2.81)$$

$$D2UR = \frac{(R_b+R)^\alpha}{2(m_3+m_2)}[\left(Q_3\frac{Z_kU_\ell-R_kV_\ell}{U_\ell^2+V_\ell^2}V_\ell\right)_{at.point.3}$$

$$-\left(Q_2\frac{Z_kU_\ell-R_kV_\ell}{U_\ell^2+V_\ell^2}V_\ell\right)_{at.point.2}], \qquad (2.82)$$

or

$$D2UR = \frac{(R_b+R)^\alpha}{2(m_3+m_2)}[(Z_{k,\ell}-Z_{k-1,\ell})(U_{k,\ell+1}-U_{k,\ell})$$

$$-(R_{k,\ell}-R_{k-1,\ell})(V_{k,\ell+1}-V_{k,\ell})].$$

$$\frac{Q_3(V_{k,\ell+1}-V_{k,\ell})}{(U_{k,\ell+1}-U_{k,\ell})^2+(V_{k,\ell+1}-V_{k,\ell})^2}$$

$$-\frac{(R_b+R)^\alpha}{2(m_3+m_2)}[(Z_{k,\ell}-Z_{k-1,\ell})(U_{k,\ell}-U_{k,\ell-1})$$

$$-(R_{k,\ell}-R_{k-1,\ell})(V_{k,\ell}-V_{k,\ell-1})].$$

$$\frac{Q_2(V_{k,\ell}-V_{k,\ell-1})}{(U_{k,\ell}-U_{k,\ell-1})^2+(V_{k,\ell}-V_{k,\ell-1})^2}. \qquad (2.83)$$

In comparing the first terms appeared on the right-hand side of Eqs. (2.58) and (2.59), we know that $Z_c - Z$ is replaced by

$$\frac{Z_\ell U_k - R_\ell V_k}{U_k^2 + V_k^2}U_k. \qquad (2.84)$$

In Eq. (2.84), if $Z_\ell \approx R_\ell$, then, we can combine Z_ℓ and R_ℓ with Q and call it Q'. Therefore, Eq. (2.84) becomes

$$\frac{U_k - V_k}{U_k^2 + V_k^2}U_k. \qquad (2.85)$$

Rewrite the above equation as

$$\frac{U_k - V_k}{\sqrt{U_k^2+V_k^2}}\frac{U_k}{\sqrt{U_k^2+V_k^2}}. \qquad (2.86)$$

Using the notations shown in Fig. 2.2, one gets

$$\cos\theta = \frac{U_k}{\sqrt{U_k^2+V_k^2}}, \qquad (2.87)$$

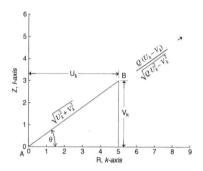

Fig. 2.2 The force due to the artificial viscosity Q is acting in the direction of the unit vector as indicated by $Q\frac{U_k-V_k}{\sqrt{U_k^2+V_k^2}}$. The unit vector formed an angle θ from the R-axis.

and the artificial viscosity Q' (where $Q' = Q \cdot Z_\ell$) is acting in the direction of $\overline{AB}$ which is the shock wave moving direction. Therefore, the component of the artificial viscosity in the R direction is

$$Q'\frac{U_k - V_k}{\sqrt{U_k^2 + V_k^2}} \cos\theta, \tag{2.88}$$

$$= Q'\frac{U_k - V_k}{\sqrt{U_k^2 + V_k^2}}\frac{U_k}{\sqrt{U_k^2 + V_k^2}}, \tag{2.89}$$

$$= Q'\frac{U_k - V_k}{U_k^2 + V_k^2} U_k. \tag{2.90}$$

When $U_k = 1$ and $V_k = 0$, Eq. (2.90) becomes

$$Q' = Z_\ell Q = (Z_c - Z)Q. \tag{2.91}$$

Equation (2.91) is exactly the same as the first term on the right-hand side of Eq. (2.58). On the other hand, if $V_k = 1$, and $U_k = 0$, then, the first term on the right-hand side of Eq. (2.76) becomes

$$Q_3 \frac{Z_\ell U_k - R_\ell V_k}{U_k^2 + V_k^2} V_k, \tag{2.92}$$

$$= Q_3 \frac{Z_\ell(0) - R_\ell(1)}{0^2 + 1^2} \cdot 1, \tag{2.93}$$

$$= -Q_3 R_\ell = -Q_3(R_c - R). \tag{2.94}$$

The above equation is the same as in Eq. (2.75). If $Z_\ell \neq R_\ell$, one may introduce small error in Eqs. (2.84) and (2.92), but, Q is the artificial viscosity which is some kind of smoothing function any way. Therefore, the small error introduced can be ignored. This is why artificial viscosity is modified to look like in Eqs. (2.59) and (2.76).

During the calculation of the U component of the momentum equation, one should always use Eq. (2.84) instead of using Z_ℓ or $(Z_c - Z)$ as described in Eq. (2.54) which may produce some errors for most cases. For example, when $U_k = V_k > 0$, we have the shock wave running at 45° from the R-axis. By using QZ_ℓ for the momentum equation, one will get a larger velocity U than it should be. If one uses $Q\frac{Z_\ell U_k - R_\ell V_k}{U_k^2 + V_k^2} U_k$ for calculating the U component, then, the calculated U will be much smaller with the shock wave running in the correct direction. Consequently, the computed results of the grids are much better and the aspect ratios of the cells are more uniform. This means that there are less long-thin zones and more near square zones.

In the standard way, we assume that all quantities on the right-hand side of Eqs. (2.51) and (2.68) are known at some time

$$t^n = \sum_{i=1}^{n} i\Delta t, \qquad (2.95)$$

so that

$$\frac{\Delta U}{\Delta t} = \frac{U^{n+\frac{1}{2}} - U^{n-\frac{1}{2}}}{t^{n+\frac{1}{2}} - t^{n-\frac{1}{2}}}, \qquad (2.96)$$

and

$$\frac{\Delta V}{\Delta t} = \frac{V^{n+\frac{1}{2}} - V^{n-\frac{1}{2}}}{t^{n+\frac{1}{2}} - t^{n-\frac{1}{2}}}. \qquad (2.97)$$

Position at t^{n+1} are then calculated from

$$R^{n+1} - R^n = U^{n+\frac{1}{2}}(t^{n+1} - t^n), \qquad (2.98)$$

and

$$Z^{n+1} - Z^n = V^{n+\frac{1}{2}}(t^{n+1} - t^n). \qquad (2.99)$$

2.3 Calculation of Stress Deviators at t^{n+1}

Rather than differencing the energy equation at this point, we first calculate the stress deviators at t^{n+1} because they will be needed in the energy equation. For the elastic, perfectly plastic model we will employ the Prandtl–Reuss theory as discussed in Ref. [2.3]. Given the stress deviators $(S^{RR})^n$, $(S^{\theta\theta})^n$, $(S^{ZZ})^n$ and $(S^{RZ})^n$, we first assume that during $\Delta t = t^{n+1} - t^n$ the medium remains within the elastic range and, consequently, we use Hookes' law in current form

$$\frac{DS^{RR}}{Dt} = 2G\left(\frac{\partial U}{\partial R} - \frac{1}{3v}\frac{Dv}{Dt}\right), \qquad (2.100)$$

$$\frac{DS^{ZZ}}{Dt} = 2G\left(\frac{\partial V}{\partial Z} - \frac{1}{3v}\frac{Dv}{Dt}\right), \qquad (2.101)$$

$$\frac{DS^{RZ}}{Dt} = G\left(\frac{\partial U}{\partial Z} + \frac{\partial V}{\partial R}\right), \qquad (2.102)$$

$$\frac{DS^{\theta\theta}}{Dt} = 2G\left(\frac{\alpha U}{R} - \frac{1}{3v}\frac{Dv}{Dt}\right) = \frac{D}{Dt}\left(-S^{RR} - S^{ZZ}\right), \qquad (2.103)$$

where G is the modulus of rigidity.

To difference these equations, we use Eq. (2.100) as a prototype. Transforming Eq. (2.100) to Lagrangian coordinate we have

$$\frac{DS^{RR}}{Dt} = 2G(j^{-1})\left(Z_\ell U_k - Z_k U_\ell - \frac{1}{3v}\frac{Dv}{Dt}\right). \qquad (2.104)$$

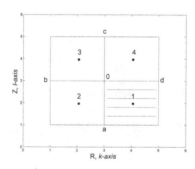

Fig. 2.3 The arrangement of the Lagrangian cells, cell 1 is correlated with the point (k, ℓ) relabeled 0 for convenience.

Because S^{RR} is consider a cell-centered variable, we should difference the quantities Z_ℓ, U_k, etc., symmetrically from values of Z, U, etc., at the four corner points of a cell. Further, all terms in Eq. (2.104) not differentiated with respect to t should be centered at $t^{n+\frac{1}{2}}$. Figure 2.3 again illustrates the arrangement of the Lagrangian cells in the code. Cell 1 in Fig. 2.3 is logically correlated with the point (k, ℓ) relabeled 0 for convenience. Three passes in the code are made through the mesh in each time cycle. In the first pass, accelerations are calculated at t^n and the components of the velocity are advanced to $t^{n+\frac{1}{2}}$. In the second pass, R and Z are advanced to t^{n+1}. The calculation of these stress deviators will be made in the third pass, therefore, we will have the following information available

$$R^{n+1}, Z^{n+1}, R^n, Z^n, U^{n+\frac{1}{2}}, V^{n+\frac{1}{2}}, \text{ at points } 0, d, f, a, \qquad (2.105)$$

and

$$(S^{ij})^n, P^n, v^{n+1}, \text{ at points } 1. \qquad (2.106)$$

From Fig. 2.3 we difference the terms $Z_\ell U_k, -Z_k U_\ell$ in Eq. (2.104) as

$$[Z_\ell U_k]^{n+\frac{1}{2}} = \frac{1}{2}\left[\frac{Z_0 - Z_a}{\ell_0 - \ell_a} + \frac{Z_d - Z_f}{\ell_d - \ell_f}\right]^{n+\frac{1}{2}} \frac{1}{2}\left[\frac{U_d - U_0}{k_d - k_0} + \frac{U_f - U_a}{k_f - k_a}\right]^{n+\frac{1}{2}}, \qquad (2.107)$$

and

$$[Z_k U_\ell]^{n+\frac{1}{2}} = \frac{1}{2}\left[\frac{Z_d - Z_0}{k_d - k_0} + \frac{Z_f - Z_a}{k_f - k_a}\right]^{n+\frac{1}{2}} \frac{1}{2}\left[\frac{U_0 - U_a}{\ell_0 - \ell_a} + \frac{U_d - U_f}{\ell_d - \ell_f}\right]^{n+\frac{1}{2}}, \qquad (2.108)$$

where

$$(Z_i)^{n+\frac{1}{2}} = \frac{1}{2}[(Z_i)^{n+1} + (Z_i)^n], \text{ and } i = 0, d, f, a. \qquad (2.109)$$

The product of any two terms in Eqs. (2.107) and (2.108) will yield expressions in the denominator, such as $(\ell_0 - \ell_a)(k_d - k_0)$, $(\ell_0 - \ell_a)(k_f - k_a)$, etc. Thus, if the mesh is initially rectangular, this common denominator will be the initial area of cell 1, which we denote by A_0. However, by Eq. (2.30) as applied to cell 1, we have

$$j = \frac{\rho_0 k^\alpha}{\rho R^\alpha} = \frac{m_1/A_0}{m_1/A} = \frac{A}{A_0}. \text{ (all at n)}. \qquad (2.110)$$

Therefore,

$$[j^{-1}(Z_\ell U_k - Z_k U_\ell)]^{n+\frac{1}{2}} = \frac{1}{\frac{1}{2}(A^{n+1}+A^n)}[\frac{1}{4}(Z_0 - Z_a + Z_d - Z_f)$$

$$\times (U_d - U_0 + U_f - U_a) - \frac{1}{4}(Z_d - Z_0 + Z_f - Z_a)(U_0 - U_a + U_d - U_f)]^{n+\frac{1}{2}}. \quad (2.111)$$

For convenience, let $\frac{\overline{\Delta U}}{\Delta R}$ denote the left-hand side of Eq. (2.111), and after multiplication and cancellation on the right-hand side, this reduces to

$$\frac{\overline{\Delta U}}{\Delta R} = \frac{1}{(A^{n+1} + A^n)}[(Z_0 - Z_f)(U_d - U_a) + (Z_d - Z_a)(U_f - U_0)]^{n+\frac{1}{2}}. \quad (2.112)$$

Then, following this method, the velocity gradients needed in Eqs. (2.101) and (2.102) difference as

$$\frac{\overline{\Delta V}}{\Delta Z} = \frac{1}{(A^{n+1} + A^n)}[(R_d - R_a)(V_0 - V_f) + (R_f - R_0)(V_d - V_a)]^{n+\frac{1}{2}}, \quad (2.113)$$

and

$$\frac{\overline{\Delta U}}{\Delta Z} + \frac{\overline{\Delta V}}{\Delta R} = \frac{1}{(A^{n+1}+A^n)}[(R_d - R_a)(U_0 - U_f) + (R_f - R_0)(U_d - U_a)$$

$$+ (Z_0 - Z_f)(V_d - V_a) + (Z_d - Z_a)(V_f - V_0)]^{n+\frac{1}{2}}. \quad (2.114)$$

Equations (2.100) through (2.103) are in final form as

$$(S^{RR})^{n+1} - (S^{RR})^n = 2G\left[\frac{\overline{\Delta U}}{\Delta R}\Delta t - \frac{2}{3}\frac{v^{n+1} - v^n}{v^{n+1} + v^n}\right], \quad (2.115)$$

$$(S^{ZZ})^{n+1} - (S^{ZZ})^n = 2G\left[\frac{\overline{\Delta V}}{\Delta Z}\Delta t - \frac{2}{3}\frac{v^{n+1} - v^n}{v^{n+1} + v^n}\right], \quad (2.116)$$

$$(S^{RZ})^{n+1} - (S^{RZ})^n = G\left[\frac{\overline{\Delta U}}{\Delta Z} + \frac{\overline{\Delta V}}{\Delta R}\right]\Delta t, \quad (2.117)$$

and

$$(S^{\theta\theta})^{n+1} = -(S^{RR})^{n+1} - (S^{ZZ})^{n+1}. \quad (2.118)$$

With these values of the stress deviators at t^{n+1}, we must test to see whether the medium in cell 1 has actually remained within the elastic range.

Therefore, we form the second invariant of the stress deviator tensor, ϕ is defined as

$$\phi = [(S^{RR})^2 + (S^{ZZ})^2 + (S^{\theta\theta})^2 + 2(S^{RZ})^2]^{n+1}, \quad (2.119)$$

and by using the von-Mises yield criterion, we test to see whether

$$\phi \leq \frac{2}{3}(Y^0)^2. \quad (2.120)$$

Y^0 is the yield in simple tension. If condition (2.120) holds, then the values for $(S^{ij})^{n+1}$ calculated in Eqs. (2.115) through (2.118) are the true values for t^{n+1}, except for a correction due to rotation of cell 1 during Δt. This correction will be discussed later. If Eq. (2.120) is violated, we know that at some unknown time within Δt the medium has begun to flow plastically and Hookes' law no longer applies. It is presumed that we have already computed Δt by the stability criterion used in the code. As a first attempt, we take as a model (especially when this situation is encountered for the first time for a given cell) that the medium remains elastic over an increment, Δt_1, and then flows plastically over an increment, Δt_2, where

$$\Delta t = \Delta t_1 + \Delta t_2. \quad (2.121)$$

For a first guess for Δt_1, we suggest using Eq. (2.120) as an indicator and take

$$\Delta t_1 = \Delta t\{1 - (\frac{2/3(Y^0)^2}{\phi^{n+1}})^{\frac{1}{2}}\}. \quad (2.122)$$

Now with Δt_1 we return to Eqs. (2.115) through (2.118) and compute new values of S^{ij} at Δt_1, and a new value of ϕ at Δt_1 from Eq. (2.119). Again, we test whether or not at Δt_1

$$\phi_{\Delta t_1} = \frac{2}{3}(Y^0)^2. \quad (2.123)$$

It may take several iterations before the equality demanded in Eq. (2.123) is met. A possible alternate method is to subdivide Δt into five or ten equal parts and through a do-loop find that subdivision closest to matching the equality. For convenience let us call the final values for the stresses, which meet the yield condition in Eq. (2.123), $(S^{ij})^*$. For the remainder of $\Delta t(\Delta t_2)$, we must difference the Prandtl–Reuss equation for plastic flow.

$$\dot{S}^{ij} = 2G\dot{e}^{ij} - \frac{3GW}{(Y^0)^2}S^{ij}, \quad (2.124)$$

where e^{ij} is the total strain rate deviator and for this model, e^{ij} will just equal the plastic strain rate deviator over the increment Δt_2. W is the rate at which work is being done per unit volume on the mass in cell 1 changing its shape. In Eulerian coordinates, this quantity is

$$\dot{W} = (\frac{\partial U}{\partial R})S^{RR} + (\frac{\partial U}{\partial Z} + \frac{\partial V}{\partial R})S^{RZ} + (\frac{\partial V}{\partial Z})S^{ZZ} + (\frac{\alpha U}{R})S^{\theta\theta}. \quad (2.125)$$

Presumably, W is positive or zero during this plastic phase. Transforming Eqs. (2.124) and (2.125) to Lagrangian coordinates and differencing we have

$$(S^{RR})^{n+1} - (S^{RR})^* = 2G\left[\frac{\overline{\Delta U}}{\Delta R}\Delta t_2 - \frac{2}{3}\frac{v^{n+1}-v^*}{v^{n+1}+v^*}\right]$$

$$-\frac{3GW_t\Delta t_2}{2(Y^0)^2}[(S^{RR})^{n+1} + (S^{RR})^*], \quad (2.126)$$

$$(S^{ZZ})^{n+1} - (S^{ZZ})^* = 2G\left[\frac{\overline{\Delta V}}{\Delta Z}\Delta t_2 - \frac{2}{3}\frac{v^{n+1}-v^*}{v^{n+1}+v^*}\right]$$

$$-\frac{3GW_t\Delta t_2}{2(Y^0)^2}[(S^{ZZ})^{n+1} + (S^{ZZ})^*], \quad (2.127)$$

$$(S^{RZ})^{n+1} - (S^{RZ})^* = 2G\left[\frac{\overline{\Delta U}}{\Delta Z} + \frac{\overline{\Delta V}}{\Delta R}\right]\Delta t_2$$

$$-\frac{3GW_t\Delta t_2}{2(Y^0)^2}[(S^{RZ})^{n+1} + (S^{RZ})^*], \quad (2.128)$$

and

$$W_t = \tfrac{1}{2}\{[(S^{RR})^{n+1} + (S^{RR})^*]\tfrac{\overline{\Delta U}}{\Delta R} + [(S^{\theta\theta})^{n+1} + (S^{\theta\theta})^*]\tfrac{\alpha U}{R}$$

$$+[(S^{ZZ})^{n+1} + (S^{ZZ})^*]\frac{\overline{\Delta V}}{\Delta Z} + [(S^{RZ})^{n+1} + (S^{RZ})^*][\frac{\overline{\Delta U}}{\Delta Z} + \frac{\overline{\Delta V}}{\Delta R}]\}. \quad (2.129)$$

Equations (2.126) through (2.129) constitute an implicit set of three equations in our three unknown: S^{RR}, S^{ZZ}, and S^{RZ} at t^{n+1} (Eq. (2.118) may be used to eliminate $S^{\theta\theta}$ in Eq. (2.129)). Note also that the volume at Δt_1 indicated as v^* must be computed. To begin an iteration toward a solution of Eqs. (2.126) through (2.129), we suggest to set

$$(S^{ij})^{n+1} = (S^{ij})^*, \qquad (2.130)$$

in Eq. (2.129) and then each of the Eqs. (2.126) through (2.129) may be explicitly solved for its value of $(S^{ij})^{n+1}$. After the final values of $(S^{ij})^{n+1}$ are found, it would certainly be prudent to test the von-Mises equality with these values as well as the sign of W_t.

2.4 Correction of Stresses for Rigid Body Rotation During Δt

Finally, there is yet another adjustment for $(S^{ij})^{n+1}$ if, during $\Delta t = t^{n+1} - t^n$, the mass in cell 1 has rotated appreciably as a rigid body. Under these conditions, Hookes' law demands

$$(S^{ij})^{n+1} = (S^{ij})^n, \qquad (2.131)$$

because there is no dilatation or deformation during Δt. However, the force caused by stresses acting on the faces are rotated as illustrated in Fig. 2.4. Figure 2.4 shows a mass cell at t^n under tension (the dotted lines represent its original volume) in the R direction. The four corner points are moving such as to rotate the cell as a rigid body. We have translated the cell at t^{n+1} only for clarification in Fig. 2.4. Now the equations of motion contain the spatial derivatives of the stresses at a point and these stresses are referred to the fixed R, and Z, axis. We must, therefore, refer $(S^{ij})^{n+1}$ to this axis. (In the transformation to Lagrangian coordinates, we did not transfer (S^{ij}) as a tensor and then differentiate (S^{ij}).) Another way to say this is that in its original form, Hookes' law is a static relationship of strain to stress. We are extending Hookes' relationship to a dynamic situation through time differentiation and in so doing we must account for rigid body motion in this ad hoc manner because Hookes' law "know" nothing about rigid body rotation. Attempts to formally extend Hookes' law, as well as other relations, have given rise in the literature to many so-called objective stress derivatives [2.3], which derivatives are not unique. We prefer to relate

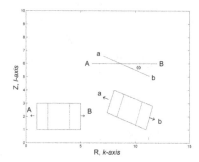

Fig. 2.4 A rigid body rotated an angle of ω during $\Delta t = t^{n+1} - t^n$. The angle ω is the smaller angle formed by lines $\overline{AB}$ and $\overline{ab}$.

$(S^{ij})^{n+1}$ to the fixed frame by the ordinary tensor transformations.

$$S_0^{RR} = S^{RR} \cos^2 \omega + S^{ZZ} \sin^2 \omega + 2 S^{RZ} \sin \omega \cos \omega, \tag{2.132}$$

$$S_0^{ZZ} = S^{RR} \sin^2 \omega + S^{ZZ} \cos^2 \omega - 2 S^{RZ} \sin \omega \cos \omega, \tag{2.133}$$

and

$$S_0^{RZ} = S^{RZ}(\cos^2 \omega - \sin^2 \omega) - (S^{RR} - S^{ZZ}) \sin \omega \cos \omega, \tag{2.134}$$

where

$$\sin \omega = \frac{\Delta t}{2}\left(\frac{\partial U}{\partial Z} - \frac{\partial V}{\partial R}\right) = \frac{\Delta t}{2(A^{n+1} + A^n)}[(R_d - R_a)(U_0 - U_f)$$

$$+ (R_f - R_0)(U_d - U_a) - (Z_0 - Z_f)(V_d - V_a) - (Z_d - Z_a)(V_f - V_0)]^{n+\frac{1}{2}}. \tag{2.135}$$

These equations are derived in Section 3.2.2.7. The value of (S^{ij}), to be used on the right-hand side of Eqs. (2.132) through (2.134), are those calculated in the previous section. ω is the angle of clockwise rotation of the mass cell in the Z and R plane as indicated in Fig. 2.4.

2.5 Calculation of P^{n+1} and ϵ^{n+1}

Rather than to use Eq. (2.33) for the energy equation, it will be easier to first introduce the deviators into Eq. (2.4), then, we have

$$\rho \frac{D\epsilon}{Dt} = -P \nabla \cdot \vec{u} + S^{RR}\frac{\partial U}{\partial R} + \alpha S^{\theta\theta}\frac{U}{R} + S^{ZZ}\frac{\partial V}{\partial Z} + S^{RZ}\left(\frac{\partial U}{\partial Z} + \frac{\partial V}{\partial R}\right). \tag{2.136}$$

By Eq. (2.125), this can be more easily written as

$$\rho \frac{D\epsilon}{Dt} = -P \nabla \cdot \vec{u} + \dot{W}. \tag{2.137}$$

Then by Eqs. (2.26) and (2.27) where we identify

$$\nabla \cdot \vec{u} = \frac{j'_t}{j'} = \frac{v_t}{v}. \tag{2.138}$$

Equation (2.137) becomes in Lagrangian coordinates,

$$\rho \frac{D\epsilon}{Dt} = -P \frac{v_t}{v} + W_t. \tag{2.139}$$

Multiplying through by v, we have

$$\frac{D\epsilon}{Dt} = -Pv_t + vW_t, \tag{2.140}$$

which in difference form is

$$\epsilon^{n+1} - \epsilon^n = -\frac{1}{2}(P^{n+1} + P^n)(v^{n+1} - v^n) + \frac{1}{2}(v^{n+1} + v^n)(W_t)^{n+\frac{1}{2}} \delta t. \tag{2.141}$$

$(W_t)^{n+\frac{1}{2}}$ in difference form is expressed in Eq. (2.129) wherein the final values of $(S^{ij})^{n+1}$ are used as computed in the previous section. In addition, we have an equation of state in the form

$$P = f(\epsilon, v). \tag{2.142}$$

A good model of the equation of state is the quadratic form polynomial that calculates the pressure as a function of density ratio and internal energy as discussed in Section 4.3 of Ref. [2.4]. The expression is

$$P(Mbar) = \frac{A_1\mu + A_2\mu|\mu| + (B_0 + B_1\mu + B_2\mu^2)\epsilon + (C_0 + C_1\mu)\epsilon^2}{\epsilon + D_0}, \tag{2.143}$$

where

$$\mu = \frac{\rho}{\rho_0} - 1. \tag{2.144}$$

In Eq. (2.143), ρ is the density (g/cm^3), ρ_0 is the initial density (g/cm^3), ϵ is the specific internal energy $(Mbar - cm^3/g)$, $A_1, A_2, B_0, B_1, B_2, C_0, C_1$

and D_0 are coefficients that are material dependent. The coefficients for some material are given in Table 4.2 of Ref. [2.4].

Formally, we substitute Eq. (2.142) into Eq. (2.141) to get

$$\epsilon^{n+1} - \epsilon^n = -\frac{1}{2}[f(\epsilon^{n+1}, v^{n+1}) + P^n](v^{n+1} - v^n) + \frac{1}{2}(v^{n+1} + v^n)(W_t)^{n+\frac{1}{2}}\Delta t. \tag{2.145}$$

Depending upon the form of Eq. (2.142), Eq. (2.145) can be solved for ϵ^{n+1} explicitly or by iterative methods. After ϵ^{n+1} has been found, Eq. (2.142) is used to calculate

$$P^{n+1} = f(\epsilon^{n+1}, v^{n+1}). \tag{2.146}$$

Throughout this section we have used ϵ and v, respectively, as the specific internal energy and the specific volume. In the code the internal energy per initial cell volume and the cell volume per initial cell volume are used. We denote these quantities with a subscript m and we have

$$v_m = \rho_0 v, \tag{2.147}$$

and

$$E_m = \rho_0 \epsilon. \tag{2.148}$$

If one examines the equations in this section wherein v or ϵ were used and substitutes v_m/ρ_0 and E_m/ρ_0, respectively, he will see that the factor ρ_0 will cancel. Thus, these equations may be immediately interpreted in terms of the code variables through

$$v = v_m, \tag{2.149}$$

and

$$\epsilon = E_m. \tag{2.150}$$

2.6 Calculation of Longitudinal Sound Speed

Another effect of an elastic-plastic medium is in the calculation of the sound speed needed in the stability criterion used in the code. The code uses the hydrodynamic or so called "bulk" sound speed for a fluid medium and by definition this is

$$c_B^2 = (\frac{\partial P}{\partial \rho})_s, \quad (2.151)$$

here s is the entropy. In the elastic range, there is no bulk wave, physically, because an elastic medium admits only longitudinal and transverse waves with respective speeds

$$c_\ell^2 = \frac{\lambda + 2G}{\rho}, \quad (2.152)$$

and

$$c_t^2 = \frac{G}{\rho}. \quad (2.153)$$

However, for a solid obeying Hookes' law we may inquire what is $(\frac{\partial P}{\partial \rho})_s$? Because Hookes' law states

$$P = -K\frac{\Delta v}{v} = K\frac{\Delta \rho}{\rho}, \quad (2.154)$$

where

$$K = \lambda + \frac{2}{3}G, \quad (2.155)$$

then

$$(\frac{\partial P}{\partial \rho})_T = \frac{K}{\rho}. \quad (2.156)$$

The partial derivative in Eq. (2.156) is at constant temperature because K in Eq. (2.154) is usually determined from ultrasonics isothermal conditions. Then from the thermodynamic identity,

$$(\frac{\partial P}{\partial \rho})_s = (\frac{\partial P}{\partial \rho})_T + \frac{(\beta K)^2 T}{\rho^2 c_v}, \quad (2.157)$$

one has

$$(\frac{\partial P}{\partial \rho})_s = \frac{K}{\rho}\left[1 + \frac{\beta^2 K T}{\rho^2 c_v}\right]. \quad (2.158)$$

For metals over the elastic range, the second term on the right-hand side is like 0.05, so to a good approximation for an elastic solid

$$c_B^2 = \frac{K}{\rho}. \tag{2.159}$$

Then, eliminating λ between Eqs. (2.152) and (2.155) we have

$$c_\ell^2 = \frac{K + \frac{4}{3}G}{\rho}, \tag{2.160}$$

and by Eq. (2.159), we have

$$c_\ell^2 = c_B^2 + \frac{4}{3}\frac{G}{\rho}. \tag{2.161}$$

Equation (2.161) permit us to find c_ℓ^2 by adding $\frac{4}{3}\frac{G}{\rho}$ to the "hydro" sound speed squared.

2.7 Artificial Viscosity Used in the Two-Dimensional Lagrangian Code

In the code calculation Q is computed only when the cell is compressing, for example, a running shock wave. The artificial viscosity used in the present code calculation is

$$Q = 1.4\rho A \left(\frac{\tau^n - \tau^{n-1}}{\tau^n \Delta t}\right)^2, \tag{2.162}$$

where ρ is the density, A the area, $\tau = \frac{1}{\rho}$ and t the time.

This chapter is a modified version of Blewett's Report [2.5].

Bibliography

2.1 Herrmann, W. (1964) *Comparison of finite difference expressions used in Lagrangian fluid flow calculations*, Air Force Weapons Laboratory report WL-TR-64-104.
2.2 Prager, W and Hodge, PG. (1951). *Theory of perfectly plastic solids*, John Wiley and Sons, Inc., New York.
2.3 Eringen, AC. (1962). *Nonlinear theory of continuous media*, McGraw-Hill, Inc., New York.
2.4 Lee, WH. (2006). *Computer simulation of shaped charge problems*, World Scientific Publishing Co., Singapore.
2.5 Blewett, PJ. (1964). *Stress calculation in 2D Lagrangian code*, Los Alamos National Laboratory report LA-64-104.

Chapter 3

Two-Dimensional Eulerian Method

Notations

E specific internal energy per unit volume ($\frac{Mbar-cm^3}{cm^3}$)
E the specific total energy as used in Eq. (3.34) ($\frac{joule}{g}$)
e equivalent plastic strain (no unit)
G shear modulus of elasticity ($Mbar$)
I specific internal energy per unit mass ($\frac{Mbar-cm^3}{g}$)
M cell mass (g)
P pressure ($Mbar$)
R, r radial coordinate (cm)
S^{ij} stress deviator tensor ($Mbar$)
$S^{rr}, S^{zz}, S^{rz}, S^{\theta\theta}$ stress deviator components ($Mbar$)
s distance (cm)
t time (μsec)
Δt time step (μsec)
Δt_{max} maximum allowed time step (μsec)
U velocity in R direction ($\frac{cm}{\mu sec}$)
$\overline{U}$ average velocity defined by Eq. (3.56) ($\frac{cm}{\mu sec}$)
$|U|$ in Eq. (3.57), may be $|U|$ or $|V|$ dependent on the direction of calculation ($\frac{cm}{\mu sec}$)
u_R displacement in the R direction (cm)
V velocity in Z direction ($\frac{cm}{\mu sec}$)
$\overline{V}$ average velocity defined by Eq. (3.56) ($\frac{cm}{\mu sec}$)
u_Z displacement in the Z direction (cm)
$\dot{W}$ rate of energy source due to work hardening ($\frac{Mbar-cm^3}{cm^3-\mu sec}$)
Y flow stress of elasticity ($Mbar$)
Z, z axial coordinate (cm)

Greek letters

δ_{ij} Kronecker delta (no unit)
ϵ specific internal energy per mass ($\frac{Mbar-cm^3}{g}$)
γ ratio of the specific heat, i.e., $\gamma = C_P/C_V$ (no unit)
η normalized density = $\frac{\rho}{\rho_0}$ (no unit)
θ angular coordinate
λ half of the grid size in R direction, i.e., $\lambda = \Delta R/2$ (cm)
μ normalized density minus one ($=\eta - 1 = \frac{\rho}{\rho_0} - 1$)(no unit)
ρ density ($\frac{g}{cm^3}$)
σ^{ij} stress tensor ($Mbar$)
$\sigma^{rr}, \sigma^{zz}, \sigma^{rz}, \sigma^{\theta\theta}$ stress components ($Mbar$)
τ specific volume ($\frac{cm^3}{g}$)

Subscripts

0 initial value
e elastic regime
kk normal strains, i.e., $e_{rr}, e_{zz}, e_{\theta\theta}$
p plastic regime
R derivative with respect to R coordinate
t derivative with respect to time
Z derivative with respect to Z coordinate

Superscripts

n time at n time-step, i.e., $t^n = t_0 + n \cdot \Delta t$
kk normal stresses, i.e., $\sigma^{rr}, \sigma^{zz}, \sigma^{\theta\theta}$

3.1 Introduction

There is considerable interest in solving multi-material compressible flow problems with material interface. A Lagrangian approach would be a quite natural choice — for example HEMP [3.1], TOODY [3.2], MAGEE [3.3 and 3.4]. However, for a large material distortion, the Lagrangian calculations can no longer be continued, so an Eulerian or a combined Lagrangian/Eulerian-type scheme has been applied.

Although many schemes have been invented in the various Eulerian code developments, materials are treated in only two ways, namely the particle-in-cell (PIC) method [3.5], where materials are represented by discrete mass points called Particle, and the continuous Eulerian methods, as in SOIL [3.6], HELP [3.7], and CSQ [3.8]. Computing economy is gained in the continuous method; however, the Lagrangian-type capability of the PIC method is lost and subsequently is replaced by various interface treatment.

In development of the present code, the goal is to retain PIC capability while improving the accuracy and computing economy. Recently, Clark [3.9] proved the accuracy of the standard PIC method to be a first-order scheme in hydrodynamic computation of multi-material problems and proposed a second-order scheme. Here a similar second-order scheme is used to compute a multi-material elastic-plastic flow, including phase transition and spall. The truncation error, stability analysis, and other details of computing are presented in the following section.

3.2 General Description of Physical Formulation

3.2.1 The Conservation Equation for a Stress-Supporting Medium

The conservation equation in a two-dimensional cylindrical or plane Eulerian coordinate can be written as follows:

Mass
$$\frac{D\rho}{Dt} + \rho(\frac{\partial U}{\partial R} + \alpha\frac{U}{R} + \frac{\partial V}{\partial Z}) = 0, \quad (3.1)$$

Momentum
$$\rho\frac{DU}{Dt} = \frac{\partial \sigma^{RR}}{\partial R} + \frac{\partial \sigma^{RZ}}{\partial Z} + \frac{\alpha}{R}(\sigma^{RR} - \sigma^{\theta\theta}), \quad (3.2)$$

$$\rho\frac{DV}{Dt} = \frac{\partial \sigma^{ZZ}}{\partial Z} + \frac{\partial \sigma^{RZ}}{\partial R} + \frac{\alpha}{R}\sigma^{RZ}, \quad (3.3)$$

Energy
$$\rho\frac{D\epsilon}{Dt} = \sigma^{RR}\frac{\partial U}{\partial R} + \alpha\sigma^{\theta\theta}\frac{U}{R} + \sigma^{ZZ}\frac{\partial V}{\partial Z} + \sigma^{RZ}(\frac{\partial U}{\partial Z} + \frac{\partial V}{\partial R}), \quad (3.4)$$

where $\alpha = 0$ for plane geometry and $\alpha = 1$ for cylindrical geometry with the following definitions

$$\frac{D}{Dt} = \frac{\partial}{\partial t} + U\frac{\partial}{\partial R} + V\frac{\partial}{\partial Z}, \quad (3.5)$$

$$\sigma^{ij} = S^{ij} - P\delta_{ij}, \quad (3.6)$$

where σ^{ij} is the stress tensor and δ_{ij} the Kronecker delta. For compression, P is negative of the normal stresses, therefore

$$P = -\frac{1}{3}\sigma^{kk} = -\frac{1}{3}(\sigma^{RR} + \sigma^{ZZ} + \sigma^{\theta\theta}), \quad (3.7)$$

$$S^{ij} = \text{stress deviator}, \quad (3.8)$$

and
$$\epsilon = \text{specific internal energy}. \quad (3.9)$$

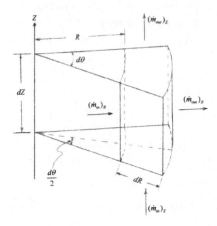

Fig. 3.1 The control volume and the mass flows through it.

3.2.1.1 The Derivation of the Mass Conservation Equation

Figure 3.1 shows the control volume bounded by $dR, dZ, Rd\theta$ and $(R + dR)d\theta$. The accumulation of the mass in the control volume can be obtained from $(\dot{m}_{in})_R, (\dot{m}_{out})_R, (\dot{m}_{in})_Z$ and $(\dot{m}_{out})_Z$. Since

$$(\dot{m}_{in})_R = (\rho U - \frac{\partial \rho U}{\partial R}\frac{dR}{2})(Rd\theta)(dZ), \qquad (3.10)$$

$$(\dot{m}_{out})_R = (\rho U + \frac{\partial \rho U}{\partial R}\frac{dR}{2})(R+dR)(d\theta)(dZ), \qquad (3.11)$$

where ρU is defined at the center of the control volume which is located at $R = R + \frac{dR}{2}$ and $Z = \frac{dZ}{2}$. Therefore, the accumulation of the mass in the R direction is

$$(\dot{m}_{in})_R - (\dot{m}_{out})_R = (\rho U - \frac{\partial \rho U}{\partial R}\frac{dR}{2})(Rd\theta)(dZ)$$

$$-(\rho U + \frac{\partial \rho U}{\partial R}\frac{dR}{2})(R+dR)(d\theta)(dZ), \qquad (3.12)$$

$$= -\frac{\partial \rho U}{\partial R}(dR)(Rd\theta)(dZ) - (\rho U)(dR)(R+dR)(d\theta)(dZ). \qquad (3.13)$$

Accumulation of mass in the Z direction for the control volume is

$$(\dot{m}_{in})_Z - (\dot{m}_{out})_Z = (\rho V - \frac{\partial \rho V}{\partial Z}\frac{dZ}{2})(R+\frac{dR}{2})(d\theta)(dR)$$

$$-(\rho V + \frac{\partial \rho V}{\partial Z}\frac{dZ}{2})(R+\frac{dR}{2})(d\theta)(dR), \qquad (3.14)$$

$$= -\frac{\partial \rho V}{\partial Z}(dZ)(Rd\theta)(dR). \tag{3.15}$$

In the derivation of Eq. (3.15) we assume $(dR)^2 \ll dR$. Since the mass increases in the control volume is

$$\dot{m} = \frac{dm}{dt} = \frac{d}{dt}(\rho Rd\theta dRdZ) = \frac{d\rho}{dt}(Rd\theta)(dR)(d\theta). \tag{3.16}$$

From Eqs. (3.13), (3.15) and (3.16), we get

$$\frac{d\rho}{dt}(Rd\theta)(dR)(d\theta) = -\frac{\partial \rho U}{\partial R}(dR)(Rd\theta)(dZ)$$

$$-(\rho U)(dR)(R+dR)(d\theta)(dZ) - \frac{\partial \rho V}{\partial Z}(dZ)(Rd\theta)(dR). \tag{3.17}$$

Equation (3.17) can be simplified as

$$\frac{d\rho}{dt} = -\frac{\partial \rho U}{\partial R} - \frac{\partial \rho V}{\partial Z} - \frac{\rho U}{R}, \tag{3.18}$$

or

$$\frac{d\rho}{dt} + \frac{\partial \rho U}{\partial R} + \frac{\partial \rho V}{\partial Z} + \frac{\rho U}{R} = 0, \tag{3.19}$$

or

$$\frac{d\rho}{dt} + U\frac{\partial \rho}{\partial R} + V\frac{\partial \rho}{\partial Z} + \rho(\frac{\partial U}{\partial R} + \frac{\partial V}{\partial Z} + \frac{U}{R}) = 0, \tag{3.20}$$

or

$$\frac{D\rho}{Dt} + \rho(\frac{\partial U}{\partial R} + \frac{\partial V}{\partial Z} + \frac{U}{R}) = 0. \tag{3.21}$$

For cylindrical coordinate, $\alpha = 1$, one gets Eq. (3.1) as

$$\frac{D\rho}{Dt} + \rho(\frac{\partial U}{\partial R} + \frac{\partial V}{\partial Z} + \frac{U}{R}) = 0. \tag{3.22}$$

3.2.1.2 The Derivation of the Momentum Conservation Equations

The momentum conservation equation can be derived from the Newton's second law, that is $F = ma$. In fluid mechanic, the pressure dominates the momentum balance. But in solid mechanic, one does not see the pressure, since the pressure is embedded inside the stress tensor. Therefore, we calculate the force from the stress components and use $F = ma = m\frac{DU}{Dt}$ to get the velocity component U in the R direction. As shown in Fig. 3.2, to derive the momentum equation in the R direction, one has to consider

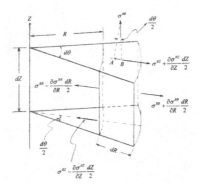

Fig. 3.2 The control volume and the stresses acting on it.

the stress components $\sigma^{\theta\theta}, \sigma^{RR}$, and σ^{RZ}. In solid mechanics, the particle motion has the Lagrangian character which will take certain amount of the material to move with the particle, therefore, the particle acceleration is represented by $\frac{DU}{Dt}$ and not by $\frac{dU}{dt}$ which is the Eulerian approximation. For the momentum in the R direction, we have

$$F_R = ma_R = m\frac{DU}{Dt}, \qquad (3.23)$$

where F_R is the force in the R direction and a_R is the acceleration. The mass is

$$m = \rho d(Volume) = \rho(Rd\theta)(dR)(dZ). \qquad (3.24)$$

The force due to $\sigma^{RR}, \sigma^{\theta\theta}$ and σ^{RZ} is

$$F_R = (\sigma^{RR} + \tfrac{\partial \sigma^{RR}}{\partial R}\tfrac{dR}{2})[dZ(R+dR)d\theta] - (\sigma^{RR} - \tfrac{\partial \sigma^{RR}}{\partial R}\tfrac{dR}{2})[dZ(Rd\theta)]$$

$$+ (\sigma^{RZ} + \tfrac{\partial \sigma^{RZ}}{\partial Z}\tfrac{dZ}{2})\{\tfrac{1}{2}[Rd\theta + (R+dR)d\theta]dR\}$$

$$- (\sigma^{RZ} - \tfrac{\partial \sigma^{RZ}}{\partial Z}\tfrac{dZ}{2})\{\tfrac{1}{2}[Rd\theta + (R+dR)d\theta]dR\}$$

$$-2\sigma^{\theta\theta}\frac{d\theta}{2}(dR)(dZ). \qquad (3.25)$$

With some algebra Eq. (3.25) becomes

$$F_R = \sigma^{RR}(dR)(dZ)(d\theta) + \frac{\partial \sigma^{RR}}{\partial R}(dR)(dZ)(d\theta)(R)$$

$$+ \frac{\partial \sigma^{RZ}}{\partial Z}(dR)(dZ)(d\theta)(R) - \sigma^{\theta\theta}(d\theta)(dR)(dZ). \qquad (3.26)$$

Substituting Eqs. (3.26) and (3.24) into Eq. (3.23), it follows

$$\rho(Rd\theta)(dR)(dZ)\frac{DU}{Dt} = \sigma^{RR}(dR)(dZ)(d\theta) + \frac{\partial \sigma^{RR}}{\partial R}(dR)(dZ)(d\theta)(R)$$

$$+\frac{\partial \sigma^{RZ}}{\partial Z}(dR)(dZ)(d\theta)(R) - \sigma^{\theta\theta}(d\theta)(dR)(dZ). \qquad (3.27)$$

Dividing the above equation by $(Rd\theta)(dR)(dZ)$, one gets

$$\rho\frac{DU}{Dt} = [\sigma_R^{RR} + \sigma_Z^{ZZ} + \frac{1}{R}(\sigma^{RR} - \sigma^{\theta\theta})], \qquad (3.28)$$

which is the same as Eq. (3.2) when one sets $\alpha = 1$ for cylindrical coordinate system.

To derive the momentum equation in the Z direction, one has to consider the stress components σ^{ZZ} and σ^{RZ} as shown in Fig. 3.3. The force due to σ^{ZZ}, and σ^{RZ} is

$$F_Z = (\sigma^{ZZ} + \frac{\partial \sigma^{ZZ}}{\partial Z}\frac{dZ}{2})(R + \frac{dR}{2})(d\theta)(dR)$$

$$-(\sigma^{ZZ} - \frac{\partial \sigma^{ZZ}}{\partial Z}\frac{dZ}{2})(R + \frac{dR}{2})(d\theta)(dR)$$

$$+(\sigma^{RZ} + \frac{\partial \sigma^{RZ}}{\partial R}\frac{dR}{2})(R + dR)(d\theta)(dZ)$$

$$-(\sigma^{RZ} - \frac{\partial \sigma^{RZ}}{\partial R}\frac{dR}{2})(R)(d\theta)(dZ). \qquad (3.29)$$

With some algebra Eq. (3.29) becomes

$$F_Z = \frac{\partial \sigma^{ZZ}}{\partial Z}(dR)(dZ)(d\theta)(R + \frac{dR}{2}) + \frac{\partial \sigma^{RZ}}{\partial R}(dR)(dZ)(d\theta)(R)$$

$$+\sigma^{RZ}(d\theta)(dR)(dZ) + \frac{\partial \sigma^{RZ}}{\partial R}\frac{dR}{2}(dR)(dZ)(d\theta)(R). \qquad (3.30)$$

From the Newton's second law, we have

$$F_Z = ma_Z = m\frac{DV}{Dt}. \qquad (3.31)$$

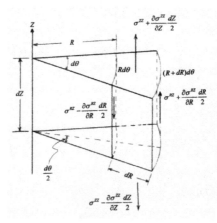

Fig. 3.3 The control volume and the stresses acting on it.

Substituting Eqs. (3.24) and (3.30) into Eq. (3.31) with the fact that $(dR)^2 \ll dR.$, it follows

$$\rho(Rd\theta)(dR)(dZ)\frac{DV}{Dt} = \frac{\partial \sigma^{ZZ}}{\partial Z}(dR)(dZ)(d\theta)(R)$$
$$+ \frac{\partial \sigma^{RZ}}{\partial R}(dR)(dZ)(d\theta)(R) + \sigma^{RZ}(d\theta)(dR)(dZ).$$
(3.32)

Dividing the above equation by $(Rd\theta)(dR)(dZ)$, one gets

$$\rho\frac{DV}{Dt} = \sigma^{ZZ}_Z + \sigma^{RZ}_R + \frac{1}{R}(\sigma^{RZ}),$$
(3.33)

which is identical to Eq. (3.3) when one sets $\alpha = 1$ for cylindrical coordinate system.

3.2.1.3 *The Derivation of the Energy Conservation Equations*

From the first law of thermodynamics without any source, heat conduction, and heat convection, then, we have

$$E = \epsilon + K.E.,$$
(3.34)

where
E is the specific total energy $(\frac{joule}{g})$ or $(\frac{j}{g})$,
ϵ is the specific internal energy $(\frac{j}{g})$,
$K.E.$ is the specific kinetic energy $(\frac{j}{g})$.

Two-Dimensional Eulerian Method 45

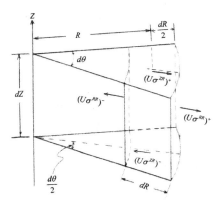

Fig. 3.4 The control volume and the stresses acting on it.

The rate of the increase in energy of a body is equal to the rate at which the applied forces do the work on the body. Therefore

$$\rho \frac{DE}{Dt} = \frac{d}{dt}[\sum \text{work done by stress}]/\text{Volume}, \qquad (3.35)$$

or

$$\rho \frac{DE}{Dt} = \frac{d}{dt}[\sum \text{Stress} \times \text{Area} \times \text{Distance}]/\text{Volume}, \qquad (3.36)$$

or

$$\rho \frac{DE}{Dt} = [\sum \text{Stress} \times \text{Area} \times \frac{ds}{dt}]/\text{Volume}, \qquad (3.37)$$

or

$$\rho \frac{DE}{Dt} = [\sum \text{Stress} \times \text{Area} \times \text{Velocity}]/\text{Volume}. \qquad (3.38)$$

Sometime, we call the result of this multiplication (Stress x Area x Velocity) the stress power since it has the unit of $\frac{j}{sec}$. The right hand side of Eq. (3.38) has the unit of $\frac{j}{sec*cm^3}$ which is the same as the left hand side of Eq. (3.38).

Figure 3.4 shows the (Stress x Velocity) components which contribute to the stress power in the R direction for the control volume. As shown in Fig. 3.4, one can write

$$(U\sigma^{RR})^+ = (U\sigma^{RR}) + \frac{\partial(U\sigma^{RR})}{\partial R}\frac{dR}{2}, \qquad (3.39)$$

$$(U\sigma^{RR})^- = (U\sigma^{RR}) - \frac{\partial(U\sigma^{RR})}{\partial R}\frac{dR}{2}. \qquad (3.40)$$

For two-dimensional cylindrical problems, the velocity in the θ direction is zero, therefore, $(w\sigma^{\theta\theta}) = 0$. For $(U\sigma^{RZ})$, we have

$$(U\sigma^{RZ})^+ = (U\sigma^{RZ}) + \frac{\partial(U\sigma^{RZ})}{\partial Z}\frac{dZ}{2}, \qquad (3.41)$$

$$(U\sigma^{RZ})^- = (U\sigma^{RZ}) - \frac{\partial(U\sigma^{RZ})}{\partial Z}\frac{dZ}{2}. \qquad (3.42)$$

The stress power resulted from $(U\sigma^{RR})$ is

$$[(U\sigma^{RR})^+(R+\tfrac{dR}{2}) - (U\sigma^{RR})^-(R-\tfrac{dR}{2})](d\theta)(dZ)$$

$$= [(U\sigma^{RR})^+ - (U\sigma^{RR})^-]R(d\theta)(dZ)$$

$$+[(U\sigma^{RR})^+ + (U\sigma^{RR})^-]\frac{(dR)(d\theta)(dZ)}{2}, \qquad (3.43)$$

$$= (\frac{\partial U\sigma^{RR}}{\partial R})(dR)(Rd\theta)(dZ) + U\sigma^{RR}(dR)(d\theta)(dZ). \qquad (3.44)$$

The stress power resulted from $(U\sigma^{RZ})$ is

$$[(U\sigma^{RZ})^+ - (U\sigma^{RZ})^-]R(d\theta)(dR)$$

$$= (\frac{\partial U\sigma^{RZ}}{\partial Z})(dR)(Rd\theta)(dZ). \qquad (3.45)$$

Therefore, the total stress power in the R direction is the summation of Eqs. (3.44) and (3.45) which follows

$$(\frac{\partial U\sigma^{RR}}{\partial R})(dR)(Rd\theta)(dZ) + U\sigma^{RR}(dR)(d\theta)(dZ)$$

$$+(\frac{\partial U\sigma^{RZ}}{\partial Z})(dR)(Rd\theta)(dZ). \qquad (3.46)$$

Figure 3.5 shows the (Stress x Velocity) components which contribute to the stress power in the Z direction for the control volume. As shown in Fig. 3.5, one can write

$$(V\sigma^{ZZ})^+ = (V\sigma^{ZZ}) + \frac{\partial(V\sigma^{ZZ})}{\partial Z}\frac{dZ}{2}, \qquad (3.47)$$

$$(V\sigma^{ZZ})^- = (V\sigma^{ZZ}) - \frac{\partial(V\sigma^{ZZ})}{\partial Z}\frac{dZ}{2}. \qquad (3.48)$$

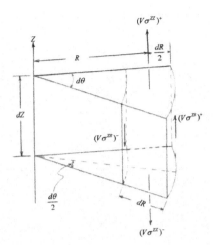

Fig. 3.5 The control volume and the stresses acting on it.

For $(V\sigma^{ZR})$, we have

$$(V\sigma^{ZR})^+ = (V\sigma^{ZR}) + \frac{\partial(V\sigma^{ZR})}{\partial R}\frac{dR}{2}, \qquad (3.49)$$

$$(V\sigma^{ZR})^- = (V\sigma^{ZR}) - \frac{\partial(V\sigma^{ZR})}{\partial R}\frac{dR}{2}. \qquad (3.50)$$

The stress power resulted from $(V\sigma^{ZZ})$ is

$$[(V\sigma^{ZZ})^+ - (V\sigma^{ZZ})^-]R(d\theta)(dR)$$

$$= (\frac{\partial V\sigma^{ZZ}}{\partial Z})(dR)(Rd\theta)(dZ). \qquad (3.51)$$

The stress power resulted from $(V\sigma^{ZR})$ is

$$[(V\sigma^{ZR})^+(R+\tfrac{dR}{2}) - (V\sigma^{ZR})^-(R-\tfrac{dR}{2})](d\theta)(dZ)$$

$$= [(V\sigma^{ZR})^+ - (V\sigma^{ZR})^-]R(d\theta)(dZ)$$

$$+ [(V\sigma^{ZR})^+ + (V\sigma^{ZR})^-](\frac{dR}{2})(d\theta)(dZ), \qquad (3.52)$$

$$= \frac{\partial V\sigma^{ZR}}{\partial R}(dR)(Rd\theta)(dZ) + (V\sigma^{ZR})(d\theta)(dZ)(dR). \qquad (3.53)$$

Therefore, the total stress power in the Z direction is the summation of Eqs. (3.51) and (3.53) which follows

$$\frac{\partial V\sigma^{ZZ}}{\partial Z}(dR)(Rd\theta)(dZ) + \frac{\partial V\sigma^{ZR}}{\partial R}(dR)(Rd\theta)(dZ) + (V\sigma^{ZR})(d\theta)(dZ)(dR). \tag{3.54}$$

For kinetic energy, we know that

$$K.E. = \sum(\text{Velocity x Momentum}). \tag{3.55}$$

Since the momentum equations in two-dimensional cylindrical coordinate are

$$\rho\frac{DU}{Dt} = \frac{\partial \sigma^{RR}}{\partial R} + \frac{\partial \sigma^{RZ}}{\partial Z} + \frac{1}{R}(\sigma^{RR} - \sigma^{\theta\theta}), \tag{3.56}$$

$$\rho\frac{DV}{Dt} = \frac{\partial \sigma^{ZZ}}{\partial Z} + \frac{\partial \sigma^{RZ}}{\partial R} + \frac{1}{R}\sigma^{RZ}. \tag{3.57}$$

The rate of change in kinetic energy is

$$\rho\frac{D(K.E.)}{Dt} = U(\rho\frac{DU}{Dt}) + V(\rho\frac{DV}{Dt}), \tag{3.58}$$

or

$$\rho\frac{D(K.E.)}{Dt} = \rho\frac{D(\frac{U^2}{2})}{Dt} + \rho\frac{D(\frac{V^2}{2})}{Dt}. \tag{3.59}$$

From the above equation, one obtains

$$K.E. = \frac{U^2}{2} + \frac{V^2}{2}. \tag{3.60}$$

Substituting Eqs. (3.56) and (3.57) into the right-hand side of Eq. (3.58), we have

$$\rho\frac{D(K.E.)}{Dt} = U[\frac{\partial \sigma^{RR}}{\partial R} + \frac{\partial \sigma^{RZ}}{\partial Z} + \frac{1}{R}(\sigma^{RR} - \sigma^{\theta\theta})]$$

$$+ V[\frac{\partial \sigma^{ZZ}}{\partial Z} + \frac{\partial \sigma^{RZ}}{\partial R} + \frac{1}{R}\sigma^{RZ}]. \tag{3.61}$$

From Eq. (3.34), we can write the specific internal energy as

$$\epsilon = E - K.E. \tag{3.62}$$

The above equation can be written as

$$\rho\frac{D(\epsilon)}{Dt} = \rho\frac{D(E)}{Dt} - \rho\frac{D(K.E.)}{Dt}. \tag{3.63}$$

Since
$$\rho \frac{D(E)}{Dt} = [\text{Eq. (3.46)} + \text{Eq. (3.54)}]/\text{Volume}, \qquad (3.64)$$

or

$$\rho \frac{D(E)}{Dt} = \frac{\partial U \sigma^{RR}}{\partial R} + \frac{U \sigma^{RR}}{R} + \frac{\partial U \sigma^{RZ}}{\partial Z}$$
$$+ \frac{\partial V \sigma^{ZZ}}{\partial Z} + \frac{\partial V \sigma^{RZ}}{\partial R} + \frac{V \sigma^{ZR}}{R}, \qquad (3.65)$$

or

$$\rho \frac{D(E)}{Dt} = U \frac{\partial \sigma^{RR}}{\partial R} + \sigma^{RR} \frac{\partial U}{\partial R} + \frac{U \sigma^{RR}}{R} + U \frac{\partial \sigma^{RZ}}{\partial Z} + \sigma^{RZ} \frac{\partial U}{\partial Z}$$
$$+ V \frac{\partial \sigma^{ZZ}}{\partial Z} + \sigma^{ZZ} \frac{\partial V}{\partial Z} + V \frac{\partial \sigma^{ZR}}{\partial R}$$
$$+ \sigma^{RR} \frac{\partial V}{\partial R} + \frac{V \sigma^{ZR}}{R}. \qquad (3.66)$$

The internal energy equation is

$$\rho \frac{D(\epsilon)}{Dt} = \rho \frac{D(E)}{Dt} - \rho \frac{D(K.E.)}{Dt}, \qquad (3.67)$$

$$= \sigma^{RR} \frac{\partial U}{\partial R} + \sigma^{RZ} \frac{\partial U}{\partial Z} + \sigma^{ZZ} \frac{\partial V}{\partial Z} + \sigma^{ZR} \frac{\partial V}{\partial R} + \frac{U \sigma^{\theta\theta}}{R}, \qquad (3.68)$$

or

$$\rho \frac{D(\epsilon)}{Dt} = \sigma^{RR} \frac{\partial U}{\partial R} + \frac{U \sigma^{\theta\theta}}{R} + \sigma^{ZZ} \frac{\partial V}{\partial Z} + \sigma^{RZ} \left(\frac{\partial U}{\partial Z} + \frac{\partial V}{\partial R} \right). \qquad (3.69)$$

The above equation is the internal energy equation for a two-dimensional cylindrical coordinate system.

3.2.2 Equation of State

A stress-supporting medium flowing under a wide range of stresses exhibits a variety of different physical characteristics. Depending on its retention of elastic character, the flow may be elastic or plastic. Here a straightforward approach is taken, yet there does not seem to exist any better model.

3.2.2.1 Stress in Elastic Regime

when a material flows elastically, Hooke's law in current form can be written as

$$\frac{DS^{RR}}{Dt} = 2G\left(\frac{\partial U}{\partial R} + \frac{1}{3\rho}\frac{D\rho}{Dt}\right) + \delta^{RR}, \tag{3.70}$$

$$\frac{DS^{ZZ}}{Dt} = 2G\left(\frac{\partial V}{\partial Z} + \frac{1}{3\rho}\frac{D\rho}{Dt}\right) + \delta^{ZZ}, \tag{3.71}$$

$$\frac{DS^{RZ}}{Dt} = G\left(\frac{\partial U}{\partial Z} + \frac{\partial V}{\partial R}\right) + \delta^{RZ}, \tag{3.72}$$

$$\frac{DS^{\theta\theta}}{Dt} = 2G\left(\frac{\alpha U}{R} + \frac{1}{3\rho}\frac{D\rho}{Dt}\right) = \frac{D}{Dt}(-S^{RR} - S^{ZZ}), \tag{3.73}$$

and

$$S^{RR} + S^{ZZ} + S^{\theta\theta} = 0, \tag{3.74}$$

where

$$G = \text{modulus of elasticity in shear} = f(\rho, \epsilon, \text{material}),$$

and

$$\delta^{ij} = \text{correction for rigid body rotation}.$$

In tensor form

$$\frac{DS^{ij}}{Dt} = 2G\dot{e}_{ij} + \delta^{ij}, \tag{3.75}$$

where

$\dot{e}_{ij}$ = strain rate deviator

$$= \begin{bmatrix} \frac{\partial U}{\partial R} + \frac{1}{3\rho}\frac{D\rho}{Dt} & \frac{1}{2}\left(\frac{\partial U}{\partial Z} + \frac{\partial V}{\partial R}\right) & 0 \\ \frac{1}{2}\left(\frac{\partial U}{\partial Z} + \frac{\partial V}{\partial R}\right) & \frac{\partial V}{\partial Z} + \frac{1}{3\rho}\frac{D\rho}{Dt} & 0 \\ 0 & 0 & \alpha\frac{U}{R} + \frac{1}{3\rho}\frac{D\rho}{Dt} \end{bmatrix}. \tag{3.76}$$

3.2.2.2 The Derivation of the Hooke's Law with Deviatoric Stress

This section will show how to derive the Hooke's law for deviatoric stress S^{RR} from the deviatoric strain, e'_{RR}. If e_{ij} is the strain and e'_{ij} is the deviatoric strain (or strain deviation), then

$$e'_{ij} = e_{ij} - \frac{1}{3}e_{kk}\delta_{ij}. \tag{3.77}$$

The relationship between the deviatoric stress, S^{ij}, and the deviatoric strain, e'_{ij}, can be obtained from page 216 of Ref. [3.12] which gives

$$S^{ij} = 2Ge'_{ij}. \tag{3.78}$$

For $i = 1$ and $j = 1$, Eq. (3.77) becomes

$$e'_{11} = e_{11} - \frac{1}{3}(e_{11} + e_{22} + e_{33}), \tag{3.79}$$

or

$$e'_{RR} = e_{RR} - \frac{1}{3}(e_{RR} + e_{\theta\theta} + e_{ZZ}). \tag{3.80}$$

Now, applying $\frac{D}{Dt}$ to Eq. (3.80), one obtains

$$\frac{De'_{RR}}{Dt} = \dot{e}_{RR} - \frac{1}{3}(\dot{e}_{RR} + \dot{e}_{\theta\theta} + \dot{e}_{ZZ}). \tag{3.81}$$

For two-dimensional cylindrical coordinate system, the strains are

$$e_{RR} = \frac{\partial u_R}{\partial R}, \tag{3.82}$$

$$e_{\theta\theta} = \frac{u_R}{R}, \tag{3.83}$$

$$e_{ZZ} = \frac{\partial u_Z}{\partial Z}, \tag{3.84}$$

where u_R and u_Z are the displacements in the R and Z directions respectively. Substituting Eqs. (3.82)-(3.84) into the right hand side of Eq. (3.81), we have

$$\frac{De'_{RR}}{Dt} = \frac{\partial \dot{u}_R}{\partial R} - \frac{1}{3}(\frac{\partial \dot{u}_R}{\partial R} + \frac{\dot{u}_R}{R} + \frac{\partial \dot{u}_Z}{\partial Z}). \tag{3.85}$$

The above equation is equivalent to

$$\frac{De'_{RR}}{Dt} = \frac{\partial U}{\partial R} - \frac{1}{3}(\frac{\partial U}{\partial R} + \frac{U}{R} + \frac{\partial V}{\partial Z}). \tag{3.86}$$

For $\alpha = 1$ as in cylindrical coordinate system, the continuity equation, i.e. Eq. (3.1), can be written as

$$\frac{D\rho}{Dt} = -\rho(\frac{\partial U}{\partial R} + \frac{U}{R} + \frac{\partial V}{\partial Z}). \tag{3.87}$$

Substituting Eq. (3.87) into Eq. (3.86), we get

$$\frac{De'_{RR}}{Dt} = \frac{\partial U}{\partial R} + \frac{1}{3\rho}\frac{D\rho}{Dt}. \tag{3.88}$$

Applying $\frac{D}{Dt}$ to Eq. (3.78), one gets

$$\frac{DS^{ij}}{Dt} = 2G\frac{De'_{ij}}{Dt}. \tag{3.89}$$

Substituting Eq. (3.88) into the right hand side of Eq. (3.89) and adding the rigid body rotation correction term, it results

$$\frac{DS^{RR}}{Dt} = 2G(\frac{\partial U}{\partial R} + \frac{1}{3\rho}\frac{D\rho}{Dt}) + \delta^{RR}. \tag{3.90}$$

Equation (3.90) is the same as Eq. (3.70). Using the same procedure in deriving Eq. (3.70), one can obtain Eqs. (3.71), (3.72) and (3.73).

3.2.2.3 Stresses in Plastic Regime

For plastic flows, Prandtl and Reuss consider both plastic and elastic strain simultaneously and arrived at the following flow equation

$$\dot{S}^{ij} = 2G\dot{e}_{ij} - \frac{G\dot{W}}{\frac{1}{3}(Y^0)^2}S^{ij}, \tag{3.91}$$

where $\dot{W} = \sum_{ij} S^{ij}\dot{e}_{ij}$ is the plastic work/unit volume and Y^0 = yield stress in simple tension.

Expanding Eq. (3.91), we get

$$\frac{DS^{RR}}{Dt} = 2G(\frac{\partial U}{\partial R} + \frac{1}{3\rho}\frac{D\rho}{Dt}) - \frac{G\dot{W}}{\frac{1}{3}(Y^0)^2}S^{RR} + \delta^{RR}, \tag{3.92}$$

$$\frac{DS^{ZZ}}{Dt} = 2G(\frac{\partial V}{\partial Z} + \frac{1}{3\rho}\frac{D\rho}{Dt}) - \frac{G\dot{W}}{\frac{1}{3}(Y^0)^2}S^{ZZ} + \delta^{ZZ}, \tag{3.93}$$

and

$$\frac{DS^{RZ}}{Dt} = G(\frac{\partial U}{\partial Z} + \frac{\partial V}{\partial R}) - \frac{G\dot{W}}{\frac{1}{3}(Y^0)^2}S^{RZ} + \delta^{RZ}. \tag{3.94}$$

cylinder is called the Tresca's hexagonal yield surface. Now, let's suppose that a small time step, Δt, is chosen such that the Prandtl–Reuss relation is valid. During the time increment Δt, a material can be in an elastic period Δt_1, then in a plastic period Δt_2, that is, $\Delta t = \Delta t_1 + \Delta t_2$. Then for a rigorous computation of S^{ij}, Δt_1 has to be obtained such that

$$\phi_{\Delta t_1} = \frac{2}{3}(Y^0)^2, \quad (3.104)$$

is satisfied. An iterative procedure might be considered. However, considering the accuracy of the yield criterion itself and plastic flow model, a simple method seems practical. An experience with a Lagrangian computation shows that the following scheme proposed by Wilkins [3.1] is adequate in computing stress deviators on yield surface, $(S^{ij})^*$, such that

$$(S^{ij})^* = (S^{ij})\frac{\sqrt{\frac{2}{3}}Y^0}{\phi}. \quad (3.105)$$

Therefore, if the flow becomes plastic during Δt period, S^{ij} is first calculated based on elastic assumptions. Then, the new stress deviators on yield surface $(S^{ij})^*$ are approximated by Eq. (3.105).

3.2.2.6 Stress Correction for Rigid Body Rotation

The magnitude of stresses does not change during a rigid body rotation, and new stresses, after rotation as well as after deformation, have to be expressed in terms of the original coordinate system. Hookes' law is a static relationship of strain and stress. To extend Hookes' relationship to a dynamic situation through time differentiation, we must account for rigid body motion. Since the equations of motion contain the spatial derivatives of the stresses at a point and these stresses are referred to the fixed R and Z-axis. We must, therefore, refer $(S^{ij})^{n+1}$ to this axis. The corrections for rigid body rotation before and after deformation have to be expressed in terms of the original coordinate system. The correction for rigid body rotation is relatively straightforward. However, there seems to be no unique way to correct stresses for deformation. One possibility is to include this effect into governing differential equation (e.g., see Swegle [3.2]). Even though technically interesting, the inclusion of deformation correction does not appear to be very significant for practical purposes.

Suppose S^{ij} is rotated by ω during a time increment, Δt. Then, the new stress deviator in the original coordinate system, S_0^{ij}, can be written as

$$S_0^{RR} = S^{RR}\cos^2\omega + S^{ZZ}\sin^2\omega + 2S^{RZ}\sin\omega\cos\omega, \quad (3.106)$$

$$S_0^{ZZ} = S^{RR}\sin^2\omega + S^{ZZ}\cos^2\omega - 2S^{RZ}\sin\omega\cos\omega, \tag{3.107}$$

and

$$S_0^{RZ} = S^{RZ}(\cos^2\omega - \sin^2\omega) - (S^{RR} - S^{ZZ})\sin\omega\cos\omega, \tag{3.108}$$

where

$$\sin\omega = \frac{\Delta t}{2}\left(\frac{\partial U}{\partial Z} - \frac{\partial V}{\partial R}\right). \tag{3.109}$$

Instead of including δ^{ij} to differential equation as in Eqs. (3.70)–(3.73) and (3.92)–(3.94), S^{ij} correction can be done to the resulting stresses at the end of each computational time step. Thus, it is convenient to define

$$\delta_0^{ij} = \Delta t \dot\delta^{ij} = S_0^{ij} - S^{ij}, \tag{3.110}$$

or

$$S_0^{ij} = S^{ij} + \delta_0^{ij}. \tag{3.111}$$

Then

$$\delta_0^{RR} = \frac{1}{2}(S^{RR} - S^{ZZ})(\cos 2\omega - 1) + S^{RZ}\sin 2\omega, \tag{3.112}$$

$$\delta_0^{ZZ} = -\delta_0^{RR}, \tag{3.113}$$

and

$$\delta_0^{RZ} = S^{RZ}(\cos 2\omega - 1) - \frac{1}{2}(S^{RR} - S^{ZZ})\sin 2\omega. \tag{3.114}$$

For a small ω, we may further simplify Eqs. (3.112)–(3.114) by

$$\sin 2\omega = 2\sin\omega = \Delta t\left(\frac{\partial U}{\partial Z} - \frac{\partial V}{\partial R}\right). \tag{3.115}$$

3.2.2.7 Derivation of Stresses Subjected to Rigid Body Rotation

In the previous section, the new deviatoric stresses after the rigid body rotation are given by Eqs. (3.106)–(3.108). Here we like to derive those equations using the force equilibrium principle, that is $\sum F_{x_1} = 0$ or $\sum F_{y_1} = 0$, where x_1 and y_1 are the rectangular coordinates after rotating an angle ω from the original x and y coordinates as shown in Fig. 3.7. Base on the equation of equilibrium, $\sum F_{x_1} = 0$, neglecting $\tau_{x_1 y_1}$ since it is

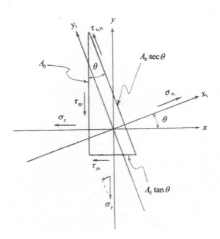

Fig. 3.7 New stresses subjected to rigid body rotation.

perpendicular to x_1, one gets

$$\sigma_{x_1} A_0 \sec\theta - \sigma_x A_0 \cos\theta - (\tau_{xy} A_0) \sec\theta$$
$$-\sigma_y(A_0 \tan\theta)\sin\theta - \tau_{yx}(A_0 \tan\theta)\cos\theta = 0. \quad (3.116)$$

For $\sum F_{y_1} = 0$, one only counts $\tau_{xy}, \tau_{x_1 y_1}, \tau_{yx}, \sigma_x$ and σ_y for the force balance excluding σ_{x_1} and it results

$$\tau_{x_1 y_1} A_0 \sec\theta + \sigma_x A_0 \sin\theta - (\tau_{xy} A_0)\cos\theta$$
$$-\sigma_y(A_0 \tan\theta)\cos\theta + \tau_{yx}(A_0 \tan\theta)\sin\theta = 0. \quad (3.117)$$

Simplifying Eqs. (3.116) and (3.117) with setting $\tau_{xy} = \tau_{yx}$, one gets

$$\sigma_{x_1} = \sigma_x \cos^2\theta + \sigma_y \sin^2\theta + 2\tau_{yx} \sin\theta\cos\theta, \quad (3.118)$$

and

$$\tau_{x_1 y_1} = -(\sigma_x - \sigma_y)\sin\theta\cos\theta + \tau_{xy}(\cos^2\theta - \sin^2\theta). \quad (3.119)$$

For deviatoric stresses, one can set $\sigma_{x_1} = S_0^{RR}$, $\sigma_x = S^{RR}$, $\sigma_y = S^{ZZ}$, $\tau_{xy} = S^{RZ}$ and $\tau_{x_1 y_1} = S_0^{RZ}$, then, Eqs. (3.118) and (3.119) become

$$S_0^{RR} = S^{RR} \cos^2\omega + S^{ZZ} \sin^2\omega + 2S^{RZ} \sin\omega\cos\omega, \quad (3.120)$$

and

$$S_0^{RZ} = S^{RZ}(\cos^2\omega - \sin^2\omega) - (S^{RR} - S^{ZZ})\sin\omega\cos\omega. \quad (3.121)$$

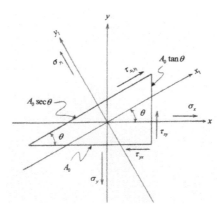

Fig. 3.8 New stresses subjected to rigid body rotation.

Equation (3.120) is identical to Eq. (3.106), while Eq. (3.121) is also the same as in Eq. (3.108). Now, we are going to derive Eq. (3.107) for S_0^{ZZ}. As shown in Fig. 3.8, the force equilibrium in the y_1 direction is $\sum F_{y_1} = 0$ which can be written as

$$\sigma_{y_1} A_0 \sec\theta - \sigma_x(A_0 \tan\theta)\sin\theta + \tau_{xy}(A_0 \tan\theta)\cos\theta$$
$$- \sigma_y(A_0 \cos\theta) + \tau_{yx}(A_0 \sin\theta) = 0. \quad (3.122)$$

Simplifying Eq. (3.122) with setting $\tau_{xy} = \tau_{yx}$, one gets

$$\sigma_{y_1} \sec\theta = \sigma_x \tan\theta \sin\theta + \sigma_y \cos\theta - \tau_{xy}\tan\theta \cos\theta - \tau_{yx}\sin\theta. \quad (3.123)$$

Dividing the above equation by $\sec\theta$, one gets

$$\sigma_{y_1} = \sigma_x \sin^2\theta + \sigma_y \cos^2\theta - 2\tau_{yx}\sin\theta \cos\theta. \quad (3.124)$$

Using the force equilibrium equation $\sum F_{x_1} = 0$, we have

$$\sigma_x A_0 \tan\theta \cos\theta - \sigma_y(A_0 \sin\theta) + \tau_{x_1 y_1}(A_0 \sec\theta)$$
$$+ \tau_{xy}(A_0 \tan\theta)\sin\theta - \tau_{yx}(A_0 \cos\theta) = 0, \quad (3.125)$$

or

$$\tau_{x_1 y_1} \sec\theta = -\sigma_x \tan\theta \cos\theta + \sigma_y \sin\theta - \tau_{xy}\tan\theta \sin\theta + \tau_{yx}\cos\theta, \quad (3.126)$$

or

$$\tau_{x_1 y_1} = -\sigma_x \sin\theta \cos\theta + \sigma_y \sin\theta \cos\theta - \tau_{xy}(\sin^2\theta - \cos^2\theta), \quad (3.127)$$

or
$$\tau_{x_1y_1} = \tau_{xy}(\cos^2\theta - \sin^2\theta) - (\sigma_x - \sigma_y)\sin\theta\cos\theta. \quad (3.128)$$
For deviatoric stresses, one can set $\sigma_{y_1} = S_0^{ZZ}$, $\sigma_x = S^{RR}$, $\sigma_y = S^{ZZ}$, $\tau_{xy} = S^{RZ}$ and $\tau_{x_1y_1} = S_0^{RZ}$, then, Eqs. (3.124) and (3.128) become
$$S_0^{ZZ} = S^{RR}\sin^2\omega + S^{ZZ}\cos^2\omega - 2S^{RZ}\sin\omega\cos\omega, \quad (3.129)$$
and
$$S_0^{RZ} = S^{RZ}(\cos^2\omega - \sin^2\omega) - (S^{RR} - S^{ZZ})\sin\omega\cos\omega. \quad (3.130)$$
Equation (3.129) is identical to Eq. (3.107), while Eq. (3.130) is also the same as in Eq. (3.108).

3.2.3 Spall

Material failure can be considered under a number of different conditions. However, under intense, short-duration tension loading, a particular type of fracture called "spall" is frequently produced. For high pressure compression problems, the material failure is likely to occur in this form. In spalling, small, independent cracks or voids are produced and usually an extensive crack propagation does not take place. The shape of void thus produced may depend on material structure. However, effects of crystallographic orientation on gross damage as well as the stress distribution relative to the shape of the void seem to be insignificant.

Considering these characteristics of spall and considering that our problems do not require much computation once a material is significantly fractured, a simplistic model seems to be adequate. Therefore, for the present code, the following criteria are used to check material failure and phase transition.

3.2.3.1 Melting

When $\epsilon >$ melt energy, the material is melted and subsequently is treated as a pure hydrodynamic fluid. Melt energy will be described in detail later.

3.2.3.2 Spall

When pressure gets down to a certain specified spall pressure ($P \leq P_{spall}$), the material is assumed to be spalled, as shown in Fig. 3.9. Once spalled, the material can not support any stress, and P and S^{ij} are set to be zero until recombined. For recombining, if $\tau \leq \tau_{spall}$, the material is regarded as recombined.

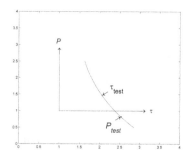

Fig. 3.9 The pressure-specific volume curve for spall and τ_{test}.

3.2.3.3 Fracture

Fracture criterion is based on tensile strength. If principal stresses, $\sigma_I >$ tensile strength, the material is assume to be fractured. A fractured cell is treated as spalled. For a more complete check, the compressive strength also has to be compared with principal stresses. However, for the class of problems we are solving, a material is probably melting before it reaches the compressive limit. Therefore, the fracture criteria based on compressive stress are not included. Computational detail for a spalled cell is given in Sections 3.3.5.2–3.3.6.3.

3.3 Computational Scheme

As briefly mentioned earlier, the second-order PIC, developed by Clark [3.9] in conjunction with hydrodynamic computation, is extended to elastic-plastic flow here.

3.3.1 General Discussion

One of the first questions in computing an elastic-plastic flow is where to define stresses. Wherever defined in a continuum, stresses are physically important. Depending on the choice of a difference scheme, a particular definition can be more convenient than others. For example, in a typical Lagrangian grid (Fig. 3.10), the velocity at point 1, 2, 3, and 4 are known. Then, it seems natural to assign S^{ij} at the cell center to solve Eqs. (3.5) or (3.7). In PIC method, however, it is debatable whether a cell-centered stress is more advantageous than other possibilities, such as defining S^{RZ} along cell boundaries. For bookkeeping purposes, a cell-centered S^{ij} would

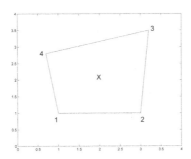

Fig. 3.10 A typical Lagrangian mesh with velocities defined at vertices 1, 2, 3, and 4.

be a very convenient choice. As will be shown, in re-mapping S^{ij} to original coordinate, a volume weighting scheme is second-order accurate, and a cell-centered S^{ij} is satisfactory.

A volume weighting of S^{ij} is essentially an interpolating scheme. Physically, it may be more sensible to balance forces on a small material element by defining S^{ij} on cell boundaries. However, since the force balance would require an interpolation scheme, conceptually it is not much different from the volume weighting.

3.3.2 Summary of Calculation Procedure

An operator splitting method is applied here. The governing set of cylindrical equation is split in radial and axial directions. Then, a separate calculation is performed in each direction using the predictor and the corrector scheme. The order of this calculation is alternated for each time advancement to maintain the accuracy of one-dimensional procedure. More detail description is given in the following Sections. A sequence of computational procedure is presented briefly below.

3.3.2.1 *Phase I: Lagrangian Phase*

Based on known quantities at nth time step (denoted by a superscript n), Lagrangian quantities at $(n+1)$th time step are computed here (denoted by a wiggle). The following steps are taken logically.

Step 1: Calculate P and S^{ij} at $(n+\frac{1}{2})$.
Step 2: Calculate $\widetilde{U}_i, \widetilde{\epsilon}_i$ using $P_i^{n+1/2}$ and $(S^{ij})_i^{n+1/2}$. Then, calculate $\widetilde{S}^{ij}$.
Step 3: Temporarily assign the momentum $(M_i U_i)$ and the internal energy $(M_i \epsilon_i)$ to cell i.

3.3.2.2 Phase II: Particle Transport and Remapping

In this phase, particle are transported with proper velocities, which will be described in Sections 3.3.5.1 and 3.3.5.2. Along with particle, portions of momentum, internal energy and stresses are transported. These transported quantities are then summed up for each cell and new quantities at $(n+1)$th time step are calculated.

Step 1: Move particles.
Step 2: Remap; the new cell mass, momentum, internal energy, and stresses are calculated; subsequently, the velocities and specific internal energy are determined.

3.3.3 Lagrangian Phase

In this Section, Lagrangian phase calculation is elaborated by equations in the radial direction after operator splitting. Similar equations can easily be obtained in the axial direction or in Cartesian coordinates by dropping radial terms.

3.3.3.1 Equations to be Solved in the Radial Direction

Mass

$$\rho_t = -\overline{U}\rho_R - \rho(U_R + \alpha\frac{U}{R}). \tag{3.131}$$

Momentum

$$U_t = -\overline{U}U_R + \frac{1}{\rho}[\sigma_R^{RR} + \frac{\alpha}{R}(\sigma^{RR} - \sigma^{\theta\theta})], \tag{3.132}$$

$$V_t = -\overline{U}V_R + \frac{1}{\rho}[\sigma_R^{RZ} + \frac{\alpha}{R}(\sigma^{RZ})]. \tag{3.133}$$

Energy

$$\epsilon_t = -\overline{U}\epsilon_R + \frac{1}{\rho}[\sigma^{RR}U_R + \frac{\alpha U}{R}(\sigma^{\theta\theta}) + \sigma^{RZ}V_R]. \tag{3.134}$$

Equation of State:
Pressure

$$P = f(\rho, \epsilon). \tag{3.135}$$

Elastic Regime

$$S_t^{RR} = -\overline{U}S_R^{RR} + 2G[U_R + \frac{1}{3\rho}(\rho_t + U\rho_R)], \tag{3.136}$$

$$= -\overline{U}S_R^{RR} + 2G[\frac{1}{3}(2U_R - \frac{\alpha U}{R})], \tag{3.137}$$

$$S_t^{ZZ} = -\overline{U}S_R^{ZZ} + 2G[\frac{1}{3\rho}(\rho_t + U\rho_R)], \tag{3.138}$$

$$= -\overline{U}S_R^{ZZ} + 2G[-\frac{1}{3}(U_R + \frac{\alpha U}{R})], \tag{3.139}$$

$$S_t^{RZ} = -\overline{U}S_R^{RZ} + GV_R, \tag{3.140}$$

and

$$S^{\theta\theta} = -(S^{RR} + S^{ZZ}). \tag{3.141}$$

Plastic Regime

$$S_t^{RR} = -\overline{U}S_R^{RR} + 2G[\frac{1}{3}(2U_R - \frac{\alpha U}{R})] - \frac{G\dot{W}}{\frac{1}{3}(Y^0)^2}S^{RR}, \tag{3.142}$$

$$S_t^{ZZ} = -\overline{U}S_R^{ZZ} + 2G[-\frac{1}{3}(U_R + \frac{\alpha U}{R})] - \frac{G\dot{W}}{\frac{1}{3}(Y^0)^2}S^{ZZ}, \tag{3.143}$$

$$S_t^{RZ} = -\overline{U}S_R^{RZ} + GV_R - \frac{G\dot{W}}{\frac{1}{3}(Y^0)^2}S^{RZ}, \tag{3.144}$$

$$S^{\theta\theta} = -(S^{RR} + S^{ZZ}), \tag{3.145}$$

and

$$\dot{W} = U_R S^{RR} + \frac{\alpha U}{R}S^{\theta\theta} + V_R S^{RZ}, \tag{3.146}$$

$$= (U_R - \frac{\alpha U}{R})S^{RR} - \frac{\alpha U}{R}S^{ZZ} + V_R S^{RZ}. \tag{3.147}$$

3.3.3.2 Derivation of the Equations Described in the Previous Section

From Eq. (3.1), we have

$$\frac{D\rho}{Dt} + \rho(\frac{\partial U}{\partial R} + \alpha\frac{U}{R} + \frac{\partial V}{\partial Z}) = 0. \qquad (3.148)$$

Expanding $\frac{D\rho}{Dt}$, one gets

$$\frac{\partial \rho}{\partial t} + U\frac{\partial \rho}{\partial R} + V\frac{\partial \rho}{\partial Z} + \rho(\frac{\partial U}{\partial R} + \alpha\frac{U}{R} + \frac{\partial V}{\partial Z}) = 0. \qquad (3.149)$$

For radial direction only, one can drop $\frac{\partial}{\partial Z}$ terms, therefore

$$\frac{\partial \rho}{\partial t} + U\frac{\partial \rho}{\partial R} + \rho(\frac{\partial U}{\partial R} + \alpha\frac{U}{R}) = 0, \qquad (3.150)$$

or

$$\rho_t = -\overline{U}\rho_R - \rho(U_R + \alpha\frac{U}{R}). \qquad (3.151)$$

The above equation is the same as Eq. (3.131).

Expanding $\frac{DU}{Dt}$ from the left-hand side of the momentum Eq. (3.2), one gets

$$\rho[\frac{\partial U}{\partial t} + U\frac{\partial U}{\partial R} + V\frac{\partial U}{\partial Z}] = \frac{\partial \sigma^{RR}}{\partial R} + \frac{\partial \sigma^{RZ}}{\partial Z} + \frac{\alpha}{R}(\sigma^{RR} - \sigma^{\theta\theta}). \qquad (3.152)$$

For the radial direction calculation only, we have

$$U_t = -\overline{U}\frac{\partial U}{\partial R} + \frac{1}{\rho}[\sigma_R^{RR} + \frac{\alpha}{R}(\sigma^{RR} - \sigma^{\theta\theta})]. \qquad (3.153)$$

Equation (3.153) is identical to Eq. (3.132). The axial direction momentum equation, i.e., Eq. (3.133), can be obtained by using the similar procedure for deriving Eq. (3.153).

For the internal energy, we start with Eq. (3.4) which is

$$\rho\frac{D\epsilon}{Dt} = \sigma^{RR}\frac{\partial U}{\partial R} + \alpha\sigma^{\theta\theta}\frac{U}{R} + \sigma^{ZZ}\frac{\partial V}{\partial Z} + \sigma^{RZ}(\frac{\partial U}{\partial Z} + \frac{\partial V}{\partial R}). \qquad (3.154)$$

By expanding the left hand side of Eq. (3.154) and discard the $\frac{\partial}{\partial Z}$ terms, we have

$$\rho(\frac{\partial \epsilon}{\partial t} + U\frac{\partial \epsilon}{\partial R}) = \sigma^{RR}\frac{\partial U}{\partial R} + \alpha\sigma^{\theta\theta}\frac{U}{R} + \sigma^{RZ}\frac{\partial V}{\partial R}, \qquad (3.155)$$

or

$$\epsilon_t = -\overline{U}\epsilon_R + \frac{1}{\rho}[\sigma^{RR}U_R + \alpha\frac{U}{R}\sigma^{\theta\theta} + \sigma^{RZ}V_R], \qquad (3.156)$$

which is the same as in Eq. (3.134).

For the deviatoric stresses in elastic regime, we start from Eq. (3.70) with expanding $\frac{DS^{RR}}{Dt}$ and set $\frac{\partial}{\partial Z} = 0$ which results in

$$\frac{\partial S^{RR}}{\partial t} + U\frac{\partial S^{RR}}{\partial R} = 2G[\frac{\partial U}{\partial R} + \frac{1}{3\rho}(\frac{\partial \rho}{\partial t} + U\frac{\partial \rho}{\partial R})] + \delta^{RR}, \qquad (3.157)$$

or

$$S_t^{RR} = -\overline{U}S_R^{RR} + 2G[U_R + \frac{1}{3\rho}(\rho_t + U\rho_R)] + \delta^{RR}. \qquad (3.158)$$

From Eq. (3.131), one can substituting $\rho_t + U\rho_R$ into Eq. (3.158) and obtains

$$S_t^{RR} = -\overline{U}S_R^{RR} + 2G[U_R + \frac{1}{3\rho}(-\rho)(U\rho_R + \alpha\frac{U}{R})] + \delta^{RR}, \qquad (3.159)$$

or

$$S_t^{RR} = -\overline{U}S_R^{RR} + 2G[\frac{1}{3}(2U_R + \alpha\frac{U}{R})] + \delta^{RR}, \qquad (3.160)$$

which is the same as Eq. (3.137). Equations (3.139) and (3.140) can be derived using similar procedure in obtaining Eq. (3.160).

For the plastic work, we start with Eq. (3.97) and set $\frac{\partial}{\partial Z} = 0$ which results in

$$\dot{W} = (\frac{\partial U}{\partial R})S^{RR} + \frac{\partial V}{\partial R}S^{RZ} + (\frac{\alpha U}{R})S^{\theta\theta}. \qquad (3.161)$$

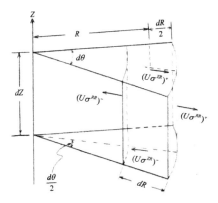

Fig. 3.11 Sign convention for normal and shear stresses.

or

$$\dot{W} = U_R S^{RR} + V_R S^{RZ} + (\frac{\alpha U}{R})(-S^{RR} - S^{ZZ}), \quad (3.162)$$

or

$$\dot{W} = (U_R - \frac{\alpha U}{R})S^{RR} - \frac{\alpha U}{R}S^{ZZ} + V_R S^{RZ}, \quad (3.163)$$

which is the same as Eq. (3.147).

3.3.3.3 Lagrangian Phase Calculation

Step 1: From Reynold's transport theorem [3.10] we calculate $\rho^{n+1/2}$ and $\epsilon^{n+1/2}$ as below

$$f^{n+1/2} = f^n + \frac{\delta t}{2}\frac{Df}{Dt} = f^n + \frac{\delta t}{2}(f_t + Uf_R), \quad (3.164)$$

and

$$P^{n+1/2} = g(\rho^{n+1/2}, \epsilon^{n+1/2}). \quad (3.165)$$

Step 2: Calculate $\widetilde{U}, \widetilde{\epsilon}$ and $\widetilde{S}^{ij}$. To visualize this Lagrangian phase calculation, let us consider the sketch of a cylindrical section as shown in Fig. (3.11). From Eq. (3.132), one gets

$$\frac{U^{n+1} - U^n}{\Delta t} = -\overline{U}U_R + \frac{1}{\rho}[\sigma_R^{RR} + \frac{\alpha}{R}(\sigma^{RR} - \sigma^{\theta\theta})], \quad (3.166)$$

or

$$U^{n+1} = U^n - \Delta t \overline{U} U_R + \frac{\Delta t}{\rho}[\sigma_R^{RR} + \frac{\alpha}{R}(\sigma^{RR} - \sigma^{\theta\theta})], \quad (3.167)$$

$$U^{n+1} = U^n - \Delta t \overline{U} U_R + \frac{\Delta t}{\rho}[\frac{(\sigma^{RR})^+ - (\sigma^{RR})^-}{\Delta R} + \frac{(\sigma^{RR})^+ + (\sigma^{RR})^-}{2R}]. \quad (3.168)$$

Therefore the contribution of σ^{RR} to U^{n+1} is

$$= \delta t\{[\frac{(\sigma^{RR})_R}{\rho}]^{n+1} + [\frac{\sigma^{RR}}{\rho R}]^{n+1}\}. \quad (3.169)$$

The contribution of σ^{RR} to $\widetilde{U}$ is

$$= \int_0^{\delta t} \frac{[(\sigma^{RR})^+(R + \frac{\Delta R}{2}) - (\sigma^{RR})^-(R - \frac{\Delta R}{2})]\Delta\theta\Delta Z}{(R\Delta R\Delta\theta\Delta Z)\rho} dt, \quad (3.170)$$

$$= \int_0^{\delta t} \{\frac{[(\sigma^{RR})^+ - (\sigma^{RR})^-]}{\rho \Delta R} + \frac{[(\sigma^{RR})^+ + (\sigma^{RR})^-]}{2\rho R}\} dt, \qquad (3.171)$$

$$= \delta t \{[\frac{(\sigma^{RR})_R}{\rho}]^{n+1/2} + [\frac{\sigma^{RR}}{\rho R}]^{n+1/2}\}. \qquad (3.172)$$

$\sigma^{\theta\theta}$ contribution to $\widetilde{U}$

$$= -\int_0^{\delta t} \frac{[(\sigma^{\theta\theta})^+ + (\sigma^{\theta\theta})^-]\Delta\theta \Delta R \Delta Z}{2(R\Delta R \Delta\theta \Delta Z)\rho} dt, \qquad (3.173)$$

$$= -\delta t [\frac{(\sigma^{\theta\theta})}{\rho R}]^{n+1/2}. \qquad (3.174)$$

Adding Eqs. (3.172) and (3.174), we get

$$\widetilde{U} = U^n + \delta t \{[\frac{(\sigma^{RR})_R}{\rho}]^{n+1/2} + \alpha[\frac{\sigma^{RR} - \sigma^{\theta\theta}}{\rho R}]^{n+1/2}\}. \qquad (3.175)$$

Now, let us consider $\widetilde{\epsilon}$.

$\sigma^{RR} U$ contribution to energy

$$= \int_0^{\delta t} \frac{[(\sigma^{RR})^+(R+\frac{\Delta R}{2})U^+ - (\sigma^{RR})^-(R-\frac{\Delta R}{2})U^-]\Delta\theta \Delta Z}{(R\Delta R \Delta\theta \Delta Z)\rho} dt, \qquad (3.176)$$

$$= \int_0^{\delta t} \{\frac{[(\sigma^{RR}U)^+ - (\sigma^{RR}U)^-]}{\rho \Delta R} + \frac{[(\sigma^{RR}U)^+ + (\sigma^{RR}U)^-]}{2\rho R}\} dt, \qquad (3.177)$$

$$= \delta t \{[\frac{(\sigma^{RR}U)_R}{\rho}]^{n+1/2} + [\frac{\sigma^{RR}U}{\rho R}]^{n+1/2}\}. \qquad (3.178)$$

The total net force due to $\sigma^{\theta\theta}$, i.e., $0 \leq \theta \leq 360°$, is zero. Therefore the energy contributed from $\sigma^{\theta\theta}$ is also zero since energy is equal to the force times the distance. Currently, we only deal with problems in the radial direction, the energy contribution from σ^{ZZ} can be discarded.

Then

$$\widetilde{\epsilon} + \frac{\widetilde{U}^2 + \widetilde{V}^2}{2} = \epsilon^n + \frac{(U^n)^2 + (V^n)^2}{2} + \delta t \{[\frac{(\sigma^{RR}U)_R}{\rho}]^{n+1/2}$$

$$+ [\frac{\sigma^{RR}U}{\rho R}]^{n+1/2} + [\frac{\sigma^{RZ}V}{\rho}]^{n+1/2}\}. \qquad (3.179)$$

Since we are doing radial-direction calculation, we can discard $\widetilde{V}^2$ and $(V^n)^2$.

Therefore Eq. (3.179) becomes

$$\tilde{\epsilon} = \epsilon^n - \frac{\tilde{U}^2 - (U^n)^2}{2} + \delta t \{[\frac{(\sigma^{RR}U)_R}{\rho}]^{n+1/2}$$

$$+ [\frac{\sigma^{RR}U}{\rho R}]^{n+1/2} + [\frac{\sigma^{RZ}V}{\rho}]^{n+1/2}\}, \qquad (3.180)$$

or

$$\tilde{\epsilon} = \epsilon^n - \frac{(\tilde{U}+U^n)(\tilde{U}-U^n)}{2} + \delta t \{[\frac{(\sigma^{RR}U)_R}{\rho}]^{n+1/2}$$

$$+ [\frac{\sigma^{RR}U}{\rho R}]^{n+1/2} + [\frac{\sigma^{RZ}V}{\rho}]^{n+1/2}\}, \qquad (3.181)$$

or

$$\tilde{\epsilon} = \epsilon^n - \overline{U}(\tilde{U} - U^n) + \delta t \{[\frac{(\sigma^{RR}U)_R}{\rho}]^{n+1/2}$$

$$+ [\frac{\sigma^{RR}U}{\rho R}]^{n+1/2} + [\frac{\sigma^{RZ}V}{\rho}]^{n+1/2}\}. \qquad (3.182)$$

Substituting $(\tilde{U} - U^n)$ from Eq. (3.175) into Eq. (3.182), one gets

$$\tilde{\epsilon} = \epsilon^n + \delta t \{[\frac{(\sigma^{RR}U)_R}{\rho}]^{n+1/2} + [\frac{\sigma^{RR}U}{\rho R}]^{n+1/2}\}$$

$$- \delta t \{[\frac{(\sigma^{RR})_R}{\rho}]^{n+1/2} + \alpha[\frac{\sigma^{RR} - \sigma^{\theta\theta}}{\rho R}]^{n+1/2}\}\overline{U} + [\frac{\sigma^{RZ}V}{\rho}]^{n+1/2}\}, \qquad (3.183)$$

or

$$\tilde{\epsilon} = \epsilon^n + \delta t \{[\frac{\sigma^{RR}(U)_R}{\rho}]^{n+1/2} + [\frac{\sigma^{\theta\theta}U}{\rho R}]^{n+1/2}\} + [\frac{\sigma^{RZ}V}{\rho}]^{n+1/2} + O(\delta t^2). \qquad (3.184)$$

Other quantities can be obtained similarly.

The Lagrangian phase calculation in the radial direction is summarized below

$$\tilde{U} = U^n + \delta t \{[\frac{(\sigma^{RR})_R}{\rho}]^{n+1/2} + \alpha[\frac{\sigma^{RR} - \sigma^{\theta\theta}}{\rho R}]^{n+1/2}\}, \qquad (3.185)$$

$$\widetilde{V} = V^n + \delta t \{ [\frac{(\sigma^{RZ})_R}{\rho}]^{n+1/2} + \alpha [\frac{\sigma^{RZ}}{\rho R}]^{n+1/2} \}, \tag{3.186}$$

and

$$\widetilde{\epsilon} = \epsilon^n + \delta t \{ [\frac{\sigma^{RR}(U)_R}{\rho}]^{n+1/2} + \alpha [\frac{\sigma^{\theta\theta} U}{\rho R}]^{n+1/2} + [\frac{\sigma^{RZ} V_R}{\rho}]^{n+1/2} \} + O(\delta t^2). \tag{3.187}$$

In the elastic regime

$$\widetilde{S}^{RR} = S^{RR} + \delta t \{ 2G[\frac{1}{3}(2U_R - \frac{\alpha U}{R})] \}^{n+1/2}, \tag{3.188}$$

$$\widetilde{S}^{ZZ} = S^{ZZ} + \delta t \{ 2G[-\frac{1}{3}(U_R + \frac{\alpha U}{R})] \}^{n+1/2}, \tag{3.189}$$

$$\widetilde{S}^{RZ} = S^{RZ} + \delta t (GV_R)^{n+1/2}. \tag{3.190}$$

In the plastic regime

$$(\widetilde{S}^{ij})_p = (\widetilde{S}^{ij})_e - \delta t [\frac{3G\dot{W}}{(Y^0)^2} S^{ij}]^{n+1/2}, \tag{3.191}$$

where $(\widetilde{S}^{ij})_e$ = elastic portion given by Eqs. (3.188)-(3.190). Equation (3.191) is not straightforward; more details will be given later. An alternative form of Eqs. (3.185)- (3.190) is convenient for programming. From Eq. (3.164), we have

$$(\frac{\sigma_R}{\rho})^{n+1/2} = (\frac{\sigma_R}{\rho})^n + \frac{\delta t}{2} \frac{D}{Dt}(\frac{\sigma_R}{\rho}), \tag{3.192}$$

and

$$\frac{[(\sigma)^{n+1/2}]_R}{\rho^n} = \frac{1}{\rho^n}(\sigma^n + \frac{\delta t}{2}\frac{D\sigma}{Dt})_R. \tag{3.193}$$

Subtracting Eq. (3.193) from Eq. (3.192), one gets

$$(\frac{\sigma_R}{\rho})^{n+1/2} - \frac{[(\sigma)^{n+1/2}]_R}{\rho^n}$$

$$= (\frac{\sigma_R}{\rho})^n + \frac{\delta t}{2}\frac{D}{Dt}(\frac{\sigma_R}{\rho}) - \frac{1}{\rho^n}(\sigma^n + \frac{\delta t}{2}\frac{D\sigma}{Dt})_R, \tag{3.194}$$

$$= \frac{\delta t}{2}[\frac{D}{Dt}(\frac{\sigma_R}{\rho}) - \frac{1}{\rho}(\frac{D\sigma}{Dt})_R], \tag{3.195}$$

$$= \frac{\delta t}{2}[(\frac{\sigma_R}{\rho})_t + U(\frac{\sigma_R}{\rho})_R - \frac{1}{\rho}(\sigma_t + U\sigma_R)_R], \quad (3.196)$$

$$= \frac{\delta t}{2}\{\frac{\sigma_R}{\rho^2}[U\rho_R + \rho(U_R + \alpha\frac{U}{R})] + \frac{\sigma_{tR}}{\rho} - \rho_R U\frac{\sigma_R}{\rho^2}$$

$$+U\frac{\sigma_{RR}}{\rho^2} - \frac{1}{\rho}(\sigma_{tR} + U_R\sigma_R + U\sigma_{RR})\}, \quad (3.197)$$

$$= \alpha\frac{\delta t}{2}(\frac{\sigma_R}{\rho}\frac{U}{R}). \quad (3.198)$$

Let $\Delta\sigma = \sigma^{RR} - \sigma^{\theta\theta}$, then

$$\alpha[(\frac{\Delta\sigma}{R\rho})^{n+1/2} - \frac{(\Delta\sigma)^{n+1/2}}{R\rho^n}]$$

$$= \alpha\{\frac{\delta t}{2}[(\frac{\Delta\sigma}{R\rho})_t + U(\frac{\Delta\sigma}{R\rho})_R] - (\frac{\delta t}{2})(\frac{1}{R\rho^n})[(\Delta\sigma)_t + U(\Delta\sigma)_R]\}, \quad (3.199)$$

$$= \alpha(\frac{\delta t}{2})\{\frac{1}{R}(\frac{\Delta\sigma}{\rho})_t + U(\frac{\Delta\sigma}{R\rho})_R - \frac{1}{R\rho}[(\Delta\sigma)_t + U(\Delta\sigma)_R]\}, \quad (3.200)$$

$$= \alpha(\frac{\delta t}{2})\{\frac{1}{R}\frac{1}{\rho^2}[\rho\frac{\partial(\Delta\sigma)}{\partial t} - (\Delta\sigma)\frac{\partial\rho}{\partial t}]$$

$$+U(\frac{1}{R\rho})^2[R\rho\frac{\partial(\Delta\sigma)}{\partial R} - (\Delta\sigma)\frac{\partial R\rho}{\partial R}] - (\frac{1}{R\rho})[\frac{\partial(\Delta\sigma)}{\partial t} + U\frac{\partial(\Delta\sigma)}{\partial R}]\}, \quad (3.201)$$

$$= \alpha(\frac{\delta t}{2})\{-\frac{1}{R\rho^2}(\Delta\sigma)\frac{\partial\rho}{\partial t} - \frac{U(\Delta\sigma)}{R^2\rho^2}[R\frac{\partial\rho}{\partial R} + \rho]\}, \quad (3.202)$$

$$= \alpha(\frac{\delta t}{2})\{-(\frac{\Delta\sigma}{R\rho^2})[\frac{\partial\rho}{\partial t} + U\frac{\partial\rho}{\partial R}] - \frac{U(\Delta\sigma)}{R^2\rho}\}, \quad (3.203)$$

$$= \alpha(\frac{\delta t}{2})(\frac{\Delta\sigma}{R\rho})\{-\frac{1}{\rho}[\frac{\partial\rho}{\partial t} + U\frac{\partial\rho}{\partial R}] - \frac{U}{R}\}, \quad (3.204)$$

$$= \alpha(\frac{\delta t}{2})(\frac{\Delta\sigma}{R\rho})\{-(\frac{1}{\rho})[-\rho(U_R + \alpha\frac{U}{R})] - \frac{U}{R}\}, \quad (3.205)$$

$$= \alpha(\frac{\delta t}{2})(\frac{\Delta\sigma}{R\rho})[(U_R + \alpha\frac{U}{R} - \frac{U}{R}). \quad (3.206)$$

For cylindrical coordinate, $\alpha = 1$, therefore, Eq. (3.206) becomes

$$= \alpha(\frac{\delta t}{2})(\frac{\Delta\sigma}{\rho R})U_R. \quad (3.207)$$

Using $\overline{U} = \frac{1}{2}(U^n + \widetilde{U}) = U^{n+1/2}$, one gets

$$(\frac{\sigma U_R}{\rho})^{n+1/2} - \frac{\sigma U^{n+1/2}\overline{U}_R}{\rho^n}$$

$$= \sigma^{n+1/2}\{(\frac{U_R}{\rho})^n + \frac{\delta t}{2}[(\frac{U_R}{\rho})_t + U(\frac{U_R}{\rho})_R] - (\frac{U_R}{\rho})^n - \frac{1}{\rho}\frac{\delta t}{2}(U_{Rt} + UU_R)\}, \quad (3.208)$$

$$= \sigma^{n+1/2}(\frac{\delta t}{2})\{\frac{1}{\rho^2}[\rho(\frac{\partial U_R}{\partial t}) - U_R(\frac{\partial \rho}{\partial t})]$$

$$+U(\frac{1}{\rho^2})[\rho\frac{\partial U_R}{\partial R} - U_R\frac{\partial \rho}{\partial R}] - \frac{1}{\rho^n}[\frac{\partial U_R}{\partial t} + U\frac{\partial U_R}{\partial R}]\}, \quad (3.209)$$

$$= \sigma^{n+1/2}(\frac{\delta t}{2})\{-\frac{U_R}{\rho^2}[\frac{\partial \rho}{\partial t} + U\frac{\partial \rho}{\partial R}]$$

$$+\frac{1}{\rho}[\frac{\partial U_R}{\partial t} + U\frac{\partial U_R}{\partial R}] - \frac{1}{\rho^n}[\frac{\partial U_R}{\partial t} + U\frac{\partial U_R}{\partial R}]\}. \quad (3.210)$$

Substituting $\frac{\partial \rho}{\partial t} + U\frac{\partial \rho}{\partial R}$ from Eq. (3.131) into Eq. (3.210) which becomes

$$= \sigma^{n+1/2}(\frac{\delta t}{2})\{(-\frac{U_R}{\rho^2})[-\rho(U_R + \alpha\frac{U}{R})]\}, \quad (3.211)$$

or

$$= \sigma^{n+1/2}(\frac{\delta t}{2})[\frac{\alpha U(U_R)}{\rho R}]. \quad (3.212)$$

In deriving Eq. (3.212), we use the fact that $U_R >> (U_R)^2$.
For the energy Eq. (3.187), we have

$$(\frac{\sigma U}{\rho R})^{n+1/2} - \frac{\sigma^{n+1/2}\overline{U}}{\rho R}$$

$$= \sigma^{n+1/2}U^{n+1/2}[(\frac{1}{\rho R})^{n+1/2} - (\frac{1}{\rho R})^n], \quad (3.213)$$

$$= \sigma^{n+1/2}U^{n+1/2}\{(\frac{1}{\rho R})^n + \frac{\delta t}{2}[(\frac{1}{\rho R})_t + U(\frac{1}{\rho R})_R] - (\frac{1}{\rho R})^n\}, \quad (3.214)$$

$$= \sigma^{n+1/2}U^{n+1/2}(\frac{\delta t}{2})\{(\frac{1}{\rho R})^2[-\frac{\partial(\rho R)}{\partial t}] + U(\frac{1}{\rho R})^2[-\frac{\partial(\rho R)}{\partial R}]\}, \quad (3.215)$$

$$= \sigma^{n+1/2} U^{n+1/2} (\frac{\delta t}{2})(\frac{-1}{(\rho R)^2})[\rho \frac{\partial R}{\partial t} + R\frac{\partial \rho}{\partial t} + U(\rho \frac{\partial R}{\partial R} + R\frac{\partial \rho}{\partial R})], \quad (3.216)$$

$$= -\sigma^{n+1/2} U^{n+1/2} (\frac{\delta t}{2})(\frac{1}{(\rho R)^2})[R(\frac{\partial \rho}{\partial t} + U\frac{\partial \rho}{\partial R}) + \rho(\frac{\partial R}{\partial t} + U)], \quad (3.217)$$

$$= -\sigma^{n+1/2} U^{n+1/2} (\frac{\delta t}{2})(\frac{1}{(\rho R)^2})[-\rho R(U_R + \alpha \frac{U}{R}) + 2U\rho], \quad (3.218)$$

$$= \sigma^{n+1/2} U^{n+1/2} (\frac{\delta t}{2})[-\frac{2U}{\rho R^2} + \frac{1}{\rho R}(U_R + \alpha \frac{U}{R})], \quad (3.219)$$

$$= \sigma^{n+1/2} U^{n+1/2} (\frac{\delta t}{2})(-\frac{2U}{\rho R^2} + \frac{\alpha U}{\rho R^2} + \frac{U_R}{\rho R}). \quad (3.220)$$

For cylindrical coordinate, $\alpha = 1$, therefore, Eq. (3.220) becomes

$$= \sigma^{n+1/2} U^{n+1/2} (\frac{\delta t}{2})(-\frac{U}{\rho R^2} + \frac{U_R}{\rho R}), \quad (3.221)$$

$$= \sigma^{n+1/2} U^{n+1/2} (\frac{\delta t}{2})(\frac{1}{\rho R})(U_R - \frac{U}{R}). \quad (3.222)$$

The shear stress term in Eq. (3.187) with setting $S = \sigma^{RZ}$ is

$$(\frac{SV_R}{\rho})^{n+1/2} - \frac{S^{n+1/2}\overline{V}_R}{\rho}$$

$$= (S)^{n+1/2}[(\frac{V_R}{\rho})^{n+1/2} - \frac{1}{\rho^n}V_R^{n+1/2}], \quad (3.223)$$

$$= (S)^{n+1/2}\{(\frac{V_R}{\rho})^n + (\frac{\delta t}{2})[(\frac{V_R}{\rho})_t + U(\frac{V_R}{\rho})_R]$$

$$- \frac{1}{\rho^n}[(V_R)^n + (\frac{\delta t}{2})(\frac{\partial V_R}{\partial t} + U\frac{\partial V_R}{\partial R})]\}, \quad (3.224)$$

$$= (S)^{n+1/2}(\frac{\delta t}{2})[(\frac{V_R}{\rho})_t + U(\frac{V_R}{\rho})_R - \frac{1}{\rho^n}(\frac{\partial V_R}{\partial t} + U\frac{\partial V_R}{\partial R})], \quad (3.225)$$

$$= (S)^{n+1/2}[(\frac{\delta t}{2})[(\frac{1}{\rho^2})(\rho \frac{V_R}{t} - V_R \frac{\partial \rho}{\partial t})$$

$$+ U(\frac{1}{\rho^2})(\rho \frac{\partial V_R}{\partial R} - V_R \frac{\partial \rho}{\partial R}) - \frac{1}{\rho}(\frac{\partial V_R}{\partial t} + U\frac{\partial V_R}{\partial R})], \quad (3.226)$$

$$= (S)^{n+1/2}(\frac{\delta t}{2})[(-\frac{V_R}{\rho^2})(\frac{\partial \rho}{\partial t} + U\frac{\partial \rho}{\partial R}) + (\frac{U}{\rho})(\frac{\partial V_R}{\partial R}) - (\frac{U}{\rho})(\frac{\partial V_R}{\partial R})], \quad (3.227)$$

$$= (S)^{n+1/2}(\frac{\delta t}{2})[(\frac{V_R}{\rho})(U_R + \alpha\frac{U}{R})], \qquad (3.228)$$

since $(\delta t)(U_R) << (\delta t)(\alpha\frac{U}{R})$, Equation (3.228) becomes

$$= \alpha\frac{\delta t}{2}\frac{(SU)V_R}{\rho R}. \qquad (3.229)$$

For the deviatoric stresses in the elastic regime, we have

$$(GU_R)^{n+1/2} - \overline{GU}_R$$

$$= (GU_R)^n + (\tfrac{\delta t}{2})[(GU_R)_t + U(GU_R)_R]$$

$$-[G^n + (\frac{\delta t}{2})(G_t + UG_R)][U^n + (\frac{\delta t}{2})(U_t + UU_R)]_R, \qquad (3.230)$$

$$= (\tfrac{\delta t}{2})[(GU_R)_t + U(GU_R)_R]$$

$$-[G^n(\frac{\delta t}{2})(U_t + UU_R) + U^n(\frac{\delta t}{2})(G_t + UG_R)], \qquad (3.231)$$

$$= (\frac{\delta t}{2})[(GU_R)_t + U(GU_R)_R] - [G(U_t + UU_R) + U(G_t + UG_R)]. \qquad (3.232)$$

Since $\frac{\partial G}{\partial t} = 0$ and $\frac{\partial G}{\partial R} = 0$, also neglecting the high order terms such as $\frac{\partial^2}{\partial t \partial R}$, therefore, Eq. (3.232) becomes

$$= (\frac{\delta t}{2})(-GU_RU_R) = -(\frac{\delta t}{2})G(U_R)^2 + O(\Delta^2). \qquad (3.233)$$

$$(\tfrac{GU}{R})^{n+1/2} - \overline{\tfrac{GU}{R}}$$

$$= (\tfrac{GU}{R}) + (\tfrac{\delta t}{2})[(\tfrac{GU}{R})_t + U((\tfrac{GU}{R})_R]$$

$$-(\frac{1}{R})[G + (\frac{\delta t}{2})(G_t + UG_R)][U + (\frac{\delta t}{2})(U_t + UU_R)], \qquad (3.234)$$

$$= (\tfrac{GU}{R}) + (\tfrac{\delta t}{2})[(\tfrac{GU}{R})_t + U(\tfrac{GU}{R})_R]$$

$$-(\frac{1}{R})(GU) - (\frac{1}{R})[G(\frac{\delta t}{2})(U_t + UU_R) + (\frac{\delta t}{2})(G_t + UG_R)(U)], \qquad (3.235)$$

$$= (\frac{\delta t}{2})[(\tfrac{GU}{R})_t + U(\tfrac{GU}{R})_R - (\frac{1}{R})G(U_t + UU_R) - (\frac{1}{R})U(G_t + UG_R)], \qquad (3.236)$$

$$= (\frac{\delta t}{2})[(\frac{GU}{R})_t + U(\frac{GU}{R})_R - (\frac{1}{R})(UG_t + U^2 G_R + GU_t + GUU_R), \quad (3.237)$$

$$= (\tfrac{\delta t}{2})[(\tfrac{GU}{R})_t + \tfrac{U}{R^2}(R\tfrac{\partial GU}{\partial R} - GU)$$

$$-(\frac{1}{R})(UG_t + U^2 G_R + GU_t + GUU_R). \quad (3.238)$$

In Eq. (3.238), the only important term is $-GU$, other terms are all in high order derivatives, therefore, it becomes

$$= (\frac{\delta t}{2})(-\frac{GU^2}{R^2}) = -(\frac{\delta t}{2})(\frac{GU^2}{R^2}) + O(\Delta^2). \quad (3.239)$$

From Eq. (3.185) we have

$$\tilde{U} = U^n + \delta t \{ [\frac{(\sigma^{RR})_R}{\rho}]^{n+1/2} + \alpha [\frac{\sigma^{RR} - \sigma^{\theta\theta}}{\rho R}]^{n+1/2} \}. \quad (3.240)$$

On the right hand side of Eq. (3.240), substituting $[\frac{(\sigma^{RR})_R}{\rho}]^{n+1/2}$ by Eq. (3.198) and $\alpha[\frac{\sigma^{RR}-\sigma^{\theta\theta}}{\rho R}]^{n+1/2}$ by Eq. (3.207), one gets

$$\tilde{U} = U^n + \tfrac{\delta t}{\rho^n}[(\sigma^{RR})_R^{n+1/2} + \tfrac{\alpha}{R}(\Delta\sigma)^{n+1/2}]$$

$$+\frac{\delta t}{\rho^n}[\frac{\alpha \delta t}{2}(\sigma_R^{RR}\frac{U}{R} + \frac{\Delta\sigma U_R}{R})]. \quad (3.241)$$

From Eq. (3.186) we have

$$\tilde{V} = V^n + \delta t \{ [\frac{(\sigma^{RZ})_R}{\rho}]^{n+1/2} + \alpha [\frac{\sigma^{RZ}}{\rho R}]^{n+1/2} \}. \quad (3.242)$$

On the right hand side of Eq. (3.242), substituting $[\frac{(\sigma^{RZ})_R}{\rho}]^{n+1/2}$ by Eq. (3.198) and $\alpha[\frac{\sigma^{RZ}}{\rho R}]^{n+1/2}$ by Eq. (3.207), one gets

$$\tilde{V} = V^n + \tfrac{\delta t}{\rho^n}[(\sigma^{RZ})_R^{n+1/2} + \tfrac{\alpha}{R}(\sigma^{RZ})^{n+1/2}]$$

$$+\frac{\delta t}{\rho^n}[\frac{\alpha \delta t}{2}(\sigma_R^{RZ}\frac{U}{R} + \frac{\sigma^{RZ} U_R}{R})]. \quad (3.243)$$

From Eq. (3.187) we have

$$\tilde{\epsilon} = \epsilon^n + \delta t \{ [\frac{\sigma^{RR}(U)_R}{\rho}]^{n+1/2} + \alpha[\frac{\sigma^{\theta\theta}U}{\rho R}]^{n+1/2} + [\frac{\sigma^{RZ}V_R}{\rho}]^{n+1/2} \} + O(\delta t^2). \quad (3.244)$$

On the right hand side of Eq. (3.244), substituting $[\frac{\sigma^{RR}(U)_R}{\rho}]^{n+1/2}$ by Eq. (3.212), $\alpha[\frac{\sigma^{\theta\theta}U}{\rho R}]^{n+1/2}$ by Eq. (3.222) and $[\frac{\sigma^{RZ}V_R}{\rho}]^{n+1/2}$ by Eq. (3.229), one gets

$$\widetilde{\epsilon} = \epsilon^n + \frac{\delta t}{\rho^n}[(\sigma^{RR})^{n+1/2}\overline{U}_R + \frac{\alpha}{R}(\sigma^{\theta\theta})^{n+1/2}\overline{U} + (\sigma^{RZ})^{n+1/2}\overline{V}_R]$$

$$+\frac{\delta t}{\rho^n}[\frac{\alpha\delta t}{2}(\sigma^{RR}U_R\frac{U}{R} + \frac{\sigma^{\theta\theta}U}{R}(U_R - \frac{U}{R}) + \frac{\sigma^{RZ}V_R U}{R})], \qquad (3.245)$$

or

$$\widetilde{\epsilon} = \epsilon^n + \frac{\delta t}{\rho^n}[(\sigma^{RR})^{n+1/2}\frac{\partial(R\overline{U})}{R\partial R} - \frac{\overline{U}}{R}(\sigma^{RR} - \sigma^{\theta\theta})^{n+1/2} + (\sigma^{RZ})^{n+1/2}\overline{V}_R]$$

$$+\frac{\delta t}{\rho^n}[\frac{\alpha\delta t}{2}(\sigma^{RR}U_R\frac{U}{R} + \frac{\sigma^{\theta\theta}U}{R}(U_R - \frac{U}{R}) + \frac{\sigma^{RZ}V_R U}{R})]. \qquad (3.246)$$

In the elastic regime, from Eq. (3.188), we have

$$\widetilde{S}^{RR} = S^{RR} + \delta t\{2G[\frac{1}{3}(2U_R - \frac{\alpha U}{R})]\}^{n+1/2}, \qquad (3.247)$$

or

$$\widetilde{S}^{RR} = S^{RR} + \delta t\{\frac{2}{3}[2GU_R - \frac{\alpha GU}{R}]\}^{n+1/2}. \qquad (3.248)$$

From Eq. (3.233), we have

$$(GU_R)^{n+1/2} = -(\frac{\delta t}{2})G(U_R)^2 + \overline{GU}_R. \qquad (3.249)$$

From Eq. (3.239), we have

$$(\frac{GU}{R})^{n+1/2} = -(\frac{\delta t}{2})(\frac{GU^2}{R^2}) + \frac{\overline{GU}}{R}. \qquad (3.250)$$

Substituting Eqs. (3.249) and (3.250) into (3.248), one gets

$$\widetilde{S}^{RR} = S^{RR} + \frac{2}{3}\delta t[-\delta t G(U_R)^2 + 2\overline{GU}_R + \frac{\alpha(\delta t)GU^2}{2R^2} - \frac{\alpha\overline{GU}}{R}]. \qquad (3.251)$$

Sine $G^n = \overline{G}$, Eq. (3.251) becomes

$$\widetilde{S}^{RR} = S^{RR} + \frac{2}{3}\delta t\overline{G}[-\delta t(U_R)^2 + 2\overline{U}_R + \frac{\alpha(\delta t)U^2}{R^2} - \frac{\alpha\overline{U}}{R}], \qquad (3.252)$$

or

$$\widetilde{S}^{RR} = S^{RR} + (\delta t)\{\frac{2\overline{G}}{3}[2\overline{U}_R(1 - \frac{U_R\delta t}{2}) - (\frac{\alpha\overline{U}}{R})(1 - \frac{U\delta t}{2R})]\}. \qquad (3.253)$$

In deriving Eq. (3.253), we assume that $\overline{U}_R = U_R$ and $U^2 = U \cdot \overline{U}$.
From Eq. (3.189), we have

$$\widetilde{S}^{ZZ} = S^{ZZ} + \delta t \{2G[-\frac{1}{3}(U_R + \frac{\alpha U}{R})]\}^{n+1/2}. \qquad (3.254)$$

Substituting $(GU_R)^{n+1/2}$ from Eq. (3.249) and $(\frac{GU}{R})^{n+1/2}$ from Eq. (3.250) into the right hand side of Eq. (3.254), we have

$$\widetilde{S}^{ZZ} = S^{ZZ} + \frac{2}{3}(\delta t)[(\frac{\delta t}{2})(G)(U_R)^2 - \overline{GU}_R - (\frac{\alpha \delta t}{2})(\frac{GU^2}{R^2}) + \frac{\alpha \overline{GU}}{R}], \qquad (3.255)$$

$$= S^{ZZ} + (\delta t)\{-\frac{2}{3}\overline{G}[-(\frac{\delta t}{2})(U_R)^2 + \overline{U}_R + (\frac{\alpha \delta t}{2})(\frac{U^2}{R^2}) - \frac{\alpha \overline{GU}}{R}]\}, \qquad (3.256)$$

$$= S^{ZZ} + (\delta t)\{-\frac{2}{3}\overline{G}[\overline{U}_R(1 - \frac{U_R \delta t}{2}) + \frac{\alpha \overline{U}}{R}(1 - \frac{U \delta t}{2R})]\}. \qquad (3.257)$$

In deriving Eq. (3.257), we assume that $\overline{U}_R^2 = (U_R)(\overline{U}_R)$ and $U^2 = U \cdot \overline{U}$.
From Eq. (3.190), we have

$$\widetilde{S}^{RZ} = S^{RZ} + \delta t (GV_R)^{n+1/2}, \qquad (3.258)$$

and from Eq. (3.249), we have

$$(GV_R)^{n+1/2} = -(\frac{\delta t}{2})G(V_R)^2 + \overline{GV}_R. \qquad (3.259)$$

By substituting Eq. (3.259) into Eq. (3.258), one gets

$$\widetilde{S}^{RZ} = S^{RZ} + \delta t \{\overline{GV}_R[1 - \frac{(\delta t)V_R}{2}]\}. \qquad (3.260)$$

Again, in deriving Eq. (3.260), we assume that $(\overline{V}_R^n)^2 = (V_R^n)(\overline{V}_R)$ and $G^n = \overline{G}$.

3.3.3.4 Summary of Equations in the Axial Direction

Equations to be solved:
Mass

$$\rho_t = -\overline{V}\rho_Z - \rho V_Z. \qquad (3.261)$$

Momentum

$$U_t = -\overline{V}U_Z + \frac{\sigma_Z^{RZ}}{\rho}, \qquad (3.262)$$

$$V_t = -\overline{V}V_Z + \frac{\sigma_Z^{ZZ}}{\rho}. \qquad (3.263)$$

Energy

$$\epsilon_t = -\overline{V}\epsilon_Z + \frac{1}{\rho}(\sigma^{ZZ}V_Z + \sigma^{RZ}U_Z). \quad (3.264)$$

Equation of State:
Pressure

$$P = f(\rho, \epsilon). \quad (3.265)$$

Elastic Regime

$$S_t^{RR} = -\overline{V}S_Z^{RR} + 2G[\frac{1}{3\rho}(\rho_t + V\rho_Z)], \quad (3.266)$$

$$= -\overline{V}S_Z^{RR} + (\frac{2GV_Z}{3}), \quad (3.267)$$

$$S_t^{ZZ} = -\overline{V}S_Z^{ZZ} + 2G[V_Z + \frac{1}{3\rho}(\rho_t + V\rho_Z)], \quad (3.268)$$

$$= -\overline{V}S_Z^{ZZ} + \frac{4GV_Z}{3}, \quad (3.269)$$

$$S_t^{RZ} = -\overline{V}S_Z^{RZ} + GU_Z, \quad (3.270)$$

and

$$S^{\theta\theta} = -(S^{RR} + S^{ZZ}). \quad (3.271)$$

Plastic Regime

$$(S^{ij})_p = (S^{ij})_e - [\frac{3G\dot{W}}{(Y^0)^2}S^{ij}], \quad (3.272)$$

where $(S^{ij})_e$ = elastic component.

3.3.3.5 Derivation of the Equations Described in the Previous Section

From the mass conservation Eq. (3.1), we have

$$\frac{D\rho}{Dt} + \rho(\frac{\partial U}{\partial R} + \alpha\frac{U}{R} + \frac{\partial V}{\partial Z}) = 0, \quad (3.273)$$

or

$$\frac{\partial \rho}{\partial t} + U\frac{\partial \rho}{\partial R} + V\frac{\partial \rho}{\partial Z} + \rho(\frac{\partial U}{\partial R} + \alpha\frac{U}{R} + \frac{\partial V}{\partial Z}) = 0, \quad (3.274)$$

therefore, in the Z-direction only, we have

$$\frac{\partial \rho}{\partial t} + V\frac{\partial \rho}{\partial Z} + \rho\frac{\partial V}{\partial Z} = 0, \quad (3.275)$$

or
$$\rho_t = -\overline{V}\rho_Z - \rho V_Z. \tag{3.276}$$

From Eqs. (3.2) and (3.5), one gets

$$\rho(\frac{\partial U}{\partial t} + U\frac{\partial U}{\partial R} + V\frac{\partial U}{\partial Z}) = \frac{\partial \sigma^{RR}}{\partial R} + \frac{\partial \sigma^{RZ}}{\partial Z} + \frac{\alpha}{R}(\sigma^{RR} - \sigma^{\theta\theta}). \tag{3.277}$$

From Eq. (3.132), one gets

$$\rho U \frac{\partial U}{\partial R} = \frac{\partial \sigma^{RR}}{\partial R} + \frac{\alpha}{R}(\sigma^{RR} - \sigma^{\theta\theta}). \tag{3.278}$$

Subtracting Eq. (3.278) from Eq. (3.277), we have

$$U_t = -\overline{V}U_Z + \frac{(\sigma^{RZ})_Z}{\rho}, \tag{3.279}$$

From Eq. (3.3) and (3.5), one gets

$$\rho(\frac{\partial V}{\partial t} + U\frac{\partial V}{\partial R} + V\frac{\partial V}{\partial Z}) = \frac{\partial \sigma^{ZZ}}{\partial Z} + \frac{\partial \sigma^{RZ}}{\partial R} + \frac{\alpha}{R}\sigma^{RZ}. \tag{3.280}$$

From Eq. (3.133), we have

$$\rho U \frac{\partial V}{\partial R} = \frac{\partial \sigma^{RZ}}{\partial R} + \frac{\alpha}{R}(\sigma^{RZ}). \tag{3.281}$$

Subtracting Eq. (3.281) from Eq. (3.280), one gets

$$V_t = -\overline{V}V_Z + \frac{(\sigma^{ZZ})_Z}{\rho}. \tag{3.282}$$

From Eqs. (3.4) and (3.5), we have

$$\rho(\frac{\partial \epsilon}{\partial t} + U\frac{\partial \epsilon}{\partial R} + V\frac{\partial \epsilon}{\partial Z}) = \sigma^{RR}\frac{\partial U}{\partial R} + \alpha\sigma^{\theta\theta}\frac{U}{R} + \sigma^{ZZ}\frac{\partial V}{\partial Z} + \sigma^{RZ}(\frac{\partial U}{\partial Z} + \frac{\partial V}{\partial R}), \tag{3.283}$$

From Eq. (3.134), one gets

$$\rho(U\frac{\partial \epsilon}{\partial R}) = \sigma^{RR}\frac{\partial U}{\partial R} + \alpha\sigma^{\theta\theta}\frac{U}{R} + \sigma^{RZ}(\frac{\partial V}{\partial R}), \tag{3.284}$$

Subtracting Eq. (3.284) from Eq. (3.283), one gets

$$\epsilon_t = -\overline{V}\epsilon_Z + \frac{1}{\rho}(\sigma^{ZZ}V_Z + \sigma^{RZ}U_Z). \tag{3.285}$$

For the Hooke's law in the elastic regime, we start with Eq. (3.70) which is

$$\frac{\partial S^{RR}}{\partial t} + U\frac{\partial S^{RR}}{\partial R} + V\frac{\partial S^{RR}}{\partial Z} = 2G(\frac{\partial U}{\partial R} + \frac{1}{3\rho}\frac{D\rho}{Dt}) + \delta^{RR}. \tag{3.286}$$

From Eq. (3.136), we can write

$$U\frac{\partial S^{RR}}{\partial R} = 2G(\frac{\partial U}{\partial R}) + (2G)(\frac{1}{3\rho})(\rho_t + U\frac{\partial \rho}{\partial R}). \quad (3.287)$$

Subtracting Eq. (3.287) from Eq. (3.286), we get

$$\frac{\partial S^{RR}}{\partial t} = -V\frac{\partial S^{RR}}{\partial Z} + \frac{2G}{3\rho}(\rho_t + V\frac{\partial \rho}{\partial Z}). \quad (3.288)$$

Substituting Eq. (3.276) into Eq. (3.288), we have

$$S_t^{RR} = -\overline{V}S_Z^{RR} - \frac{2G}{3}(V_Z). \quad (3.289)$$

Using the same procedure with Eqs. (3.71), (3.138) and (3.276), we will get

$$S_t^{ZZ} = -\overline{V}S_Z^{ZZ} + \frac{4G}{3}(V_Z). \quad (3.290)$$

Again, by using the same procedure with Eqs. (3.72) and (3.140), one will get

$$S_t^{RZ} = -\overline{V}S_Z^{RZ} + G(U_Z), \quad (3.291)$$

and from Eq. (3.74), we have

$$S^{\theta\theta} = -(S^{RR} + S^{ZZ}). \quad (3.292)$$

Plastic Regime

$$(S^{ij})_p = (S^{ij})_e - \delta t[\frac{3G\dot{W}}{(Y^0)^2}S^{ij}], \quad (3.293)$$

where $(S^{ij})_e$ = elastic component.

3.3.3.6 Lagrangian Phase in the Axial Direction

In this section, we describe the Lagrangian phase, that means discarding the advection terms, for solving the pertinent equations in the axial direction. The momentum Eq. (3.262) becomes

$$\tilde{U} = U^n + \delta t(\frac{\sigma_Z^{RZ}}{\rho})^{n+1/2}, \quad (3.294)$$

and from Eq. (3.198), we have

$$(\frac{\sigma_R}{\rho})^{n+1/2} = \alpha\frac{\delta t}{2}(\frac{\sigma_R}{\rho}\frac{U}{R}) + \frac{[(\sigma)^{n+1/2}]_R}{\rho^n}. \quad (3.295)$$

Since we are dealing with axial direction, there is no motion in the radial direction, therefore, $U = 0$, and Eq. (3.295) becomes

$$(\frac{\sigma_R}{\rho})^{n+1/2} = \frac{[(\sigma)^{n+1/2}]_R}{\rho^n}. \qquad (3.296)$$

Substituting Eq. (3.296) into Eq. (3.294), one gets

$$\tilde{U} = U^n + \delta t \frac{(\sigma_Z^{RZ})^{n+1/2}}{\rho^n}. \qquad (3.297)$$

For the same reason, we have

$$\tilde{V} = V^n + \delta t \frac{(\sigma_Z^{ZZ})^{n+1/2}}{\rho^n}. \qquad (3.298)$$

Now, for the internal energy, Eq. (3.264) can be written

$$\tilde{\epsilon} = \epsilon^n + \delta t (\frac{\sigma^{ZZ} V_Z}{\rho} + \frac{\sigma^{RZ} U_Z}{\rho})^{n+1/2}, \qquad (3.299)$$

or

$$\tilde{\epsilon} = \epsilon^n + \delta t [\overline{V}_Z (\frac{\sigma^{ZZ}}{\rho})^{n+1/2} + \overline{U}_Z (\frac{\sigma^{RZ}}{\rho})^{n+1/2}], \qquad (3.300)$$

or

$$\tilde{\epsilon} = \epsilon^n + \delta t [\frac{\overline{V}_Z}{\rho^n} (\sigma^{ZZ})^{n+1/2} + \frac{\overline{U}_Z}{\rho^n} (\sigma^{RZ})^{n+1/2}], \qquad (3.301)$$

or

$$\tilde{\epsilon} = \epsilon^n + \frac{\delta t}{\rho^n} [\overline{V}_Z (\sigma^{ZZ})^{n+1/2} + \overline{U}_Z (\sigma^{RZ})^{n+1/2}], \qquad (3.302)$$

and the pressure is

$$P^{n+1/2} = g(\rho^{n+1/2}, \epsilon^{n+1/2}). \qquad (3.303)$$

In the elastic regime, Eq. (3.267) can be written as

$$\tilde{S}^{RR} = (S^{RR})^n - \frac{2\delta t}{3} (GV_Z)^{n+1/2}, \qquad (3.304)$$

and from Eq. (3.233), we have

$$(GV_Z)^{n+1/2} - \overline{GV}_Z = -(\frac{\delta t}{2}) \overline{G} (\overline{V}_Z)^2, \qquad (3.305)$$

or

$$(GV_Z)^{n+1/2} = \overline{GV}_Z - (\frac{\delta t}{2}) \overline{G} (\overline{V}_Z)^2, \qquad (3.306)$$

or
$$(GV_Z)^{n+1/2} = \overline{GV}_Z(1 - \frac{\delta t}{2}\overline{V}_Z). \tag{3.307}$$
Substituting Eq. (3.307) into Eq. (3.304), one gets
$$\widetilde{S}^{RR} = (S^{RR})^n - \frac{2\delta t}{3}\overline{GV}_Z(1 - \frac{\delta t}{2}\overline{V}_Z). \tag{3.308}$$
Applying the same procedure in obtaining Eq. (3.308) to Eq. (3.269), one can have
$$\widetilde{S}^{ZZ} = (S^{ZZ})^n + \frac{4\delta t}{3}(GV_Z)^{n+1/2}, \tag{3.309}$$
or
$$\widetilde{S}^{ZZ} = (S^{ZZ})^n + \frac{4\delta t}{3}\overline{GV}_Z(1 - \frac{\delta t}{2}\overline{V}_Z). \tag{3.310}$$
Applying the same procedure in obtaining Eq. (3.308) to Eq. (3.270), we have
$$\widetilde{S}^{RZ} = (S^{RZ})^n + (\delta t)(GU_Z)^{n+1/2}, \tag{3.311}$$
or
$$\widetilde{S}^{RZ} = (S^{RZ})^n + (\delta t)\overline{GU}_Z(1 - \frac{\delta t}{2}\overline{U}_Z). \tag{3.312}$$
In the plastic regime
$$(\widetilde{S}^{ij})_p = (\widetilde{S}^{ij})_e - \delta t[\frac{3G\dot{W}}{(Y^0)^2}S^{ij}]^{n+1/2}, \tag{3.313}$$
where $(\widetilde{S}^{ij})_e$ = elastic portion given by Eqs. (3.266)-(3.271).

Since we are doing the axial-direction calculation, therefore, $U = 0$, and we can have
$$(\frac{\sigma_Z}{\rho})^{n+1/2} = \frac{(\sigma)_Z^{n+1/2}}{\rho^n}, \tag{3.314}$$
also
$$(\frac{\sigma V_Z}{\rho})^{n+1/2} = \frac{(\sigma V_Z)^{n+1/2}}{\rho^n}. \tag{3.315}$$
From Eq. (3.233) we can write

$$(GU_Z)^{n+1/2} - \overline{GU}_Z$$
$$= (GU_Z)^n + (\tfrac{\delta t}{2})[(GU_Z)_t + V(GU_Z)_Z]$$
$$- [G^n + (\frac{\delta t}{2})(G_t + VG_Z)][U^n + (\frac{\delta t}{2})(U_t + VU_Z)]_Z, \tag{3.316}$$
or
$$(GU_Z)^{n+1/2} - \overline{GU}_Z = -(\frac{\delta t}{2})(GU_ZV_Z) + O(\Delta^2). \tag{3.317}$$
Using the same procedure in obtaining Eq. (3.317), one can get
$$(GV_Z)^{n+1/2} - \overline{GV}_Z = -(\frac{\delta t}{2})(GV_ZV_Z) + O(\Delta^2). \tag{3.318}$$

3.3.4 Stress Calculation in the Plastic Regime of Flow

Since the Prandtl–Reuss equation of plastic flow is implicit in stresses, the solution procedure is not straightforward. An explicit method, consistent with the present second-order PIC, is described here in some detail for Lagrangian phase computation.

3.3.4.1 Radial Direction

Rewriting Eqs. (3.142)-(3.145) in tensor form, one gets

$$S_t^{ij} = -\overline{U} S_R^{ij} + 2G(\dot{e})^{ij} - A\dot{W} S^{ij}, \quad (3.319)$$

where $\dot{W} = S^{ij}(\dot{e})^{ij}$, and

$$(\dot{e})^{ij} = \begin{bmatrix} (\dot{e})^{RR} & (\dot{e})^{RZ} & (\dot{e})^{R\theta} \\ (\dot{e})^{ZR} & (\dot{e})^{ZZ} & (\dot{e})^{Z\theta} \\ (\dot{e})^{\theta R} & (\dot{e})^{\theta Z} & (\dot{e})^{\theta\theta} \end{bmatrix}, \quad (3.320)$$

or

$$(\dot{e})^{ij} = \begin{bmatrix} U_R + \frac{1}{3\rho}(\rho_t + U\rho_R) & \frac{V_R}{2} & 0 \\ \frac{V_R}{2} & \frac{1}{3\rho}(\rho_t + U\rho_R) & 0 \\ 0 & 0 & \frac{\alpha U}{R} + \frac{1}{3\rho}(\rho_t + U\rho_R) \end{bmatrix}, \quad (3.321)$$

or

$$(\dot{e})^{ij} = \begin{bmatrix} \frac{1}{3}(2U_R - \frac{\alpha U}{R}) & \frac{V_R}{2} & 0 \\ \frac{V_R}{2} & -\frac{1}{3}(U_R - \frac{\alpha U}{R}) & 0 \\ 0 & 0 & -\frac{1}{3}(U_R - \frac{\alpha U}{R}) \end{bmatrix}, \quad (3.322)$$

and

$$A = \frac{3G}{(Y^0)^2}. \quad (3.323)$$

From Eq. (3.76), we have

$$(\dot{e})^{RR} = \frac{\partial U}{\partial R} + \frac{1}{3\rho}\frac{D\rho}{Dt}. \quad (3.324)$$

Since we are dealing with radial direction only, Eq. (3.324) becomes

$$(\dot{e})^{RR} = U_R + \frac{1}{3\rho}(\rho_t + U\rho_R), \qquad (3.325)$$

which is the same as $(\dot{e})^{RR}$ term in Eq. (3.321).

From Eq. (3.76), we have

$$(\dot{e})^{ZZ} = \frac{\partial V}{\partial Z} + \frac{1}{3\rho}\frac{D\rho}{Dt}. \qquad (3.326)$$

Again, for radial direction only, we can set $V = 0$, therefore, Eq. (3.326) becomes

$$(\dot{e})^{ZZ} = \frac{1}{3\rho}(\rho_t + U\rho_R), \qquad (3.327)$$

which is identical to $(\dot{e})^{ZZ}$ term in Eq. (3.321).

For the strain rate deviator in the θ direction, we have

$$(\dot{e})^{\theta\theta} = -[(\dot{e})^{RR} + (\dot{e})^{ZZ}], \qquad (3.328)$$

or

$$(\dot{e})^{\theta\theta} = -[U_R + \frac{1}{3\rho}(\rho_t + U\rho_R) + \frac{1}{3\rho}(\rho_t + U\rho_R)], \qquad (3.329)$$

or

$$(\dot{e})^{\theta\theta} = -[U_R + \frac{2}{3\rho}(\rho_t + U\rho_R)]. \qquad (3.330)$$

Now, from Eq. (3.131), we have

$$\rho_t + U\rho_R = \rho(U_R + \frac{\alpha U}{R}). \qquad (3.331)$$

By substituting Eq. (3.331) into Eq. (3.330), one gets

$$(\dot{e})^{\theta\theta} = -[U_R + \frac{3}{3\rho}(\rho_t + U\rho_R) - \frac{1}{3\rho}(\rho_t + U\rho_R)], \qquad (3.332)$$

or

$$(\dot{e})^{\theta\theta} = \frac{\alpha U}{R} + \frac{1}{3\rho}(\rho_t + U\rho_R), \qquad (3.333)$$

which is identical to $(\dot{e})^{\theta\theta}$ term in Eq. (3.321). Also, by substituting Eq. (3.331) into Eq. (3.321), one will get Eq. (3.322). At this point, the derivation of Eqs. (3.321) and (3.322) is finished.

When the above equation is encountered, $(S^{ij})^n$ and $(\dot{e}^{ij})^{n+1/2}$ are known. Therefore, for Lagrangian phase calculation, the following set of equations is to be solved for the stress deviator, $\widetilde{S}^{ij}$.

$$\widetilde{S}^{ij} = (S^{ij})^n + \delta t(2G\dot{e}^{ij} - A\dot{W}S^{ij})^{n+1/2}, \qquad (3.334)$$

where
$$(\dot{W})^{n+1/2} = (\dot{e}^{ij} S^{ij})^{n+1/2}, \qquad (3.335)$$
and
$$(S^{ij})^{n+1/2} = \frac{1}{2}[\widetilde{S}^{ij} + (S^{ij})^n]. \qquad (3.336)$$

This looks messy but can be solved without iteration as shown below. From Taylor series expansion, one gets
$$\frac{1}{(1+z)^m} = (1+z)^{-m} = 1 - mz + \frac{m(m+1)}{2!}z^2 - \ldots, \qquad (3.337)$$
therefore
$$[1 + \frac{\delta t}{2}(A\dot{W})^{n+1/2}]^{-1} = 1 - \frac{\delta t}{2}(A\dot{W})^{n+1/2} + \ldots. \qquad (3.338)$$

Substituting $\widetilde{S}^{ij}$ from Eq. (3.334) into the right hand side of Eq. (3.336), we have
$$(S^{ij})^{n+1/2} = \frac{1}{2}[(S^{ij})^n + (S^{ij})^n + \delta t(2G\dot{e}^{ij} - A\dot{W}S^{ij})^{n+1/2}], \qquad (3.339)$$
or
$$(S^{ij})^{n+1/2} = (S^{ij})^n + \frac{\delta t}{2}(2G\dot{e}^{ij} - A\dot{W}S^{ij})^{n+1/2}, \qquad (3.340)$$
or
$$(S^{ij})^{n+1/2} + \frac{\delta t}{2}(A\dot{W}S^{ij})^{n+1/2} = (S^{ij})^n + \frac{\delta t}{2}(2G\dot{e}^{ij})^{n+1/2}, \qquad (3.341)$$
or
$$(S^{ij})^{n+1/2}[1 + \frac{\delta t}{2}(A\dot{W})^{n+1/2}] = (S^{ij})^n + \delta t(G\dot{e}^{ij})^{n+1/2}, \qquad (3.342)$$
or
$$(S^{ij})^{n+1/2} = \frac{(S^{ij})^n + \delta t(G\dot{e}^{ij})^{n+1/2}}{[1 + \frac{\delta t}{2}(A\dot{W})^{n+1/2}]}. \qquad (3.343)$$

By substituting Eq. (3.338) into the right hand side of Eq. (3.343), one gets
$$(S^{ij})^{n+1/2} = [(S^{ij})^n + \delta t(G\dot{e}^{ij})^{n+1/2}][1 - \frac{\delta t}{2}(A\dot{W})^{n+1/2}] + O(\Delta^2), \qquad (3.344)$$
or
$$(S^{ij})^{n+1/2} = [(S^{ij})^n + \delta t(G\dot{e}^{ij})^{n+1/2}][1 - \frac{\delta t}{2}(A\dot{W})^{n+1/2}] + O(\Delta^2), \qquad (3.345)$$

or

$$(S^{ij})^{n+1/2} = (S^{ij})^n - \tfrac{\delta t}{2}(A\dot{W})^{n+1/2}(S^{ij})^n + \delta t(G\dot{e}^{ij})^{n+1/2}$$

$$-\frac{\delta t}{2}(\delta t)(A\dot{W}G\dot{e}^{ij})^{n+1/2}. \qquad (3.346)$$

For the approximation of $O(\Delta^2)$, we can set $(\delta t)(\delta t) \approx 0$, therefore, Eq. (3.346) becomes

$$(S^{ij})^{n+1/2} = (S^{ij})^n - \frac{\delta t}{2}(A\dot{W})^{n+1/2}(S^{ij})^n + \delta t(G\dot{e}^{ij})^{n+1/2}, \qquad (3.347)$$

or

$$(S^{ij})^{n+1/2} = (S^{ij})^n + (\delta t)[(G\dot{e}^{ij})^{n+1/2} - \tfrac{1}{2}(A\dot{W})^{n+1/2}(S^{ij})^n]. \qquad (3.348)$$

Substituting Eq. (3.348) into Eq. (3.335), we have

$$(\dot{W})^{n+1/2} = \dot{e}^{ij}\{(S^{ij})^n + (\delta t)[(G\dot{e}^{ij})^{n+1/2} - \tfrac{1}{2}(A\dot{W})^{n+1/2}(S^{ij})^n]\} + O(\Delta^2). \qquad (3.349)$$

Rearranging the above equation, one gets

$$[1 + \tfrac{\delta t}{2}(A\dot{e}^{ij})^{n+1/2}(S^{ij})^n](\dot{W})^{n+1/2} = (\dot{e}^{ij})^{n+1/2}(S^{ij})^n$$

$$+(\delta t)(G\dot{e}^{ij}\dot{e}^{ij})^{n+1/2} + O(\Delta^2). \qquad (3.350)$$

From Eq. (3.349), we have

$$(\dot{W})^{n+1/2} = (\dot{e}^{ij})^{n+1/2}(S^{ij})^n + (\delta t)[(G\dot{e}^{ij}\dot{e}^{ij})^{n+1/2}$$

$$-\tfrac{1}{2}(A\dot{W}\dot{e}^{ij})^{n+1/2}(S^{ij})^n], \qquad (3.351)$$

or $(\dot{W})^{n+1/2} = (\dot{e}^{ij})^{n+1/2}(S^{ij})^n + (\delta t)(G\dot{e}^{ij}\dot{e}^{ij})^{n+1/2}$

$$-\tfrac{1}{2}(\delta t)(A\dot{W}\dot{e}^{ij})^{n+1/2}(S^{ij})^n. \qquad (3.352)$$

Using Eq. (3.335), we can write the last term in Eq. (3.352) as

$$-\tfrac{1}{2}(\delta t)(A\dot{W}\dot{e}^{ij})^{n+1/2}(S^{ij})^n = -\tfrac{1}{2}(\delta t)(A\dot{e}^{ij}S^{ij}\dot{e}^{ij})^{n+1/2}(S^{ij})^n, \qquad (3.353)$$

or

$$-\tfrac{1}{2}(\delta t)(A\dot{W}\dot{e}^{ij})^{n+1/2}(S^{ij})^n = -\tfrac{1}{2}(\delta t)(A\dot{e}^{ij}\dot{e}^{ij})^{n+1/2}(S^{ij})^n(S^{ij})^{n+1/2}. \qquad (3.354)$$

Substituting $(S^{ij})^{n+1/2}$ from Eq. (3.348) into Eq. (3.354), the RHS of Eq. (3.354) becomes

$$= -\frac{1}{2}(\delta t)(A\dot{e}^{ij}\dot{e}^{ij})^{n+1/2}(S^{ij})^n\{(S^{ij})^n + (\delta t)[(G\dot{e}^{ij})^{n+1/2}$$

$$-\frac{1}{2}(A\dot{W})^{n+1/2}(S^{ij})^n]\}. \qquad (3.355)$$

For the approximation of $O(\Delta^2)$, we can assume that $(\delta t)(\delta t) \approx 0$, therefore, Eq. (3.355) becomes

$$-\frac{1}{2}(\delta t)(A\dot{W}\dot{e}^{ij})^{n+1/2}(S^{ij})^n$$

$$= -\frac{1}{2}(\delta t)(A\dot{e}^{ij}\dot{e}^{ij})^{n+1/2}(S^{ij})^n(S^{ij})^n + O(\Delta^2). \qquad (3.356)$$

By substituting Eq. (3.356) into Eq. (3.351), one gets

$$(\dot{W})^{n+1/2} = (\dot{e}^{ij})^{n+1/2}(S^{ij})^n + (\delta t)[(G\dot{e}^{ij}\dot{e}^{ij})^{n+1/2}$$

$$-\frac{1}{2}(A\dot{e}^{ij}\dot{e}^{ij})^{n+1/2}(S^{ij})^n(S^{ij})^n] + O(\Delta^2). \qquad (3.357)$$

Since $G^{n+1/2}, A^{n+1/2}$ and $(\dot{e}^{ij})^{n+1/2}$ are known, $(\dot{W})^{n+1/2}$ can be calculated by Eq. (3.357). Substituting the result into Eq. (3.348), then from Eq. (3.334), $\widetilde{S}^{ij}$ can be obtained. For programming convenience, we use

$$(\dot{e})^{ij}(\dot{e})^{ij} = \begin{bmatrix} \frac{1}{3}(2U_R - \frac{\alpha U}{R}) & \frac{V_R}{2} & 0 \\ \frac{V_R}{2} & -\frac{1}{3}(U_R - \frac{\alpha U}{R}) & 0 \\ 0 & 0 & -\frac{1}{3}(U_R - \frac{\alpha U}{R}) \end{bmatrix}$$

$$\times \begin{bmatrix} \frac{1}{3}(2U_R - \frac{\alpha U}{R}) & \frac{V_R}{2} & 0 \\ \frac{V_R}{2} & -\frac{1}{3}(U_R - \frac{\alpha U}{R}) & 0 \\ 0 & 0 & -\frac{1}{3}(U_R - \frac{\alpha U}{R}) \end{bmatrix},$$

$$= \frac{1}{9}(2U_R - \frac{\alpha U}{R})^2 + \frac{1}{9}(U_R + \frac{\alpha U}{R})^2$$

$$+\frac{1}{9}(-U_R + \frac{2\alpha U}{R})^2 + (\frac{V_R}{2})^2 + (\frac{V_R}{2})^2, \quad (3.358)$$

$$= \frac{2}{3}(U_R^2 - \frac{\alpha U U_R}{R} + \frac{\alpha U^2}{R^2}) + \frac{V_R^2}{2}. \quad (3.359)$$

Also

$$\dot{e}^{ij}S^{ij} = \dot{e}^{RR}S^{RR} + \dot{e}^{ZZ}S^{ZZ} + \dot{e}^{\theta\theta}S^{\theta\theta} + \frac{1}{2}V_R S^{RZ} + \frac{1}{2}V_R S^{RZ}, \quad (3.360)$$

or

$$\dot{e}^{ij}S^{ij} = \dot{e}^{RR}S^{RR} + \dot{e}^{ZZ}S^{ZZ} + \dot{e}^{\theta\theta}S^{\theta\theta} + V_R S^{RZ}, \quad (3.361)$$

or

$$\dot{e}^{ij}S^{ij} = \tfrac{1}{3}(2U_R - \tfrac{\alpha U}{R})S^{RR} + V_R S^{RZ} - \tfrac{1}{3}(U_R + \tfrac{\alpha U}{R})S^{ZZ}$$

$$+ \frac{1}{3}(-U_R + \frac{\alpha U}{R})S^{\theta\theta}, \quad (3.362)$$

or

$$\dot{e}^{ij}S^{ij} = (U_R - \frac{\alpha U}{R})S^{RR} + V_R S^{RZ} - \frac{\alpha U}{R}S^{ZZ}, \quad (3.363)$$

and

$$(\frac{U}{R})^{n+1/2} = (\frac{U}{R})^n + \frac{\delta t}{2}[(\frac{U}{R})_t + U(\frac{U}{R})_R], \quad (3.364)$$

$$= (\frac{U}{R})^n + \frac{\delta t}{2}[\frac{(\frac{U}{R})^{n+1} - (\frac{U}{R})^n}{\delta t} + U\frac{\partial}{\partial R}(\frac{U}{R})], \quad (3.365)$$

$$= (\frac{U}{R})^n + \frac{1}{2}[(\frac{U}{R})^{n+1} - (\frac{U}{R})^n] + \frac{\delta t}{2}U\frac{\partial}{\partial R}(\frac{U}{R}), \quad (3.366)$$

$$= \frac{1}{2}[(\frac{U}{R})^{n+1} + (\frac{U}{R})^n] + \frac{\delta t}{2}U\frac{\partial}{\partial R}(\frac{U}{R}), \quad (3.367)$$

$$= \frac{\overline{U}}{R} - \frac{\delta t}{2}(\frac{U}{R})^2 + \frac{\delta t}{2}(UR)\frac{\partial U}{\partial R}, \quad (3.368)$$

$$= \frac{\overline{U}}{R} - \frac{\delta t}{2}(\frac{U}{R})^2 + O(\Delta^2). \quad (3.369)$$

If one makes the approximation of $\overline{U} \approx U$, then, Eq. (3.369) can be written as

$$(\frac{U}{R})^{n+1/2} = (\frac{\overline{U}}{R})(1 - \frac{\delta t}{2}\frac{U}{R}) + O(\Delta^2). \quad (3.370)$$

3.3.4.2 Axial Direction

The calculation of the plastic flow in the axial direction can be done similarly. From Eq. (3.319), one gets

$$S_t^{ij} = -\overline{V}S_Z^{ij} + 2G(\dot{e})^{ij} - A\dot{W}S^{ij}, \qquad (3.371)$$

where $\dot{W} = S^{ij}(\dot{e})^{ij}$, and

$$(\dot{e})^{ij} = \begin{bmatrix} (\dot{e})^{RR} & (\dot{e})^{RZ} & (\dot{e})^{R\theta} \\ (\dot{e})^{ZR} & (\dot{e})^{ZZ} & (\dot{e})^{Z\theta} \\ (\dot{e})^{\theta R} & (\dot{e})^{\theta Z} & (\dot{e})^{\theta\theta} \end{bmatrix}. \qquad (3.372)$$

The deviatoric strain rate for the axial direction can be derived from Eq. (3.76), therefore

$$(\dot{e})^{ij} = \begin{bmatrix} \frac{1}{3\rho}(\rho_t + V\rho_Z) & \frac{U_Z}{2} & 0 \\ \frac{U_Z}{2} & V_Z + \frac{1}{3\rho}(\rho_t + V\rho_Z) & 0 \\ 0 & 0 & \frac{1}{3\rho}(\rho_t + V\rho_Z) \end{bmatrix}. \qquad (3.373)$$

From Eq. (3.261), we have

$$\rho_t + V\rho_Z = -\rho V_Z, \qquad (3.374)$$

therefore

$$\frac{1}{3\rho}(\rho_t + V\rho_Z) = -\frac{1}{3\rho}\rho V_Z = -\frac{V_Z}{3}, \qquad (3.375)$$

and

$$V_Z + \frac{1}{3\rho}(\rho_t + V\rho_Z) = V_Z - \frac{1}{3\rho}\rho V_Z = \frac{2V_Z}{3}. \qquad (3.376)$$

Substituting Eqs. (3.375) and (3.376) into Eq. (3.373), one gets

$$(\dot{e})^{ij} = \begin{bmatrix} -\frac{1}{3}V_Z & \frac{U_Z}{2} & 0 \\ \frac{U_Z}{2} & \frac{2}{3}V_Z & 0 \\ 0 & 0 & -\frac{1}{3}V_Z \end{bmatrix}, \qquad (3.377)$$

and
$$A = \frac{3G}{(Y^0)^2}. \tag{3.378}$$

As before, Eqs. (3.357), (3.348) and (3.334) can be used with the above strain rate deviator, $(\dot{e})^{ij}$, to get $\widetilde{S}^{ij}$. Again for programming convenience, we have

$$(\dot{e})^{ij}(\dot{e})^{ij} = \begin{bmatrix} -\frac{1}{3}V_Z & \frac{U_Z}{2} & 0 \\ \frac{U_Z}{2} & \frac{2}{3}V_Z & 0 \\ 0 & 0 & -\frac{1}{3}V_Z \end{bmatrix}$$

$$\times \begin{bmatrix} -\frac{1}{3}V_Z & \frac{U_Z}{2} & 0 \\ \frac{U_Z}{2} & \frac{2}{3}V_Z & 0 \\ 0 & 0 & -\frac{1}{3}V_Z \end{bmatrix}, \tag{3.379}$$

$$= \frac{1}{9}V_Z^2 + \frac{1}{4}U_Z^2 + \frac{1}{4}U_Z^2 + \frac{4}{9}V_Z^2 + \frac{1}{9}V_Z^2, \tag{3.380}$$

$$= \frac{2}{3}V_Z^2 + \frac{1}{2}U_Z^2. \tag{3.381}$$

Also, we have

$$(\dot{e}^{ij}S^ij) = \begin{bmatrix} -\frac{1}{3}V_Z & \frac{U_Z}{2} & 0 \\ \frac{U_Z}{2} & \frac{2}{3}V_Z & 0 \\ 0 & 0 & -\frac{1}{3}V_Z \end{bmatrix}$$

$$\times \begin{bmatrix} S^{RR} & S^{RZ} & 0 \\ S^{ZR} & S^{ZZ} & 0 \\ 0 & 0 & S^{\theta\theta} \end{bmatrix}, \tag{3.382}$$

$$= (-\frac{V_Z}{3})S^{RR} + (\frac{2}{3}V_Z)S^{ZZ} - (\frac{V_Z}{3})S^{\theta\theta} + (\frac{U_Z}{2})S^{RZ} + (\frac{U_Z}{2})S^{RZ}, \tag{3.383}$$

$$= (-\frac{V_Z}{3})S^{RR} + (\frac{2}{3}V_Z)S^{ZZ} - (\frac{V_Z}{3})S^{\theta\theta} + (\frac{U_Z}{2} + \frac{U_Z}{2})S^{RZ}, \quad (3.384)$$

$$= (-\frac{V_Z}{3})S^{RR} + (\frac{2}{3}V_Z)S^{ZZ} - (\frac{V_Z}{3})(-S^{RR} - S^{ZZ}) + U_Z S^{RZ}, \quad (3.385)$$

$$= (-\frac{V_Z}{3})S^{RR} + \frac{V_Z}{3}S^{RR} + (\frac{2V_Z}{3} + \frac{V_Z}{3})S^{ZZ} + U_Z S^{RZ}, \quad (3.386)$$

$$= V_Z S^{ZZ} + U_Z S^{RZ}. \quad (3.387)$$

3.3.5 Particle Transport and Remapping

3.3.5.1 Average Velocity for the Particle

After Lagrangian phase calculation, we have $\widetilde{U}, \widetilde{V}, \widetilde{\epsilon}$ and $\widetilde{S}^{ij}$ at each cell. Consistent with these, temporary values of momentum and internal energy are assigned to cell (i) as below.

$$\text{Momentum} = M_i \widetilde{U}_i, \quad (3.388)$$

and

$$\text{Internal energy} = M_i \widetilde{\epsilon}_i, \quad (3.389)$$

where M_i = mass of cell (i).

Then, particle are transported with average velocities, that is, in the radial direction,

$$\overline{U} = \frac{1}{2}(U + \widetilde{U}). \quad (3.390)$$

Each particle carries with it a fraction of dynamic and state quantities as defined by Eqs. (3.388) and (3.389). After the particle transport, total changes in momentum, energy, and stresses are calculated for each cell. This particle transport and subsequent remapping to the original Eulerian mesh are explained in detail in conjunction with the truncation error analysis.

3.3.5.2 Particle Treatment for Void or Multi-material Cell

For each cell, we compute the location, type, and mass for each particle using the following steps:

Step 1. If a cell requires particle treatment or if the cell just spalled, then we move to the next step.

Step 2. If the particle has been previously defined, then the particle information is obtained from the particle flag words. If no particle exists in the cell, then particles will be created for each material, including void.

Two-Dimensional Eulerian Method

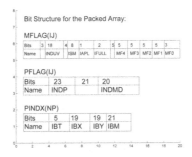

Fig. 3.12 Structure for the packed array MFLAG, PFLAG, and PINDX.

Step 3. We then move the particles using the interpolated velocities.
Step 4. If a mixed cell has a vacuum interface, then particles are used for the material region and no particle is assigned to the vacuum region.
Step 5. If a void is connected to a vacuum region, then the void region is treated as a vacuum region.
Step 6. After the particles have been moved, we save the information of the new location and the mass for each particle.
Step 7. We then are ready for the next-cycle calculation.

The PIC method is well known for its expensive CPU cost and its horrendous storage requirement. We vectorized the whole code so that it would run four times faster compared to a scalar one. We saved a large amount of storage by using three packed words that will be described below. Since we were using the CRAY-YMP machine, which contains 64 bits for each word, we divided one word into many group so that each group possessed its own information. The structures of the material flag word, MFLAG(ij), the particle flag word, PFLAG(ij), and the particle information word, PINDX(NP) are shown in Fig. 3.12. The notations and definitions used in Fig. 3.12 are:

IJ: index of cell at i (R-direction) and j (Z-direction).
NP: total number of particle in cell IJ.
INDUV: index of the velocity array, e.g., U(INDUV).
ISM: flag to indicate spallation and melting for each material (up to maximum of 4).
IAPL: for applied pressure boundary, IAPL = 1.
IFULL: 0 for empty cell, 1 for partially filled cell, 2 for full cell.

92   Computational Solid Mechanics for Oil Well Perforator Design

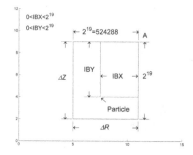

Fig. 3.13 The reference point A of any cell (IJ) is defined at the right-top corner with IBX, IBY indicating the position of the particle (Note: IBX and IBY are always positive).

MF4-MF1: material number index.
MF0: the total number of materials in the cell.
INDP: pointer for the particle array.
INDMD: index for the total (summation of all particles) mass, volume, pressure, and internal energy for the first material in cell IJ.
IBT: material type for the particle (e.g., 1–31).
IBX: normalize particle location (R-direction) in the cell (positive integer $< 2^{19}$).
IBY: normalized particle location (Z-direction) in the cell (positive integer $< 2^{19}$).
IBM: normalized particle mass fraction for one type material (positive integer $< 2^{19}$).
IBX and IBY are computed from the following (see Fig. 3.13):

```
      K = 2.0E+19
      DO   10   IX = 1, 8
      K = K - 2.0E+16
      INXT = 28 + 8.0  x   random (between 0.0 and 1.0)
               (function of IX)
      DO   10   JY = 1, 8
      IBX (JY) = K + 2.0E+13    x   random (between 0 and 7)
               (function of JY and INXT)+ 2.0E+13    x
               random (between 0 and 1.0)
      IBY(JY) = 2.0E+16   x   (JY - 1) + 2.0E+16   x
               random (between 0 and 1.0) (function of JY)
   10 continue
```

The normalized particle mass fraction IBM is defined by
IBM = (particle mass x 2^{21}) / (Total mass of the same material in the cell).
The locations of the particle, i.e., IBX and IBY, are illustrated in Fig. 3.13 with the computations:

R coordinate of particle (1) = R coordinate of point A - IBX /2^{19}.
Z coordinate of particle (1) = Z coordinate of point A - IBY /2^{19}.
Mass of particle (1) = mass of material 1 / 64.
Velocity of particle (1) = interpolated from two velocities, i.e., $U_{i+1/2}$ and $U_{i-1/2}$.
Then, particle (1) is moved to the new position. The logic for creating new particle is as follows:

Step 1. The current calculation is at cell (i).
Step 2. If cell (i) is a mixed cell, then cell (i) should have particles defined from previous cycle calculation.
Step 3. If the velocity is going to the right direction, and if the material at cell ($i + 1$) is different from cell (i), then, particles will be created for cell (i).
Step 4. If cell (i) is a pure material cell, i.e., only one material presented, but cell (i) has to spall, then, we will create one type (say blue) of particle for the material and another particle type (say red) for the void space which is defined by the local pressure (i.e., $P = 0.0$). Assign 16 particles for the void volume using random number generator to obtain the locations, i.e., the coordinates (R, Z) of each particle.

3.3.6 *Computation for Spall*

The computation on or around a spalled cell can vary widely depending on the accuracy desired on spall. Mathematically the material failure creates a discontinuity within the computational domain. Therefore, for a rigorous treatment, a proper interface condition (possibly jump condition) has to be established across a spalled cell. Basically, spall or fracture has Lagrangian characteristics and to resolve it in an Eulerian mesh system will require a special spall logic. For the present computation, a rather simple scheme is suitable for the particle method.

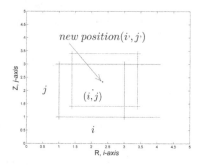

Fig. 3.14 Old and new centers of a spalled cell.

3.3.6.1 When Spalled

A void region is created and located at the cell center with a minimum number of 16 particles. Then, for subsequent time steps, this is convected with $\overline{U}$ and $\overline{V}$. P and S^{ij} are set to be zero at this point.

This is essentially the same as introducing a rectangular void at the cell center. The void region has a volume based on the equation of state initially and may grow later.

3.3.6.2 Recombination

If the center of a spalled cell (i, j), as shown in Fig. 3.14, is moved to a new location (i', j'), the specific volume for (i', j') is computed by averaging value in the dotted cell. When $\tau \leq \tau_{test}$, the dotted cell is assumed to recombine.

3.3.6.3 Propagation of Fracture Surface

The particle method in its present form will not follow a fracture surface exactly. Rather the density of the fractured cell will be lowered to behave essentially the same as an empty cell. The propagation of a crack or a material separation will be resolved in this fashion. A possible future option would be to define an interface. Particle can be created in the fractured cell and assigned a different identification number for the fractured void space. (See Fig. 3.15.)

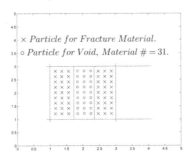

Fig. 3.15 Material number 31 is assigned to a spalled or fractured void space.

3.3.7 Time Step Control

The time step control in the present particle method is done empirically. The following condition has been applied successfully:

$$N_c \triangleq \frac{(|U|+C)\delta t}{\delta x} < 0.4, \qquad (3.391)$$

where $C \triangleq \sqrt{(\frac{\partial P}{\partial \rho})_S}$.

The above definition of bulk sound speed "C" is rather arbitrary in an elastic material. Physically in an elastic material, wave propagates in a longitudinal or transverse wave form, as shown below. In isotropic elastic media, equations of small motions in the absence of body force are

$$(\lambda + G)\frac{\partial e}{\partial x_i} + G \nabla^2 U_i - \rho \frac{\partial^2 U_i}{\partial t^2} = 0, \qquad (3.392)$$

where e = volume expansion and U_i = displacement in i-direction. When $e \doteq 0$, the deformation consists of shearing distortion and rotation only, and Eq. (3.392) becomes

$$G \nabla^2 U_i - \rho \frac{\partial^2 U_i}{\partial t^2} = 0. \qquad (3.393)$$

This is the equation for "wave of distortion" or "transverse waves." Then the wave speed is

$$C_t = \sqrt{\frac{G}{\rho}}. \qquad (3.394)$$

For the irrotational case, Eq. (3.392) becomes

$$(\lambda + 2G) \nabla^2 U_i - \rho \frac{\partial^2 U_i}{\partial t^2} = 0. \qquad (3.395)$$

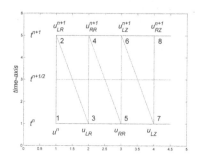

Fig. 3.16 The calculation procedure with: Lagrangian in R-direction (1-2), remap in R-direction (3-4), Lagrangian in Z-direction (5-6), remap in Z-direction (7-8).

This is the equation for "irrotational waves," "wave of dilatation," or "longitudinal waves." The wave speed is then

$$C_\ell = \sqrt{\frac{\lambda + 2G}{\rho}}. \tag{3.396}$$

From thermodynamic relations, one can write the sound speed for an elastic solid as

$$C^2 = \frac{K}{\rho}(1 + \frac{\beta^2 KT}{\rho C_v}) = \frac{K}{\rho}, \tag{3.397}$$

and

$$C_\ell^2 = C^2 + \frac{4}{3}\frac{G}{\rho}, \tag{3.398}$$

where β = thermal expansion coefficient and K = compression modulus.

3.3.8 The Logic for the Calculation Procedure

During the calculation, each cell will contain the information about its state, e.g., solid, liquid, or gas states. If the cell is in the state of solid, it will further provide the information about whether the material is in elastic or plastic regime. Therefore, there is one of the four states associated with each cell. These four states are

A. Solid in elastic regime.
B. Solid in plastic regime.
C. Liquid phase.
D. Gas phase.

In one time step, the material in the cell can change its state by one of the following seven possibilities:

Case 1. From solid elastic state to solid elastic state.
Case 2. From solid elastic state to solid plastic state.
Case 3. From solid plastic state to solid plastic state.
Case 4. From solid plastic state to liquid state.
Case 5. From liquid state to liquid state.
Case 6. From liquid state to gas state.
Case 7. From gas state to vacuum state, since the gas will be discarded.

For each case, there are certain equations should be used for calculation and they are described below:

Case 1. This is the solid elastic state at both point 1 and point 8 as shown in Fig. 3.16. The Lagrangian phase in the R-direction which goes from point 1 to point 2 uses the following equations: Eqs. (3.185), (3.186), (3.187), (3.188), (3.189) and (3.190). For the remap phase in the R-direction which goes from point 3 to point 4 uses the following equations: Eqs. (3.388), (3.389) and (3.390). For the Lagrangian phase in the Z-direction, that is from point 5 to point 6, we use Eqs. (3.262), (3.263), (3.264), (3.266), (3.269), (3.270) and (3.271). For the remap phase in the Z-direction, that is from point 7 to point 8, we use Eqs. (3.388), (3.389) and (3.390). At this point we check the spallation and fracture conditions to decide whether to set the pressure to zero or not. If there is no spallation or fracture, then, we will do the correction for the rigid body rotation by using Eqs. (3.106), (3.107) and (3.108).

Case 2. This is the solid elastic state at point 1 and solid plastic state at point 8. The whole calculation is the same as described in Case 1 except at point 8 we use Eq. (3.105) to obtain the stress deviators on the yield surface. There is no correction for rigid body rotation in this case.

Case 3. This is the solid plastic state at both point 1 and point 8. From point 1 to point 2, we use Eqs. (3.185), (3.186), (3.188), (3.189), (3.190) and (3.191). From point 3 to point 4, we use Eqs. (3.388), (3.389) and (3.390). From point 5 to point 6, we use Eqs. (3.262), (3.263), (3.264), (3.266), (3.269), (3.270), (3.271) and (3.272). At point 7, we check the yield criterion, i.e., $\phi > \frac{2}{3}(Y^0)^2$, to see if the material is still in plastic state. There is no correction for rigid body rotation in this case. From point 7 to point 8, we use Eqs. (3.388), (3.389) and (3.390).

Case 4. At point 1, the material is in solid plastic state, but after point 3, it may change to liquid state. From point 1 to point 2, we use Eqs. (3.185), (3.186), (3.187), (3.188), (3.189), (3.190) and (3.191). At point 3, we check the material temperature, if it is higher than the melting temperature of that particular material, we set both $S^{ij} = 0$ and $e^{ij} = 0$. From point 3 to point 4, we use Eqs. (3.388), (3.389) and (3.390). From point 5 to point 6, we use Eqs. (3.262), (3.263), (3.264), (3.266), (3.269), (3.270), (3.271) and (3.272). From point 7 to point 8, we use equations similar to Eqs. (3.388), (3.389) and (3.390) to do the remap in the Z-direction. There is no correction for rigid body rotation in this case.

Case 5. This is the liquid phase for both point 1 and point 8. From point 1 to point 2, we use Eqs. (3.185), (3.186) and (3.187). From point 3 to point 4, we use Eqs. (3.388), (3.389) and (3.390). From point 5 to point 6, we use Eqs. (3.262), (3.263) and (3.264). At point 7, we use Eq. (3.135) to obtain pressure. There is no correction for rigid body rotation for this case. From point 7 to point 8, we use equations similar to Eqs. (3.388), (3.389) and (3.390).

Case 6. At point 1, we have liquid phase, but after point 3, it may change to gas phase if the temperature is higher than boiling point. Once the material becomes gas phase we will discard it. From point 1 to point 2, we use Eqs. (3.185), (3.186) and (3.187). From point 3 to point 4, we use Eqs. (3.388), (3.389) and (3.390). From point 5 to point 6, we use Eqs. (3.262), (3.263) and (3.264). Again, at point 7 we will check the material temperature to see if it is higher than the boiling temperature. If it does, then, we will discard the material.

Case 7. If a cell contains a vacuum volume due to fracture, i.e., void opening, the void may be closed up when the local pressure becomes higher. If the cell material becomes gas phase due to the boiling from liquid phase, we will discard the gas material.

3.4 Truncation Error Analysis

Here we prove the truncation error of the proposed method and explain details of the particle transport and remapping. Again radial equations are considered, and for convenience, the radial velocity is assumed to be positive. The analysis is done on mass, momentum, energy, and stress equations.

3.4.1 Mass

From Eq. (3.390), a particle moves with

$$\overline{U} = \frac{1}{2}(U^n + \widetilde{U}). \tag{3.399}$$

From Eq. (3.185), the Lagrangian phase in the R-direction, we have

$$\widetilde{U} = U^n + \delta t \{ [\frac{(\sigma^{RR})_R}{\rho}]^{n+1/2} + \alpha [\frac{\sigma^{RR} - \sigma^{\theta\theta}}{\rho R}]^{n+1/2} \}. \tag{3.400}$$

Substituting Eq. (3.400) into Eq. (3.399), one gets

$$\overline{U} = U^n + \frac{\delta t}{2}(\frac{\sigma_R^{RR}}{\rho} + \alpha \frac{\Delta\sigma}{R\rho})^{n+1/2}. \tag{3.401}$$

Equation (3.399) can be written as

$$\overline{U} = \frac{1}{2}(2\widetilde{U} + U^n - \widetilde{U}), \tag{3.402}$$

or

$$\overline{U} = \widetilde{U} - \frac{1}{2}(\widetilde{U} - U^n). \tag{3.403}$$

Substituting $\widetilde{U} - U^n$ from Eq. (3.400) into Eq. (3.403), we have

$$\overline{U} = \widetilde{U} - \frac{\delta t}{2}(\frac{\sigma_R^{RR}}{\rho} + \alpha \frac{\Delta\sigma}{R\rho})^{n+1/2}, \tag{3.404}$$

where $\Delta\sigma = \sigma^{RR} - \sigma^{\theta\theta}$.

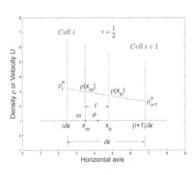

Fig. 3.17 A particle, originally at position P, can move with a maximum distance ℓ.

The maximum distance ℓ that a section of fluid can cross the cell boundary during a time step δt, is shown in Fig. 3.17. Let the $\overline{U}(R_m)$ be the average velocity at the location of $R = R_m$, then

$$\overline{U}(R_m) = \overline{U}_{i+\frac{1}{2}} - \ell(\overline{U}_R)_{i+\frac{1}{2}} = \frac{\overline{U}_i + \overline{U}_{i+1}}{2} + \ell \frac{\overline{U}_i - \overline{U}_{i+1}}{\delta R}, \tag{3.405}$$

$$\ell = \overline{U}(R_m)\delta t, \tag{3.406}$$

or

$$\ell = [\overline{U}_{i+\frac{1}{2}} - \ell(\overline{U}_R)_{i+\frac{1}{2}}]\delta t, \tag{3.407}$$

or

$$\ell = (\overline{U}_{i+\frac{1}{2}})\delta t - \ell(\overline{U}_R)_{i+\frac{1}{2}}\delta t. \tag{3.408}$$

Therefore, we have

$$\ell + \ell(\overline{U}_R)_{i+\frac{1}{2}}\delta t = (\overline{U}_{i+\frac{1}{2}})\delta t, \tag{3.409}$$

or

$$\ell[1 + (\overline{U}_R)_{i+\frac{1}{2}}\delta t] = (\overline{U}_{i+\frac{1}{2}})\delta t, \tag{3.410}$$

or

$$\ell = \frac{(\overline{U}_{i+\frac{1}{2}})\delta t}{1 + (\overline{U}_R)_{i+\frac{1}{2}}\delta t}, \tag{3.411}$$

or

$$\ell = [(\overline{U}_{i+\frac{1}{2}})\delta t][1 + (\overline{U}_R)_{i+\frac{1}{2}}\delta t]^{-1}, \tag{3.412}$$

or

$$\ell = [(\overline{U}_{i+\frac{1}{2}})\delta t][1 - (\overline{U}_R)_{i+\frac{1}{2}}\delta t] + O(\Delta^3). \tag{3.413}$$

The density at R near point P as shown in Fig. 3.17 is

$$\rho(R) = \rho_{i+\frac{1}{2}} - (R_{i+\frac{1}{2}} - R)\rho_R. \tag{3.414}$$

Then
$\delta M_{i+\frac{1}{2}}$=mass transposed from (i) to $(i+1)$ during δt, or

$$\delta M_{i+\frac{1}{2}} = 2\pi\delta Z \int_{R_{i+\frac{1}{2}}-\ell}^{R_{i+\frac{1}{2}}} \rho(R) R dR. \tag{3.415}$$

Substituting Eq. (3.414) into Eq. (3.415), one gets

$$\delta M_{i+\frac{1}{2}} = 2\pi\delta Z \int_{R_{i+\frac{1}{2}}-\ell}^{R_{i+\frac{1}{2}}} [\rho_{i+\frac{1}{2}} - (R_{i+\frac{1}{2}} - R)\rho_R] R dR, \tag{3.416}$$

$$= 2\pi\delta Z \{[\rho_{i+\frac{1}{2}} - R_{i+\frac{1}{2}}\rho_R](\frac{1}{2}R^2) + \rho_R(\frac{1}{3})R^3\}|_{R_{i+\frac{1}{2}}-\ell}^{R_{i+\frac{1}{2}}}, \tag{3.417}$$

$$= 2\pi\delta Z \{[(\frac{1}{2})(\rho_{i+\frac{1}{2}} - R_{i+\frac{1}{2}}\rho_R) + (\frac{1}{3})(\rho_R)(R)]R^2\}|_{R_{i+\frac{1}{2}}-\ell}^{R_{i+\frac{1}{2}}}, \tag{3.418}$$

$$= 2\pi\delta Z\{[(\tfrac{1}{2})(\rho_{i+\frac{1}{2}} - R_{i+\frac{1}{2}}\rho_R) + (\tfrac{1}{3})(\rho_R)(R_{i+\frac{1}{2}})](R_{i+\frac{1}{2}})^2$$

$$-[(\tfrac{1}{2})(\rho_{i+\frac{1}{2}} - R_{i+\frac{1}{2}}\rho_R) + (\tfrac{1}{3})(\rho_R)(R_{i+\frac{1}{2}} - \ell)](R_{i+\frac{1}{2}} - \ell)^2\}. \qquad (3.419)$$

In order to simplify the algebraic manipulation, we set

$$A = \frac{1}{2}(\rho_{i+\frac{1}{2}} - \rho_R R_{i+\frac{1}{2}}), \qquad (3.420)$$

$$B = \frac{1}{3}(\rho_R), \qquad (3.421)$$

$$C = 2\pi\delta Z, \qquad (3.422)$$

and

$$D = R_{i+\frac{1}{2}}. \qquad (3.423)$$

Therefore, Eq. (3.419) can be written as

$$= C\{(A + BD)(D)^2 - [A + B(D - \ell)](D - \ell)^2\}, \qquad (3.424)$$

$$= C[(A + BD)(D)^2 - (A + BD - B\ell)(D^2 - 2D\ell + \ell^2)], \qquad (3.425)$$

$$= C\{(A + BD)(D)^2 - [(A + BD)D^2 - B\ell D^2]$$

$$-(A + BD - B\ell)(-2D\ell + \ell^2)\}, \qquad (3.426)$$

$$= C[B\ell D^2 + 2AD\ell + 2BD^2\ell - 2BD\ell^2$$

$$-(A\ell^2 + BD\ell^2 - B\ell^3)], \qquad (3.427)$$

$$= C(3B\ell D^2 + 2AD\ell - 3BD\ell^2 - A\ell^2 + B\ell^3), \qquad (3.428)$$

$$= C\ell(3BD^2 + 2AD - 3BD\ell - A\ell + B\ell^2). \qquad (3.429)$$

Substituting Eqs. (3.420)-(3.423) into Eq. (3.429), we have

$$\delta M_{i+\frac{1}{2}} = C\ell(3BD^2 + 2AD - 3BD\ell - A\ell + B\ell^2), \qquad (3.430)$$

$$= C\ell\{[3(\tfrac{1}{3})(\rho_R)R_{i+\frac{1}{2}} + 2(\tfrac{1}{2})(\rho_{i+\frac{1}{2}} - \rho_R R_{i+\frac{1}{2}})]R_{i+\frac{1}{2}}$$

$$-[3(\tfrac{1}{3})(\rho_R)R_{i+\frac{1}{2}} + (\tfrac{1}{2})(\rho_{i+\frac{1}{2}} - \rho_R R_{i+\frac{1}{2}})]\ell + \tfrac{1}{3}(\rho_R)\ell^2\}, \qquad (3.431)$$

$$= C\ell\{[(\rho_R)R_{i+\frac{1}{2}} + \rho_{i+\frac{1}{2}} - \rho_R R_{i+\frac{1}{2}}]R_{i+\frac{1}{2}}$$

$$-[(\rho_R)R_{i+\frac{1}{2}} + (\frac{1}{2})(\rho_{i+\frac{1}{2}}) - (\frac{1}{2})(\rho_R R_{i+\frac{1}{2}})]\ell + \frac{1}{3}(\rho_R)\ell^2\}, \quad (3.432)$$

$$= C\ell\{(\rho_{i+\frac{1}{2}})R_{i+\frac{1}{2}} - [(\frac{1}{2})(\rho_R)R_{i+\frac{1}{2}} + (\frac{1}{2})(\rho_{i+\frac{1}{2}})]\ell$$

$$+ \frac{1}{3}(\rho_R)\ell^2\}, \quad (3.433)$$

$$= C\ell\{[\rho R - (\rho_R R + \rho)(\frac{\ell}{2})]_{i+\frac{1}{2}} + \frac{1}{3}(\rho_R)\ell^2\}, \quad (3.434)$$

$$= 2\pi\delta Z\ell\{[\rho R - (\rho_R R + \rho)(\frac{\ell}{2})]_{i+\frac{1}{2}} + \frac{1}{3}(\rho_R)\ell^2\}. \quad (3.435)$$

Substituting Eq. (3.413) into Eq. (3.435), we have

$$\delta M_{i+\frac{1}{2}} = 2\pi\delta Z\{\rho R[\overline{U}\delta t - \overline{UU}_R(\delta t)^2] - (\rho_R R + \rho)\frac{\overline{U}^2(\delta t)^2}{2}\}_{i+\frac{1}{2}} + O(\Delta^4). \quad (3.436)$$

The above equation has an accuracy of $O(\Delta^3)$, therefore, terms with $O(\Delta^4)$ are neglected, for example

$$(\delta Z)(\frac{1}{3})(\rho_R)\ell^3 = (\delta Z)(\frac{1}{3})(\rho_R)[\overline{U}\delta t - \overline{UU}_R(\delta t)^2]^3, \quad (3.437)$$

or

$$(\delta Z)(\frac{1}{3})(\rho_R)\ell^3 \approx (\delta Z)(\delta t)^3 \approx \Delta^4. \quad (3.438)$$

This is the reason we discard $(\delta Z)(\frac{1}{3})(\rho_R)\ell^3$ in deriving Eq. (3.436). Equation (3.436) can be written as

$$\delta M_{i+\frac{1}{2}} = 2\pi\delta Z(\delta t)\{\rho R\widetilde{U} - (\delta t)[\overline{UU}_R(\rho)(R) + \frac{(\rho + \rho_R R)\overline{U}^2}{2}]\}_{i+\frac{1}{2}}. \quad (3.439)$$

Let

$$A = \overline{UU}_R(\rho)(R) + \frac{(\rho + \rho_R R)\overline{U}^2}{2}, \quad (3.440)$$

and

$$B = (\overline{U})(\rho)(R). \quad (3.441)$$

For cylindrical coordinate, $\alpha = 1$, therefore, Eq. (3.404) can be written as

$$\overline{U} = \widetilde{U} - \frac{\delta t}{2}\left(\frac{\sigma_R^{RR}}{\rho} + \frac{\Delta\sigma}{R\rho}\right)^{n+1/2}. \tag{3.442}$$

Substituting Eq. (3.442) into Eq. (3.441), we have

$$B = R\rho\left[\widetilde{U} - \frac{\delta t}{2}\left(\frac{\sigma_R^{RR}}{\rho} + \frac{\Delta\sigma}{R\rho}\right)^{n+1/2}\right]. \tag{3.443}$$

Equation (3.440) can be written as

$$A = \frac{1}{2}[2\overline{UU}_R(\rho)(R) + (\rho + \rho_R R)\overline{U}^2]. \tag{3.444}$$

Substituting Eqs. (3.443) and (3.444) into Eq. (3.439), one gets
$$\delta M_{i+\frac{1}{2}} = 2\pi\delta Z(\delta t)\{\rho R\widetilde{U} - (\tfrac{\delta t}{2})[(\tfrac{\sigma_R^{RR}}{\rho} + \tfrac{\Delta\sigma}{R\rho})^{n+1/2}$$

$$+ 2\overline{UU}_R(\rho)(R) + (\rho + \rho_R R)\overline{U}^2]\}_{i+\frac{1}{2}} + O(\Delta^4). \tag{3.445}$$

The total mass change in cell (i), ΔM_i, is

$$\Delta M_i = \delta M_{i-\frac{1}{2}}(\text{in flow}) - \delta M_{i+\frac{1}{2}}(\text{outflow}), \tag{3.446}$$

$$= -[\delta M_{i+\frac{1}{2}} - \delta M_{i-\frac{1}{2}}], \tag{3.447}$$

$$= -\delta R \frac{\partial(\delta M)}{\partial R}, \tag{3.448}$$

or

$$\Delta M_i = -2\pi\delta Z(\delta t)(\delta R)\frac{\partial}{\partial R}\{\rho R\widetilde{U} - (\tfrac{\delta t}{2})[(\tfrac{\sigma_R^{RR}}{\rho} + \tfrac{\Delta\sigma}{R\rho})^{n+1/2}$$

$$+ 2\overline{UU}_R(\rho)(R) + (\rho + \rho_R R)\overline{U}^2]\}_{i+\frac{1}{2}} + O(\Delta^4). \tag{3.449}$$

In Eq. (3.449), $O(\Delta^4) = O(\delta M) \times O(\Delta^3)$.

Now the density change in cell (i) can be calculated as below.

$$\Delta\rho_i = \rho_i^{n+1} - \rho_i^n, \tag{3.450}$$

$$= \frac{\Delta M_i}{\text{Volume}}, \tag{3.451}$$

$$\Delta\rho_i = \frac{1}{2\pi R\delta R\delta Z}(-2\pi)\delta Z(\delta t)(\delta R)\frac{\partial}{\partial R}\{\rho R\widetilde{U} - (\tfrac{\delta t}{2})[(\tfrac{\sigma_R^{RR}}{\rho} + \tfrac{\Delta\sigma}{R\rho})^{n+1/2}$$

$$+ 2\overline{UU}_R(\rho)(R) + (\rho + \rho_R R)\overline{U}^2]\}_{i+\frac{1}{2}}, \tag{3.452}$$

or

$$\Delta\rho_i = -\frac{(\delta t)}{R}\{\rho R[\widetilde{U} - (\frac{\delta t}{2})(\frac{\sigma_R^{RR}}{\rho} + \frac{\Delta\sigma}{R\rho})^{n+1/2}$$

$$-\delta t[\overline{UU}_R(\rho)(R) + \frac{(\rho + \rho_R R)\overline{U}^2}{2}]\}_R + O(\Delta^3). \quad (3.453)$$

Now we will prove that Eq. (3.453) satisfies Eq. (3.131) within $O(\Delta^2)$. From the Taylor series expansion,

$$f(Z) = f(Z_0) + \frac{Z - Z_0}{1}f'(Z_0) + \frac{(Z - Z_0)^2}{2!}f''(Z_0) + \ldots + O(\Delta^2). \quad (3.454)$$

Let

$$f(Z) = \rho^{(n+1)\Delta t}, \quad (3.455)$$

$$f(Z_0) = \rho^{(n\Delta t)}, \quad (3.456)$$

and

$$Z - Z_0 = (n+1)\Delta t - (n)\Delta t = \Delta t. \quad (3.457)$$

Substituting Eqs. (3.455)-(3.457) into Eq. (3.454), we have

$$\rho^{(n+1)\Delta t} = \rho^{(n\Delta t)} + \frac{\Delta t}{1}\rho_t + \frac{(\Delta t)^2}{2}\rho_{tt} + O(\Delta^3), \quad (3.458)$$

or

$$\rho^{(n+1)\Delta t} - \rho^{(n\Delta t)} = \frac{\Delta t}{1}\rho_t + \frac{(\Delta t)^2}{2}\rho_{tt} + O(\Delta^3), \quad (3.459)$$

or

$$\frac{\rho^{(n+1)\Delta t} - \rho^{(n\Delta t)}}{\Delta t} = \rho_t + \frac{(\Delta t)}{2}\rho_{tt} + O(\Delta^2), \quad (3.460)$$

or

$$\rho_t = \frac{\rho^{(n+1)\Delta t} - \rho^{(n\Delta t)}}{\Delta t} - \frac{(\Delta t)}{2}\rho_{tt} + O(\Delta^2), \quad (3.461)$$

or

$$\rho_t = \frac{\rho^{(n+1)} - \rho^{(n)}}{\Delta t} - \frac{(\Delta t)}{2}\rho_{tt} + O(\Delta^2). \quad (3.462)$$

From Eq. (3.453), we have

$$\frac{\rho^{(n+1)} - \rho^{(n)}}{\Delta t} = -\frac{1}{R}\{\rho R[\widetilde{U} - (\frac{\delta t}{2})(\frac{\sigma_R^{RR}}{\rho} + \frac{\Delta\sigma}{R\rho})^{n+1/2}$$

$$-\delta t[\overline{UU}_R(\rho)(R) + \frac{(\rho + \rho_R R)\overline{U}^2}{2}]\}_R. \quad (3.463)$$

Applying $\frac{\partial}{\partial t}$ to Eq. (3.131) and discard the term with $-\overline{U}\rho_R$ which belongs to the remap phase only, one gets

$$\rho_{tt} = (-\rho U_R - \rho \frac{U}{R})_t. \tag{3.464}$$

Substituting Eqs. (3.463) and (3.464) into Eq. (462), we have

$$\rho_t = -\frac{1}{R}\{(\rho RU)_R - (\delta t)[(\frac{(R\sigma_R^{RR}+\Delta\sigma)}{2} - R\rho\overline{UU}_R$$

$$-\frac{(\rho+\rho_R R)\overline{U}^2}{2}]_R\} - \frac{\delta t}{2}[-\rho U_R - \rho\frac{U}{R}]_t + O(\Delta^2), \tag{3.465}$$

or

$$\rho_t = -\frac{1}{R}\{(\rho RU)_R - (\delta t)[(\frac{(R\sigma_R^{RR}+\Delta\sigma)}{2} - R\rho\overline{UU}_R$$

$$-\frac{(\rho+\rho_R R)\overline{U}^2}{2}]_R\} - \frac{\delta t}{2}[-\rho_t(U_t)_R - (\frac{\rho U}{R})_t] + O(\Delta^2). \tag{3.466}$$

If one will discard the second order terms, then, Eq. (3.466) becomes

$$\rho_t = -\frac{1}{R}(\rho RU)_R, \tag{3.467}$$

or

$$\rho_t = -\frac{1}{R}[R\frac{d(\rho U)}{dR} + (\rho U)\frac{d(R)}{dR}], \tag{3.468}$$

or

$$\rho_t = -\frac{1}{R}[R(\rho U)_R + (\rho U)], \tag{3.469}$$

or

$$\rho_t = -(\rho U)_R - \frac{\rho U}{R} + O(\Delta^2). \tag{3.470}$$

Therefore, Eq. (3.453) satisfies the conservation of mass equation to second-order accuracy.

3.4.2 Radial Momentum

New velocity is calculated based on momentum change during particle transport. The momentum transported from cell (i) to cell $(i+1)$ as shown in Fig. 3.17 can be calculated as below.

$$\delta(M\widetilde{U})_{i+\frac{1}{2}} = 2\pi\delta Z \int_{R_{i+\frac{1}{2}}-\ell}^{R_{i+\frac{1}{2}}} \rho\widetilde{U}RdR, \tag{3.471}$$

$$= 2\pi\delta Z \int_{R_{i+\frac{1}{2}}-\ell}^{R_{i+\frac{1}{2}}} \rho[\widetilde{U}_{i+\frac{1}{2}} - (R_{i+\frac{1}{2}} - R)\widetilde{U}_R]RdR, \tag{3.472}$$

$$\delta(M\widetilde{U})_{i+\frac{1}{2}} = \delta(M)_{i+\frac{1}{2}}(\widetilde{U} - \frac{\delta t}{2}\widetilde{U}\widetilde{U}_R)_{i+\frac{1}{2}} + O(\Delta^4). \tag{3.473}$$

Then the total momentum change in cell (i) is

$$\Delta(M\widetilde{U})_i = \delta(M\widetilde{U})_{i-\frac{1}{2}} - \delta(M\widetilde{U})_{i+\frac{1}{2}}, \tag{3.474}$$

$$= -\delta R \frac{\partial}{\partial R}[\delta(M\widetilde{U})] + O(\Delta^5). \tag{3.475}$$

New velocity U^{n+1} can be calculated as below

$$U^{n+1} = \frac{M\widetilde{U} + \Delta(M\widetilde{U})}{M + \Delta M} = \frac{\widetilde{U} + \Delta(M\widetilde{U})/M}{1 + \Delta M/M}. \tag{3.476}$$

Expanding Eq. (3.471) with Eq. (3.445) multiplying by $\widetilde{U}$, we get

$$(\delta M\widetilde{U})_{i+\frac{1}{2}} = 2\pi\delta Z(\delta t)\{\rho R\widetilde{U}^2 - (\tfrac{\delta t}{2})(\widetilde{U})[\sigma_R^{RR}R + \Delta\sigma$$

$$+2\overline{UU}_R(\rho)(R) + (\rho + \rho_R R)\overline{U}^2]\}_{i+\frac{1}{2}} + O(\Delta^4). \tag{3.477}$$

Now from Eq. (3.475)

$$(\Delta M\widetilde{U})_{i+\frac{1}{2}} = -2\pi\delta Z(\delta t)\{(\rho R\widetilde{U}^2)_R$$

$$-\frac{\partial}{\partial R}[\text{second term in the RHS of Eq. (3.477)}]\} + O(\Delta^5). \tag{3.478}$$

Then

$$\frac{\frac{\Delta M\widetilde{U}}{M}}{= \frac{\Delta M\widetilde{U}}{(2\pi\delta Z\delta R)(\rho R)}}, \tag{3.479}$$

$$= \underbrace{-\frac{\delta t}{\rho R}(R\rho\widetilde{U}^2)_R}_{\alpha}$$

$$+ \underbrace{\frac{\delta t^2}{2\rho R}\frac{\partial}{\partial R}}_{\alpha'}\{\text{second term in the RHS of Eq. (3.477) with }[\,]\} + O(\Delta^5).$$

$$\tag{3.480}$$

From Eqs. (3.449) and (3.445), we have

$$\frac{\Delta M}{M} = \frac{-\delta R \frac{\partial}{\partial R}(\delta M)}{M}, \tag{3.481}$$

$$= \underbrace{-\frac{\delta t}{\rho R}(R\rho \widetilde{U})_R}_{\beta}$$

$$+ \underbrace{\frac{\delta t^2}{2\rho R}\{(R\sigma_R^{RR})_R + (\Delta\sigma)_R + 2(R\rho U U_R)_R + [(R\rho_R + \rho)U^2]_R\}}_{\beta'} + O(\Delta^5). \tag{3.482}$$

From Eq. (3.476)

$$U^{n+1} = \frac{M\widetilde{U} + \Delta(M\widetilde{U})}{M + \Delta M} = \frac{\widetilde{U} + \Delta(M\widetilde{U})/M}{1 + \Delta M/M}, \tag{3.483}$$

or

$$U^{n+1} = \frac{\widetilde{U} + \alpha + \alpha'}{1 + \beta + \beta'}, \tag{3.484}$$

$$= (\widetilde{U} + \alpha + \alpha')(1 + \beta + \beta')^{-1}, \tag{3.485}$$

$$= (\widetilde{U} + \alpha + \alpha')[1 - (\beta + \beta') + (\beta + \beta')^2 - ...] + O(\Delta^3). \tag{3.486}$$

Since α and β are of order δt while α' and β' are $(\delta t)^2$, expanding Eq. (3.486) and discard terms with $(\delta t)^3$ or $(\delta t)^4$, one gets

$$U^{n+1} = \widetilde{U} + (\alpha - \beta\widetilde{U}) + (\alpha' - \alpha\beta + \beta^2\widetilde{U} - \beta'\widetilde{U}) + O(\Delta^3). \tag{3.487}$$

Also

$$(\alpha - \beta\widetilde{U}) = -\frac{\delta t}{\rho R}(R\rho\widetilde{U}^2)_R - (-\frac{\delta t}{\rho R})(R\rho\widetilde{U})_R\widetilde{U}, \tag{3.488}$$

$$= -\frac{\delta t}{\rho R}[\widetilde{U}(R\rho\widetilde{U})_R + (R\rho\widetilde{U})\widetilde{U}_R] + (\frac{\delta t}{\rho R})(R\rho\widetilde{U})_R\widetilde{U}, \tag{3.489}$$

$$= -\frac{\delta t}{\rho R}(R\rho\widetilde{U})\widetilde{U}_R, \tag{3.490}$$

$$= -\delta t \widetilde{U}\widetilde{U}_R + O(\Delta^2), \tag{3.491}$$

and

$$(\alpha' - \alpha\beta + \beta^2 \tilde{U} - \beta'\tilde{U})/\delta t^2$$

$$= \frac{1}{2R\rho}[R\sigma_R^{RR}U_R + U_R\Delta\sigma + R\rho(U^2 U_R)_R]. \tag{3.492}$$

Now rewriting Eq. (3.487) as

$$U^{n+1} = \tilde{U} + [\text{the RHS of Eq. (3.491)}] + [\text{the RHS of Eq. (3.491)}]\delta t^2. \tag{3.493}$$

From Eqs. (3.401) and (3.404), one gets

$$\tilde{U} = U^n + \delta t \left(\frac{\sigma_R^{RR}}{\rho} + \alpha\frac{\Delta\sigma}{R\rho}\right)^{n+1/2}. \tag{3.494}$$

Substituting Eq. (3.494) into Eq. (3.493), we have

$$U^{n+1} = U^n + \delta t \left(\frac{\sigma_R^{RR}}{\rho} + \alpha\frac{\Delta\sigma}{R\rho}\right)^{n+1/2} - \delta t U U_R$$

$$+ \frac{\delta t^2}{2R\rho}[R\sigma_R^{RR}U_R + U_R\Delta\sigma + R\rho(U^2 U_R)_R] + O(\Delta^3). \tag{3.495}$$

Again from the Taylor series expansion,

$$U_t = \frac{U^{(n+1)} - U^{(n)}}{\Delta t} - \frac{(\Delta t)}{2}U_{tt} + O(\Delta^2). \tag{3.496}$$

Inserting Eq. (3.495) and from differentiating Eqs. (3.132) and (3.133) into Eq. (3.496) and after some algebra, we can get

$$U_t = -UU_R + \left(\frac{\sigma_R^{RR}}{\rho} + \alpha\frac{\Delta\sigma}{R\rho}\right)^{n+1/2} + O(\Delta^2). \tag{3.497}$$

Therefore, the present particle transport and remapping scheme results in a second-order accurate radial momentum equation.

3.4.3 Axial Momentum

Now we will consider the change in axial momentum during a time advancement in radial direction. For a hydrodynamic problem, $\tilde{V} = V^n$, and the remapping of the axial momentum is straightforward. For an elastic-plastic flow, deviatoric stress component contributes to Lagrangian phase axial velocity. The truncation error analysis is similar to the radial momentum equation case.

Rewriting Eq. (3.186),

$$\widetilde{V} = V^n + \delta t\{[\frac{(\sigma^{RZ})_R}{\rho}]^{n+1/2} + \alpha[\frac{\sigma^{RZ}}{\rho R}]^{n+1/2}\}. \tag{3.498}$$

The axial momentum transported from cell (i) to cell $(i+1)$ during δt is

$$\delta(M\widetilde{V})_{i+\frac{1}{2}} = \delta(M)_{i+\frac{1}{2}}(\widetilde{V} - \frac{\delta t}{2}\widetilde{U}\widetilde{V}_R)_{i+\frac{1}{2}} + O(\Delta^4). \tag{3.499}$$

The total axial momentum change in cell (i) is

$$\Delta(M\widetilde{V})_i = \delta(M\widetilde{V})_{i-\frac{1}{2}} - \delta(M\widetilde{V})_{i+\frac{1}{2}}, \tag{3.500}$$

$$= -\delta R \frac{\partial}{\partial R}[\delta(M\widetilde{V})] + O(\Delta^5), \tag{3.501}$$

and the new axial velocity is then calculated by

$$V^{n+1} = \frac{M\widetilde{V} + \Delta(M\widetilde{V})}{M + \Delta M} = \frac{\widetilde{V} + \Delta(M\widetilde{V})/M}{1 + \Delta M/M}. \tag{3.502}$$

Some algebraic detail:

$$(\delta M\widetilde{V})_{i+\frac{1}{2}} = 2\pi\delta Z(\delta t)\{\rho R \widetilde{U}\widetilde{V} - (\frac{\delta t}{2})(\widetilde{V})[\sigma_R^{RR}R + \Delta\sigma$$

$$+2\overline{UU}_R(\rho)(R) + (\rho + \rho_R R)\overline{U}^2]\}_{i+\frac{1}{2}} + O(\Delta^4). \tag{3.503}$$

From Eq. (3.501),

$$(\Delta M\widetilde{V})_{i+\frac{1}{2}} = -2\pi\delta Z(\delta t)\{(\rho R\widetilde{U}\widetilde{V})_R$$

$$-\frac{\partial}{\partial R}[\text{second term in the RHS of Eq. (3.503)}]\} + O(\Delta^5). \tag{3.504}$$

Then

$$\frac{\Delta M\widetilde{V}}{M}$$

$$= \frac{\Delta M\widetilde{V}}{(2\pi\delta Z\delta R)(\rho R)}, \tag{3.505}$$

$$= -\underbrace{\frac{\delta t}{\rho R}(R\rho \widetilde{U}\widetilde{V})_R}_{A}$$

$$+ \underbrace{\frac{\delta t^2}{2\rho R}\frac{\partial}{\partial R}}_{A'}\{\text{second term in the RHS of Eq. (3.503) with []}\} + O(\Delta^5).$$

(3.506)

From Eq. (3.502),

$$V^{n+1} = \frac{\widetilde{V} + A + A'}{1 + \beta + \beta'}, \tag{3.507}$$

$$= (\widetilde{V} + A + A')(1 + \beta + \beta')^{-1}, \tag{3.508}$$

$$= (\widetilde{V} + A + A')[1 - (\beta + \beta') + (\beta + \beta')^2 - ...] + O(\Delta^3). \tag{3.509}$$

Since A and β are of order δt while A' and β' are $(\delta t)^2$, expanding Eq. (3.509) and discard terms with $(\delta t)^3$ or $(\delta t)^4$, one gets

$$V^{n+1} = \widetilde{V} + (A - \beta\widetilde{V}) + (A' - A\beta + \beta^2\widetilde{V} - \beta'\widetilde{V}) + O(\Delta^3). \tag{3.510}$$

Also

$$(A - \beta\widetilde{V}) = -\frac{\delta t}{\rho R}(R\rho\widetilde{U}\widetilde{V})_R - (-\frac{\delta t}{\rho R})(R\rho\widetilde{U})_R\widetilde{V}, \tag{3.511}$$

$$= -\delta t\{\widetilde{U}\widetilde{V}_R + \delta t[U(\frac{S_R^{RZ}}{\rho} + \frac{S^{RZ}}{R\rho})_R + V_R(\frac{\sigma_R^{RR}}{\rho} + \frac{\Delta\sigma}{R\rho})]\} + O(\Delta^2), \tag{3.512}$$

and

$$(A' - A\beta + \beta^2\widetilde{V} - \beta'\widetilde{V})/\delta t^2$$

$$= \frac{1}{2R\rho}[R\sigma_R^{RR}V_R + V_R\Delta\sigma + R\rho(U^2V_R)_R]. \tag{3.513}$$

Rewriting Eq. (3.510),

$$V^{n+1} = \widetilde{V} + \text{RHS of Eq. (3.512)} + \delta t^2[\text{RHS of Eq. (3.513)}]. \tag{3.514}$$

From Eq. (3.186), we have

$$\widetilde{V} = V^n + \delta t\{[\frac{(\sigma^{RZ})_R}{\rho}]^{n+1/2} + \alpha[\frac{\sigma^{RZ}}{\rho R}]^{n+1/2}\}. \tag{3.515}$$

Since $S^{RZ} = \sigma^{RZ}$, the above equation can be written as

$$\widetilde{V} = V^n + \delta t\{[\frac{(S^{RZ})_R}{\rho}]^{n+1/2} + \alpha[\frac{S^{RZ}}{\rho R}]^{n+1/2}\}. \qquad (3.516)$$

Substituting Eq. (3.516) into Eq. (3.514), it becomes

$$V^{n+1} = V^n + \delta t\{[\frac{(S^{RZ})_R}{\rho}]^{n+1/2} + [\frac{S^{RZ}}{\rho R}]^{n+1/2}\} - \delta t U V_R$$

$$-\delta t^2[U(\frac{S_R^{RZ}}{\rho} + \frac{S^{RZ}}{R\rho})_R + V_R(\frac{\sigma_R}{\rho} + \frac{\Delta \sigma}{R\rho})]$$

$$+\frac{\delta t^2}{2R\rho}[R\sigma_R^{RR}V_R + V_R\Delta\sigma + R\rho(U^2V_R)_R] + O(\Delta^3). \qquad (3.517)$$

From the Taylor series expansion,

$$V_t = \frac{V^{(n+1)} - V^{(n)}}{\Delta t} - \frac{(\Delta t)}{2}V_{tt} + O(\Delta^2). \qquad (3.518)$$

Inserting Eq. (3.517) and from differentiating Eq. (3.132) and (3.133) into Eq. (3.518) and after some simplification, we get

$$V_t = -UV_R + \frac{1}{\rho}[(S_R^{RZ})^{n+1/2} + (\frac{S^{RZ}}{R})^{n+1/2}] + O(\Delta^2). \qquad (3.519)$$

Therefore, we have a second-order accurate axial momentum equation.

3.4.4 Internal Energy

Following a procedure similar to that in Section 3.4.3 and from Eq. (3.473) we get

$$\delta(M\widetilde{\epsilon})_{i+\frac{1}{2}} = \delta(M)_{i+\frac{1}{2}}(\widetilde{\epsilon} - \frac{\delta t}{2}\widetilde{U}\widetilde{\epsilon}_R)_{i+\frac{1}{2}} + O(\Delta^4). \qquad (3.520)$$

Then the total internal energy change in cell (i) is

$$\Delta(M\widetilde{\epsilon})_i = \delta(M\widetilde{\epsilon})_{i-\frac{1}{2}} - \delta(M\widetilde{\epsilon})_{i+\frac{1}{2}}, \qquad (3.521)$$

$$= -\delta R\frac{\partial}{\partial R}[\delta(M\widetilde{\epsilon})] + O(\Delta^5). \qquad (3.522)$$

New internal energy ϵ^{n+1} can be calculated as below

$$\epsilon^{n+1} = \frac{M\widetilde{\epsilon} + \Delta(M\widetilde{\epsilon})}{M + \Delta M} = \frac{\widetilde{\epsilon} + \Delta(M\widetilde{\epsilon})/M}{1 + \Delta M/M}. \qquad (3.523)$$

Replacing $\widetilde{U}$ by $\widetilde{\epsilon}$ in Eq. (3.477), we get

$$(\delta M\widetilde{\epsilon})_{i+\frac{1}{2}} = 2\pi\delta Z(\delta t)\{\rho R\widetilde{U}\widetilde{\epsilon} - (\tfrac{\delta t}{2})(\widetilde{\epsilon})[\sigma_R^{RR}R + \Delta\sigma$$

$$+2\overline{UU}_R(\rho)(R) + (\rho + \rho_R R)\overline{U}^2]\}_{i+\frac{1}{2}} + O(\Delta^4). \tag{3.524}$$

Replacing $\widetilde{V}$ by $\widetilde{\epsilon}$ in Eq. (3.501), we get

$$\Delta(M\widetilde{\epsilon})_i = \delta(M\widetilde{\epsilon})_{i-\frac{1}{2}} - \delta(M\widetilde{\epsilon})_{i+\frac{1}{2}}, \tag{3.525}$$

$$= -\delta R\frac{\partial}{\partial R}[\delta(M\widetilde{\epsilon})] + O(\Delta^5), \tag{3.526}$$

or

$$(\Delta M\widetilde{\epsilon})_{i+\frac{1}{2}} = -2\pi\delta Z(\delta t)\{(\rho R\widetilde{U}\widetilde{\epsilon})_R$$

$$-\frac{\partial}{\partial R}[\text{second term in the RHS of Eq. (3.524)}]\} + O(\Delta^5). \tag{3.527}$$

Then

$$\frac{\frac{\Delta M\widetilde{\epsilon}}{M}}{} = \frac{\Delta M\widetilde{\epsilon}}{(2\pi\delta Z\delta R)(\rho R)}, \tag{3.528}$$

$$= \underbrace{-\frac{\delta t}{\rho R}(R\rho\widetilde{U}\widetilde{\epsilon})_R}_{C}$$

$$+\underbrace{\frac{\delta t^2 \widetilde{\epsilon}}{2\rho R}\frac{\partial}{\partial R}}_{C'}\{\text{second term in the RHS of Eq. (3.524) with [ ]}\} + O(\Delta^5).$$

$$\tag{3.529}$$

From Eq. (3.523)

$$\epsilon^{n+1} = \frac{\widetilde{\epsilon} + C + C'}{1 + \beta + \beta'}, \tag{3.530}$$

$$= (\widetilde{\epsilon} + C + C')(1 + \beta + \beta')^{-1}, \tag{3.531}$$

$$= (\widetilde{\epsilon} + C + C')[1 - (\beta + \beta') + (\beta + \beta')^2 - ...] + O(\Delta^3). \tag{3.532}$$

Since C is of order δt while C' is $(\delta t)^2$, expanding Eq. (3.532) and discard terms with $(\delta t)^3$ or $(\delta t)^4$, one gets

$$\epsilon^{n+1} = \widetilde{\epsilon} + (C - \beta\widetilde{\epsilon}) + (C' - C\beta + \beta^2\widetilde{\epsilon} - \beta'\widetilde{\epsilon}) + O(\Delta^3). \quad (3.533)$$

Also

$$(C - \beta\widetilde{\epsilon}) = -\frac{\delta t}{\rho R}(R\rho\widetilde{U}\widetilde{\epsilon})_R - (-\frac{\delta t}{\rho R})(R\rho\widetilde{U})_R\widetilde{\epsilon}, \quad (3.534)$$

$$= -\delta t\{\widetilde{U}\widetilde{\epsilon}_R + \delta t[U(\frac{S^{RZ}_R}{\rho} + \frac{S^{RZ}}{R\rho})_R + \epsilon_R(\frac{\sigma^{RR}_R}{\rho} + \frac{\Delta\sigma}{R\rho})]\} + O(\Delta^2), \quad (3.535)$$

and

$$(C' - C\beta + \beta^2\widetilde{\epsilon} - \beta'\widetilde{\epsilon})/\delta t^2$$

$$= \frac{1}{2R\rho}[R\sigma^{RR}_R\epsilon_R + \epsilon_R\Delta\sigma + R\rho(U^2\epsilon_R)_R]. \quad (3.536)$$

From Eqs. (3.533), (3.535), and (3.536), one gets

$$\epsilon^{n+1} = \widetilde{\epsilon} - \delta t\widetilde{U}\widetilde{\epsilon}_R + \frac{\delta t^2}{2R\rho}[R\sigma^{RR}_R\epsilon_R + \epsilon_R\Delta\sigma + R\rho(U^2\epsilon_R)_R] + O(\Delta^3). \quad (3.537)$$

The truncation error of this depends on the $\widetilde{\epsilon}$ expression. From the Taylor series, we have

$$\epsilon_t = \frac{\epsilon^{(n+1)} - \epsilon^{(n)}}{\delta t} - \frac{(\delta t)}{2}\epsilon_{tt} + O(\Delta^2). \quad (3.538)$$

From Eq. (3.187), we have

$$\widetilde{\epsilon} = \epsilon^n + \delta t\{[\frac{\sigma^{RR}(U)_R}{\rho}]^{n+1/2} + \alpha[\frac{\sigma^{\theta\theta}U}{\rho R}]^{n+1/2} + [\frac{\sigma^{RZ}V_R}{\rho}]^{n+1/2}\}. \quad (3.539)$$

Let

$$E = \frac{\sigma^{RR}(U)_R}{\rho}, \quad (3.540)$$

$$F = \frac{\sigma^{\theta\theta}U}{\rho R}, \quad (3.541)$$

and

$$G = \frac{\sigma^{RZ}V_R}{\rho}. \quad (3.542)$$

Substituting Eqs. (3.540)-(3.542) into Eq. (3.539) and set $\alpha = 1$, we have

$$\widetilde{\epsilon} = \epsilon^n + \delta t\{E^{n+1/2} + F^{n+1/2} + G^{n+1/2}\}. \tag{3.543}$$

From Eq. (3.401) and (3.404) with $\alpha = 1$, one gets

$$\widetilde{U} = U^n + \delta t(\frac{\sigma_R^{RR}}{\rho} + \frac{\Delta\sigma}{R\rho})^{n+1/2}. \tag{3.544}$$

Now, let

$$H = \frac{\sigma_R^{RR}}{\rho} + \frac{\Delta\sigma}{R\rho}. \tag{3.545}$$

Substituting Eq. (3.545) into Eq. (3.544), we have

$$\widetilde{U} = U^n + \delta t(H)^{n+1/2}. \tag{3.546}$$

Using Eq. (3.164), Eq. (3.546) can be written as

$$\widetilde{U} = U^n + \delta t[(H)^n + \frac{\delta t}{2}(H_t + UH_R)]. \tag{3.547}$$

Multiplying Eq. (3.547) by $\widetilde{\epsilon}_R$, we have

$$\widetilde{U}\widetilde{\epsilon}_R = U^n\widetilde{\epsilon}_R + \delta t\widetilde{\epsilon}_R[(H)^n + \frac{\delta t}{2}(H_t + UH_R)]. \tag{3.548}$$

Substituting Eq. (3.543) into Eq. (3.537), one gets

$$\frac{\epsilon^{(n+1)} - \epsilon^{(n)}}{\Delta t} = (E + F + G)^n + \frac{\delta t}{2}(E_t + UE_R + F_t + UF_R + G_t + UG_R)$$

$$-U^n\epsilon_R - \Delta t\epsilon_R[H^n + \frac{\delta t}{2}(H_t + UH_R)]$$

$$+\frac{\delta t}{2}[\frac{\sigma_R^{RR}\epsilon_R}{\rho} + \frac{\epsilon_R\Delta\sigma}{R\rho} + (U^2\epsilon_R)_R] + O(\Delta^2). \tag{3.549}$$

Substituting Eq. (3.549) into Eq. (3.538) and discard terms with δt, we have

$$\epsilon_t = -U\epsilon_R + \frac{1}{\rho}(\sigma^{RR}U_R + \sigma^{\theta\theta}\frac{U}{R} + \sigma^{RZ}V_R) + O(\Delta^2). \tag{3.550}$$

Therefore, we have a second-order scheme for ϵ.

Note that if the following expression is used instead of Eq. (3.187), we get a first-order accurate energy equation:

$$\widetilde{\epsilon} = \epsilon^n + \frac{\delta t}{\rho^n}[(\sigma^{RR})^{n+1/2}\widetilde{U}_R + (\sigma^{\theta\theta})^{n+1/2}\frac{\widetilde{U}}{R} + (\sigma^{RZ})^{n+1/2}\widetilde{V}_R) + O(\Delta). \tag{3.551}$$

Since we split governing equations in radial and axial directions in transforming the Eulerian(R, Z) to the Lagrangian(k, ℓ) coordinate system, therefore the Jacobian is

$$J = \frac{\partial(R,Z)}{\partial(k,\ell)} = \frac{\partial R}{\partial k}\frac{\partial Z}{\partial \ell} - \frac{\partial R}{\partial \ell}\frac{\partial Z}{\partial k}, \qquad (3.552)$$

where

$R = k,$

$Z = \ell,$

$\frac{\partial R}{\partial k} = 1,$

$\frac{\partial Z}{\partial \ell} = 1,$

$\frac{\partial R}{\partial \ell} = 0,$

and

$\frac{\partial Z}{\partial k} = 0.$

The Jacobian is
Jacobian = $(1)(1)-(0)(0)=1$. For R-direction calculation, one gets

$$\frac{D}{Dt} = U\frac{\partial}{\partial R} + \frac{\partial}{\partial t}. \qquad (3.553)$$

Therefore, Eq. (3.187) in the Lagrangian phase calculation is straightforward consequence.

3.4.5 Deviatoric Stresses

In the present scheme, stresses are assigned to each individual material. Therefore, the difference between volume-weighting and mass-weighting scheme in transporting stress components is contributed by the local density gradient. Both schemes result in second-order accurate results. However, for programming, the mass-weighting scheme is more convenient, and the following truncation error analysis is based on that. Similarly to Eq. (3.537), by replacing ϵ_R with S_R, ϵ^{n+1} with $(S^{ij})^{n+1}$, $\widetilde{\epsilon}$ with $\widetilde{S}^{ij}$, and $\widetilde{\epsilon}_R$ with $\widetilde{S}_R^{ij}$ we get for S^{ij} transport

$$(S^{ij})^{n+1} = \widetilde{S}^{ij} - \delta t \widetilde{U}\widetilde{S}_R^{ij} + \frac{\delta t^2}{2R\rho}[R\sigma_R^{RR}S_R + S_R\Delta\sigma + R\rho(U^2 S_R)_R] + O(\Delta^3). \qquad (3.554)$$

For the purpose of accuracy analysis, in the third term on the right hand side of Eq. (3.537), we replace ϵ_R with S_R^{RR} or just simply write S_R. Similar to Eq. (3.538), we have

$$(S^{ij})_t = \frac{(S^{ij})^{(n+1)} - (S^{ij})^{(n)}}{\delta t} - \frac{(\delta t)}{2}(S^{ij})_{tt} + O(\Delta^2). \qquad (3.555)$$

We can generalize $\widetilde{S}^{ij}$ by Eq. (3.334), that is,

$$\widetilde{S}^{ij} = (S^{ij})^n + \delta t(2G\dot{e}^{ij} - A\dot{W}S^{ij})^{n+1/2}. \qquad (3.556)$$

Also, from Eqs. (3.401) and (3.404), one gets

$$\widetilde{U} = U^n + \delta t(\frac{\sigma_R^{RR}}{\rho} + \frac{\Delta\sigma}{R\rho})^{n+1/2}. \qquad (3.557)$$

From Eqs. (3.556) and (3.557) with discarding the superscripts n and $n+1/2$, we have

$$\widetilde{U}\widetilde{S}_R^{ij} = [U + \delta t(\frac{\sigma_R^{RR}}{\rho} + \frac{\Delta\sigma}{R\rho})][(S^{ij})_R + \delta t(2G\dot{e}^{ij} - A\dot{W}S^{ij})_R], \qquad (3.558)$$

$$= US_R + \delta t[S_R(\frac{\sigma_R^{RR}}{\rho} + \frac{\Delta\sigma}{R\rho}) + U(2G\dot{e} - A\dot{W}S)_R] + O(\Delta^2). \qquad (3.559)$$

Rewriting Eq. (3.319),

$$S_t^{ij} = -\overline{U}S_R^{ij} + 2G(\dot{e})^{ij} - A\dot{W}S^{ij}. \qquad (3.560)$$

Then

$$S_{tt}^{ij} = -(\overline{U}S_R^{ij})_t + [2G(\dot{e})^{ij} - A\dot{W}S^{ij}]_t. \qquad (3.561)$$

Now from Eqs. (3.554) and (3.556), while discarding term with δt, we have

$$\frac{(S^{ij})^{n+1} - (S^{ij})^n}{\delta t} = [2G(\dot{e})^{ij} - A\dot{W}S^{ij}]^{n+1/2} - \widetilde{U}\widetilde{S}_R^{ij} + O(\Delta^2). \qquad (3.562)$$

Substituting Eqs. (3.561) and (3.562) into Eq. (3.555), while discarding term with δt or δt^2, we have

$$(S^{ij})_t = -\widetilde{U}\widetilde{S}_R^{ij} + [2G(\dot{e})^{ij} - A\dot{W}S^{ij}]^{n+1/2} + O(\Delta^2). \qquad (3.563)$$

Using the formula from Eq. (3.164), while discarding term with δt, Eq. (3.563) becomes

$$(S^{ij})_t = -\widetilde{U}\widetilde{S}_R^{ij} + [2G(\dot{e})^{ij} - A\dot{W}S^{ij}]^n + O(\Delta^2). \qquad (3.564)$$

Substituting $\widetilde{U}\widetilde{S}_R^{ij}$ from Eq. (3.559) while discarding term with δt, Eq. (3.564) becomes

$$(S^{ij})_t = -US_R^{ij} + [2G(\dot{e})^{ij} - A\dot{W}S^{ij}]^n + O(\Delta^2). \qquad (3.565)$$

Therefore, mass-weighted transport and remapping of S^{ij} result in second-order accuracy.

3.5 Equivalent Plastic Strain

Total strain rate is

$$\dot{e}^{ij} = (\dot{e}')^{ij} + (\dot{e}'')^{ij} = (\dot{e}')^{ij} + \dot{\alpha}^{ij}, \tag{3.566}$$

where $(\dot{e}')^{ij}$ is the elastic and $(\dot{e}'')^{ij}$ the plastic parts. Since Hook's law is used for elastic phase

$$(\dot{S})^{ij} = 2G(\dot{e}')^{ij}, \tag{3.567}$$

and for plastic part

$$(\dot{\alpha})^{ij} = \frac{\dot{W}}{\frac{2}{3}(Y^0)^2} S^{ij}. \tag{3.568}$$

Combining these will result in a Prandtl–Reuss equation for plastic flow as shown in Eq. (3.91), that is

$$\dot{S}^{ij} = 2G\dot{e}^{ij} - 2G\dot{\alpha}^{ij}, \tag{3.569}$$

$$= 2G\dot{e}^{ij} - \frac{G\dot{W}}{\frac{1}{3}(Y^0)^2} S^{ij}. \tag{3.570}$$

Then α_p is defined by

$$\alpha_p \triangleq (\frac{2}{3}\alpha^{ij}\alpha^{ij})^{1/2}. \tag{3.571}$$

In present method, the equivalent plastic strain, α_p, is calculated as below:

$$\alpha_p = \int (\frac{2}{3}\dot{\alpha}^{ij}\dot{\alpha}^{ij})^{1/2} dt, \tag{3.572}$$

where

$$\dot{\alpha}^{ij}\dot{\alpha}^{ij} = (\frac{\dot{W}}{\frac{2}{3}(Y^0)^2} S^{ij})^2 S^{ij}S^{ij}, \tag{3.573}$$

$$= (\frac{\dot{W}}{\frac{2}{3}(Y^0)^2} S^{ij})^2 [(S^{RR})^2 + (S^{ZZ})^2 + (S^{\theta\theta})^2 + (S^{RZ})^2], \tag{3.574}$$

$$= (\frac{\dot{W}}{\frac{2}{3}(Y^0)^2} \phi. \tag{3.575}$$

It is noted that for a perfect plastic model

$$\phi = \frac{2}{3}(Y^0)^2, \tag{3.576}$$

and then

$$\dot{\alpha}^{ij}\dot{\alpha}^{ij} = \frac{(\dot{W})^2}{\frac{2}{3}(Y^0)^2} \,. \tag{3.577}$$

Substituting Eq. (3.577) into Eq. (3.572), one gets

$$\alpha_p = \sqrt{\frac{2}{3}} \int \frac{\dot{W}}{\sqrt{\frac{2}{3}}(Y^0)} dt \,, \tag{3.578}$$

$$= \int \frac{\dot{W}}{Y^0} dt \,. \tag{3.579}$$

Applying the differentiation d to Eq. (3.572), we have

$$d\alpha_p = (\frac{2}{3}\dot{\alpha}^{ij}\dot{\alpha}^{ij})^{1/2} dt \,, \tag{3.580}$$

or

$$\frac{d\alpha_p}{dt} = (\frac{2}{3}\dot{\alpha}^{ij}\dot{\alpha}^{ij})^{1/2} \,. \tag{3.581}$$

The expression in finite difference for Eq. (3.581) is

$$\frac{\alpha_p^n - \alpha_p^{n-1}}{\Delta t} = (\frac{2}{3}\dot{\alpha}^{ij}\dot{\alpha}^{ij})^{1/2} \,. \tag{3.582}$$

Substituting $\dot{\alpha}^{ij}\dot{\alpha}^{ij}$ from Eq. (3.575), it follows

$$\alpha_p^n = \alpha_p^{n-1} + \sqrt{\frac{2}{3}}(\frac{\dot{W}}{\frac{2}{3}(Y^0)^2})\sqrt{\phi}\Delta t \,, \tag{3.583}$$

where

$$\dot{W} = (\dot{e})^{ij} S^{ij} \,. \tag{3.584}$$

From Eq. (3.97), one gets

$$\dot{W} = (\frac{\partial U}{\partial R})S^{RR} + (\frac{\partial U}{\partial Z} + \frac{\partial V}{\partial R})S^{RZ} + (\frac{\partial V}{\partial Z})S^{ZZ} + (\frac{\alpha U}{R})S^{\theta\theta} \,. \tag{3.585}$$

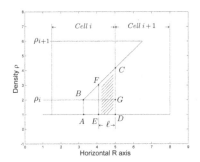

Fig. 3.18 A steep gradient of density may result in negative mass during transport.

3.6 FCT Applied to Second-Order PIC

3.6.1 *Introduction*

The FCT scheme, originally proposed by Boris and Book [3.10]. Consists of two stages, transport and flux-corrected anti-diffusion. By allowing flux correction, the FTC method generates no local extreme and maintains the positivity of mass and energy density. Thus, steep gradients and shocks can be handled well.

The two stages in FCT algorithm resemble the two computational phases of the PIC scheme, so a slightly modified FCT algorithm can easily be incorporated into the PIC method. Since the positivity of mass is automatically satisfied by PIC, the effect of anti-diffusion is utilized here to handle local extreme. For a semi-continuous version of PIC (see Clark [3.9]), the positivity of mass is maintained by a modified mass transport scheme, which is explained next.

3.6.2 *Modified Mass Transport*

Around a very steep gradient or local extreme, the mass transport scheme described in Section 3.4.1 may result in negative mass in certain cells. This situation is illustrated in Fig. 3.18.

According to Section 3.4.1, the mass represented by the shaded area will be transported from cell (i) to cell $(i+1)$. When the density gradient is steep locally, this transported mass can be larger than the total mass in cell (i). To prevent this situation, the mass transport scheme is modified as below. Let us consider a linearly interpolated density profile as given in Eq. (3.589) and shown in Fig. 3.19.

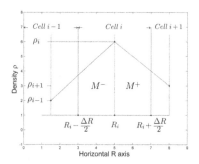

Fig. 3.19 Modified mass in cell i based on linear density profile.

Average mass in cell (i):

$$M_{avg} = 2\pi \delta Z \int_{R_i - \Delta R/2}^{R_i + \Delta R/2} \rho_i R' dR', \quad (3.586)$$

$$= (\pi \delta Z \rho_i)[(R_i + \Delta R/2)^2 - (R_i - \Delta R/2)^2], \quad (3.587)$$

$$= 2\pi \delta Z \rho_i (R_i \Delta R). \quad (3.588)$$

Mass in cell (i) based on linear density profile:

$$M_{lin} = M^- + M^+. \quad (3.589)$$

Let us denote, for convenience, that
$\rho^- = \rho_{i-1}$, $\rho^+ = \rho_{i+1}$, $R^+ = R_{i+1}$ etc.,
and
$R = R_i$, $\rho = \rho_i$, $\lambda = \frac{\Delta R}{2}$;
then, as shown in Fig. 3.20, M^- is the mass confined in the region $\overline{AGHD}$. At point A, the density is ρ_A and

$$\frac{\rho_A - \rho^-}{\rho_i - \rho^-} = \frac{\overline{EB}}{\overline{EC}} = \frac{R' - (R_i - \lambda)}{\lambda}, \quad (3.590)$$

or

$$\rho_A - \rho^- = \frac{R' - (R_i - \lambda)}{\lambda}(\rho_i - \rho^-), \quad (3.591)$$

or

$$\rho_A = \rho^- + [R' - (R_i - \lambda)]\frac{\rho_i - \rho^-}{\lambda}, \quad (3.592)$$

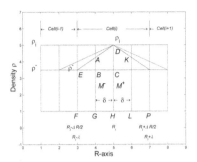

Fig. 3.20 M^- is the mass in cell i confined by the region $\overline{AGHD}$.

or

$$\rho_A = \rho^- + [R' - (R_i - \lambda)]\rho_R^-. \tag{3.593}$$

Therefore

$$M^- = 2\pi \delta Z \int_{R-\delta}^{R} \{\rho^- + [R' - (R_i - \lambda)]\rho_R^-\} R' dR', \tag{3.594}$$

$$= 2\pi \delta Z \int_{R-\delta}^{R} \{\rho^- R' + \rho_R^- (R')^2 - (R_i - \lambda)\rho_R^- R'\} dR', \tag{3.595}$$

$$= 2\pi \delta Z \{\rho^- \frac{(R')^2}{2} + \rho_R^- \frac{(R')^3}{3} - (R_i - \lambda)\rho_R^- \frac{(R')^2}{2}\}|_{R-\delta}^{R}. \tag{3.596}$$

In Eq. (3.596), we like to set $R - \delta = R - \lambda$, therefore

$$M^- = 2\pi \delta Z \{\rho_R^- \frac{(R')^3}{3} + [\rho^- - (R_i - \lambda)\rho_R^-]\frac{(R')^2}{2}\}|_{R-\lambda}^{R}, \tag{3.597}$$

or

$$M^- = 2\pi \delta Z \{\tfrac{1}{3}\rho_R^-(\lambda)(3R^2 - 3R\lambda + \lambda^2)$$

$$+ \frac{1}{2}[\rho^- - (R_i - \lambda)\rho_R^-](\lambda)(2R - \lambda)\}. \tag{3.598}$$

Since

$$\rho_R^- = \frac{\rho - \rho^-}{\lambda}. \tag{3.599}$$

Equation (3.598) becomes

$$M^- = 2\pi\delta Z\{\tfrac{1}{3}\tfrac{\rho-\rho^-}{\lambda}(\lambda)(3R^2 - 3R\lambda + \lambda^2)$$

$$+\tfrac{1}{2}[\rho^- - (R_i - \lambda)\tfrac{\rho-\rho^-}{\lambda}](\lambda)(2R - \lambda)\}, \qquad (3.600)$$

or

$$M^- = 2\pi\delta Z\{(\rho - \rho^-)(R^2 - R\lambda + \tfrac{\lambda^2}{3})$$

$$+[\tfrac{\lambda\rho^-}{2} - \tfrac{1}{2}(\rho - \rho^-)(R - \lambda)](2R - \lambda)\}, \qquad (3.601)$$

or

$$M^- = 2\pi\delta Z(\lambda)[\rho^- R - \rho^-(\tfrac{\lambda}{2}) + \tfrac{R}{2}(\rho - \rho^-) - \tfrac{\lambda}{6}(\rho - \rho^-)], \qquad (3.602)$$

or

$$M^- = 2\pi\delta Z(\lambda)[\rho(\tfrac{R}{2} - \tfrac{\lambda}{6}) + \rho^-(\tfrac{R}{2} - \tfrac{\lambda}{3})], \qquad (3.603)$$

or

$$M^- = \pi\delta Z[\rho(R\lambda - \tfrac{\lambda^2}{3}) + \rho^-(R\lambda - \tfrac{2\lambda^2}{3})]. \qquad (3.604)$$

Then, as shown in Fig. 3.20, M^+ is the mass confined in the region $\overline{DHLK}$. At point K, the density is ρ_K and

$$\rho_K = \rho + (R' - R)\rho_R^+, \qquad (3.605)$$

and

$$M^+ = 2\pi\delta Z \int_R^{R+\lambda} [\rho + (R' - R)\rho_R^+]R'dR', \qquad (3.606)$$

$$= 2\pi\delta Z \int_R^{R+\lambda} [\rho + (R' - R)(\tfrac{\rho^+ - \rho}{\lambda})]R'dR', \qquad (3.607)$$

$$= 2\pi\delta Z[\rho\tfrac{(R')^2}{2} - R(\tfrac{\rho^+ - \rho}{\lambda})\tfrac{(R')^2}{2} + (\tfrac{\rho^+ - \rho}{\lambda})\tfrac{(R')^3}{3}]|_R^{R+\lambda}, \qquad (3.608)$$

$$= 2\pi\delta Z\{\tfrac{1}{2}[\rho - R(\tfrac{\rho^+ - \rho}{\lambda})](R')^2 + \tfrac{1}{3}(\tfrac{\rho^+ - \rho}{\lambda})(R')^3\}|_R^{R+\lambda}. \qquad (3.609)$$

Let

$$X = \tfrac{1}{2}[\rho - R(\tfrac{\rho^+ - \rho}{\lambda})], \qquad (3.610)$$

and
$$Y = \frac{1}{3}(\frac{\rho^+ - \rho}{\lambda}), \tag{3.611}$$

then, Eq. (3.609) becomes

$$M^+ = 2\pi\delta Z[X \cdot (R')^2 + Y \cdot (R')^3]|_R^{R+\lambda}, \tag{3.612}$$

$$= 2\pi\delta Z\{X[(R+\lambda)^2 - R^2] + Y[(R+\lambda)^3 - R^3]\}, \tag{3.613}$$

$$= 2\pi\delta Z(\lambda)[X(2R+\lambda) + Y(3R^2 + 3R\lambda + \lambda^2)]. \tag{3.614}$$

Substituting Eqs. (3.610) and (3.611) into Eq. (3.614), one gets

$$M^+ = 2\pi\delta Z(\lambda)\{\tfrac{1}{2}[\rho - R(\tfrac{\rho^+ - \rho}{\lambda})](2R+\lambda)$$
$$+ \frac{1}{3}(\frac{\rho^+ - \rho}{\lambda})(3R^2 + 3R\lambda + \lambda^2)\}, \tag{3.615}$$

or

$$M^+ = 2\pi\delta Z\{[\rho - R(\tfrac{\rho^+ - \rho}{\lambda})](R\lambda + \tfrac{\lambda^2}{2})$$
$$+ (\rho^+ - \rho)(R^2 + R\lambda + \frac{\lambda^2}{3})\}, \tag{3.616}$$

or

$$M^+ = 2\pi\delta Z[\rho(R\lambda + \frac{\lambda^2}{2}) + (\rho^+ - \rho)(\frac{R\lambda}{2} + \frac{\lambda^2}{3})], \tag{3.617}$$

or

$$M^+ = 2\pi\delta Z[\rho(R\lambda + \frac{\lambda^2}{2}) + \frac{1}{2}R\lambda(\rho^+ - \rho) + \frac{\lambda^2}{3}(\rho^+ - \rho)]. \tag{3.618}$$

Since both of M^+ and M^- are obtained from the linear function ρ_R, therefore, both of M^+ and M^- are also linear, and

$$M_{lin} = M^- + M^+, \tag{3.619}$$

$$M_{lin} = 2\pi\delta Z[\rho(R\lambda - \tfrac{\lambda^2}{3}) + \rho^-(R\lambda - \tfrac{2\lambda^2}{3})]$$
$$+ 2\pi\delta Z[\rho(R\lambda + \frac{\lambda^2}{2}) + \frac{1}{2}R\lambda(\rho^+ - \rho) + \frac{\lambda^2}{3}(\rho^+ - \rho)], \tag{3.620}$$

or

$$M_{lin} = 2\pi\delta Z[(\frac{R\lambda}{2})\rho^- + \rho(R\lambda) + \frac{1}{2}\rho^+(R\lambda) + \frac{\lambda^2}{3}(\rho^+ - \rho^-)], \tag{3.621}$$

or
$$M_{lin} = 2\pi\delta Z[R(\frac{\Delta R}{4})(\rho^- + 2\rho + \rho^+) + \frac{(\Delta R)^2}{12}(\rho^+ - \rho^-)]. \quad (3.622)$$

Linear mass transport:

Mass transported from cell (i) to cell $(i+1)$ is, from Eq. (3.435),

$$\delta M^+ = \delta M_{i+\frac{1}{2}} = 2\pi\delta Z\ell\{[\rho R - (\rho_R R + \rho)(\frac{\ell}{2})]_{i+\frac{1}{2}} + \frac{1}{3}(\rho_R)\ell^2\}, \quad (3.623)$$

$$= 2\pi\delta Z\ell\{\rho R - (\frac{\ell}{2})[(\rho_R R + \rho) - \frac{2}{3}(\rho_R)\ell]_{i+\frac{1}{2}}\}, \quad (3.624)$$

$$= 2\pi\delta Z\ell\{\rho R - (\frac{\ell}{2})[\rho + (R - \frac{2}{3}\ell)(\rho_R)]\}_{i+\frac{1}{2}}, \quad (3.625)$$

where

$$\rho_R = \frac{\rho^+ - \rho}{\frac{\Delta R}{2}}. \quad (3.626)$$

Similarly, the mass from cell (i) to cell $(i-1)$ is

$$\delta M^- = 2\pi\delta Z\ell\{\rho R - (\frac{\ell}{2})[\rho + (R - \frac{2}{3}\ell)(\rho_R)]\}_{i-\frac{1}{2}}, \quad (3.627)$$

where

$$\rho_R = \frac{\rho - \rho^-}{\frac{\Delta R}{2}}. \quad (3.628)$$

Modified mass transport:

To avoid negative mass in a cell, Eqs. (3.625) or (3.627) are weighted as below

$$(\delta M_{i+\frac{1}{2}})_{mod} = \frac{(\delta M^+)M_{avg}}{M_{lin}}, \quad (3.629)$$

$$\frac{M_{avg}}{M_{lin}} = \frac{\rho R}{[\frac{R}{4}(\rho^- + 2\rho + \rho^+) + \frac{\Delta R}{12}(\rho^+ - \rho^-)]}, \quad (3.630)$$

$$(\delta M_{i+\frac{1}{2}})_{mod} = \frac{(2\pi\rho\Delta R\delta Z)\ell\{R_{i+\frac{1}{2}}\rho^+ - \ell[\frac{\rho^+}{2} + (R_{i+\frac{1}{2}} + \frac{2\ell}{3})\frac{(\rho^+ - \rho)}{\Delta R}]\}}{(\Delta R)^2[\frac{R}{4\Delta R}(\rho^+ + 2\rho + \rho^-) + \frac{1}{12}(\rho^+ - \rho^-)]}, \quad (3.631)$$

and

$$(\delta M_{i-\frac{1}{2}})_{mod} = \frac{(2\pi\rho\Delta R\delta Z)\ell\{R_{i-\frac{1}{2}}\rho^- - \ell[\frac{\rho^-}{2} + (R_{i-\frac{1}{2}} + \frac{2\ell}{3})\frac{(\rho - \rho^-)}{\Delta R}]\}}{(\Delta R)^2[\frac{R}{4\Delta R}(\rho^+ + 2\rho + \rho^-) + \frac{1}{12}(\rho^+ - \rho^-)]}. \quad (3.632)$$

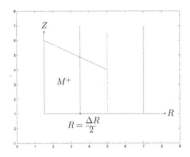

Fig. 3.21 Special treatment for the cell along z-axis.

For the center cell as shown in Fig. 3.21,

$$M_{avg} = \frac{1}{4}\pi\delta Z \rho_i (\Delta R)^2 \,, \tag{3.633}$$

$$M_{lin} = 2\pi\delta Z \int_0^{\Delta R/2} (\rho_0 + \rho_R R')R' dR' \,, \tag{3.634}$$

$$= 2\pi\delta Z(\Delta R)^2 [\frac{1}{8}(2\rho - \rho^+) + \frac{1}{6}(\rho^+ - \rho)] \,, \tag{3.635}$$

$$\delta M^+ = 2\pi\delta Z \ell \{\rho^+ R^+ - (\frac{\ell}{2})[\rho^+ + (R^+ - \frac{2}{3}\ell)(\rho_R)]\} \,, \tag{3.636}$$

$$= 2\pi\delta Z \ell [\frac{\rho^+(\Delta R - \ell)}{2} - \ell(\rho^+ - \rho)(1 - \frac{4}{3}\frac{\ell}{\Delta R})] \,, \tag{3.637}$$

and

$$(\delta M_{\frac{1}{2}})_{mod} = \frac{(2\pi\rho\delta Z)\ell[\frac{\rho^+(\Delta R - \ell)}{2} - \ell(1 - \frac{4}{3}\frac{\ell}{\Delta R})(\rho^+ - \rho_R)]}{(2\rho - \rho^+) + \frac{4}{3\Delta R}(\rho^+ - \rho)} \,. \tag{3.638}$$

3.6.3 The Modified FCT Analysis

To examine the mass transport scheme applied to the present second-order PIC method, the following simple equation is considered when advancing the density ρ in time.

$$\frac{\partial \rho}{\partial t} = -\frac{\partial}{\partial x}(\rho U) - \frac{\partial g}{\partial x} \,. \tag{3.639}$$

Assume a positive velocity for convenience, the mass transported into cell (i) from cell $(i-1)$, $\delta M_{i-1/2}$, and out of cell (i) to cell $(i+1)$, $\delta M_{i+1/2}$, is shown by crosshatched areas in Fig. 3.22.

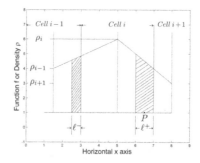

Fig. 3.22 Quantity transported from cells $i-1$ to i, and from cells i to $i+1$.

From Eq. (3.413)

$$\ell^{\pm} = [(\overline{U}_{i\pm\frac{1}{2}})\delta t][1 - (\overline{U}_x)_{i\pm\frac{1}{2}}\delta t] + O(\Delta^3), \quad (3.640)$$

then from Fig. 3.23, we have

$$\delta M_{i+\frac{1}{2}} = \frac{1}{2}[\rho(x_m) + \rho(x_b)]\ell^+ . \quad (3.641)$$

But

$$\rho(x_b) = \rho_{i+\frac{1}{2}}, \quad (3.642)$$

$$\rho(x_m) = \rho_{i+\frac{1}{2}} + \rho_{AB}, \quad (3.643)$$

$$= \rho_{i+\frac{1}{2}} + \ell^+ \cdot \frac{\rho_i - \rho_{i+1}}{\delta x}, \quad (3.644)$$

$$= \rho_{i+\frac{1}{2}} - \ell^+ \cdot \frac{\rho_{i+1} - \rho_i}{\delta x}. \quad (3.645)$$

Substituting Eqs. (3.642) and (3.645) into Eq. (3.641), yields

$$\delta M_{i+\frac{1}{2}} = \frac{1}{2}[\rho_{i+\frac{1}{2}} - \ell^+ \cdot \frac{\rho_{i+1} - \rho_i}{\delta x} + \rho_{i+\frac{1}{2}}]\ell^+, \quad (3.646)$$

or

$$\delta M_{i+\frac{1}{2}} = \frac{1}{2}[(\rho_{i+\frac{1}{2}} + \rho_{i+\frac{1}{2}}) + (\rho_i - \rho_{i+1})\frac{\ell^+}{\Delta x}]\ell^+ . \quad (3.647)$$

Using the same procedure in obtaining Eq. (3.647), one gets

$$\delta M_{i-\frac{1}{2}} = \frac{1}{2}[(\rho_{i-\frac{1}{2}} + \rho_{i-\frac{1}{2}}) + (\rho_{i-1} - \rho_i)\frac{\ell^-}{\Delta x}]\ell^-, \quad (3.648)$$

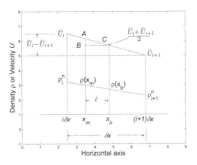

Fig. 3.23 Mass transported from cells i to $i+1$.

and

$$\Delta M_i = \text{total change of mass in cell}(i), \quad (3.649)$$

$$= \delta M_{i-\frac{1}{2}} - \delta M_{i+\frac{1}{2}}. \quad (3.650)$$

To avoid lengthy algebra, a case is considered where $g = 0$ (conservation of mass equation).

$$(M_{avg})_i = \rho_i \Delta x, \quad (3.651)$$

and

$$(M_{lin})_i = \frac{\Delta x}{2}[\rho_i + \frac{1}{4}(\rho_{i+1} + 2\rho_i + \rho_{i-1})]. \quad (3.652)$$

According to the modified mass transport scheme, the total mass change in i is

$$(\delta M_i)_{mod} = (\Delta M_i)(M_{avg})_i/(M_{lin})_i. \quad (3.653)$$

For a uniform velocity, we get simpler equations as below

$$\ell^{\pm} = U\delta t, \quad (3.654)$$

$$\Delta M_i = \frac{1}{2}[\rho_{i-1} - \rho_{i+1} + (\rho_{i-1} - 2\rho_i + \rho_{i=1})\frac{U\delta t}{\Delta x}]U\delta t, \quad (3.655)$$

and

$$(\delta M_i)_{mod} = \frac{1}{2}[\rho_{i-1} - \rho_{i+1} + (\rho_{i-1} - 2\rho_i + \rho_{i+1})\frac{U\delta t}{\Delta x}]U\delta t$$

$$\cdot \frac{\rho_i}{\frac{1}{2}[\rho_i + \frac{1}{4}(\rho_{i-1} + 2\rho_i + \rho_{i+1})]}. \quad (3.656)$$

Then we get

$$\rho_i^{n+1} = \rho_i^n + \frac{(\delta M_i)_{mod}}{\Delta x}, \quad (3.657)$$

or

$$\rho_i^{n+1} = \rho_i^n + [-\tfrac{\xi}{2}(\rho_{i-1} - \rho_{i+1}) + \tfrac{\xi^2}{2}(\rho_{i-1} - 2\rho_i + \rho_{i+1})]$$

$$\cdot \frac{2\rho_i}{[\rho_i + \tfrac{1}{4}(\rho_{i-1} + 2\rho_i + \rho_{i+1})]}, \quad (3.658)$$

or

$$\rho_i^{n+1} = \rho_i^n - \tfrac{\xi}{2}(\rho_{i-1} - \rho_{i+1}) + \tfrac{\xi^2}{2}(\rho_{i-1} - 2\rho_i + \rho_{i+1})$$

$$+ (\text{higher order terms}), \quad (3.659)$$

where $\xi = U \frac{\delta t}{\Delta x}$.

In the original FCT [3.11], the third term on the RHS of Eq. (3.659) looks like

$$(\frac{1}{8} + \frac{\xi^2}{2})(\rho_{i-1} - 2\rho_i + \rho_{i+1}). \quad (3.660)$$

The coefficient $\frac{1}{8}$ provides a strong diffusion even in the absence of convection. In our scheme, the diffusion occurs only for nonzero velocity, This term plus the additional FCT algorithm for local extremums (see Ref. [3.11]) virtually eliminates the need for artificial viscosity.

Bibliography

3.1 Wilkins, ML. (1964). Calculation of elastic-plastic flow, in *Methods in Comput. Physics*, 3, pp. 211-264.

3.2 Swegle, JW. (1978). *TOODY IV - A computer program for two-dimensional wave propagation*, Sandia National Laboratory report SAND-78-0552.

3.3 Kolsky, HG. (1955). *A method for the numerical solution of transient hydrodynamic shock problems in two space dimension*, Los Alamos Scientific Laboratory report LA-1867.

3.4 Mader, CL. (1979). *Numerical modeling of detonations*, University of California press, pp. 333-352.

3.5 Harlow, FH. (1964). The particle-in-cell computing method for fluid dynamics, in *Method in Comput. Physics*, 3, pp. 319-343.

3.6 Johnson, WE. (1982). *Modifications of OIL-type computer programs*, Ballistic Research Laboratory Contract report ARBRL-CR-0476.

3.7 Hageman, LJ, Wilkins, RT, Sedgwick,RT, and Wadell, JL. (1975). *HELP, A multi-material Eulerian program for compressible fluid and elastic-plastic in two-dimension and time*, System, Science, and Software report SSS-75-2654.

3.8 Thompson, SL. (1977). *CSQII- An Eulerian finite difference program for two-dimensional material response - Part I, material section*, Sandia National Laboratory report SAND-77-1339.

3.9 Clark, RA. (1979). *Second-order particle-in-cell computing method*, Los Alamos Scientific Laboratory report LA-UR-79-1947.

3.10 Leal, LG. (2007). *Advanced transport phenomena: Fluid mechanics and convective transport processes*, Cambridge University Press, p. 912.

3.11 Boris, JP, and Book, DL. (1973). Flux-corrected transport I, SHASTA, A fluid transport algorithm that works, *J. Comput. Physics*, 11, pp. 38-69.

3.12 Bird, RB, Stewart, WE, and Lightfoot, EN. (1960). *Transport phenomena*, John Wiley & Sons, p. 216.

Chapter 4

EOS, Constitutive Relationship and High Explosive

Notations

A pressure dependence of the shear modulus ($Mbar^{-1}$)
a coefficient of the volume dependence of gamma (no unit)
B temperature dependence of the shear modulus (K^{-1})
b coefficient of the volume dependence of gamma (no unit)
C_0 bulk sound speed ($\frac{cm}{\mu sec}$)
C_1 exponential pre-factor in rate-dependent model (μs^{-1})
C_2 coefficient of drag term in rate-dependent model ($Mber - \mu s$)
c_p lattice specific heat ($\frac{Mbar-cm^3}{g-K}$)
D_0 damage length (cm)
E specific internal energy per initial unit volume ($\frac{Mbar-cm^3}{cm_0^3}$)
e equivalent plastic strain (no unit)
G shear modulus of elasticity ($Mbar$)
G_1, G_2 Bauschinger parameters ($Mbar$)
I specific internal energy per unit mass ($\frac{Mbar-cm^3}{g}$)
k Boltzmann constant (eV/K)
M cell mass (g)
P pressure ($Mbar$)
R, r radial coordinate (cm)
S_{1-3} Hugoniot slop coefficients (no unit)
S^{ij} stress deviator tensor ($Mbar$)
$S^{rr}, S^{zz}, S^{rz}, S^{\theta\theta}$ stress deviator components ($Mbar$)
s distance (cm)
t time (μsec)
T temperature (K)

T_{m0} melting temperature at constant volume (K)
Δt time step (μsec)
Δt_{max} maximum allowed time step (μsec)
U_K activation energy (ev)
u_R displacement in the R direction (cm)
u_Z displacement in the Z direction (cm)
$\dot{W}$ rate of energy source due to work hardening ($\frac{Mbar-cm^3}{cm^3-\mu sec}$)
Y flow stress of elasticity ($Mbar$)
Y_0 flow stress at the Hugoniot elastic limit ($Mbar$)
Y_p Peierls stress ($Mbar$)
Y^*_{max} work-hardening maximum in rate-dependent model ($Mbar$)
Y_{max} work-hardening maximum ($Mbar$)
Z, z axial coordinate (cm)

Greek letters

δ_{ij} Kronecker delta (no unit)
ϵ specific internal energy per mass ($\frac{Mbar-cm^3}{g}$)
ϵ_p equivalent plastic strain (no unit)
ϵ_i initial plastic strain (no unit)
$\dot{\epsilon}_p$ strain rate (μs^{-1})
γ Grüneisen's gamma (no unit)
γ_0 initial Grüneisen's gamma (no unit)
η normalized density (or compression) = $\frac{\rho}{\rho_0}$ (no unit)
θ angular coordinate
λ half of the grid size in R direction, i.e., $\lambda = \Delta R/2$ (cm)
μ normalized density minus one (=$\eta - 1 = \frac{\rho}{\rho_0} - 1$)(no unit)
ρ density ($\frac{g}{cm^3}$)
ρ_0 initial density ($\frac{g}{cm^3}$)
σ spall stress ($Mbar$)
σ^{ij} stress tensor ($Mbar$)
$\sigma^{rr}, \sigma^{zz}, \sigma^{rz}, \sigma^{\theta\theta}$ stress components ($Mbar$)
τ specific volume ($\frac{cm^3}{g}$)

Subscripts

0 initial value
e elastic regime
kk normal strains, i.e., $e_{rr}, e_{zz}, e_{\theta\theta}$
p plastic regime

R derivative with respect to R coordinate
t derivative with respect to time
Z derivative with respect to Z coordinate

Superscripts

n time at n time-step, i.e., $t^n = t_0 + n \cdot \Delta t$
kk normal stresses, i.e., $\sigma^{rr}, \sigma^{zz}, \sigma^{\theta\theta}$

4.1 Introduction to the Equation of State

Most of the equation of state (EOS) uses pressure as a function of density and internal energy. Occasionally, there is a need for calculating temperature as a function of density and internal energy. A simple EOS model uses the u_s (shock speed) and u_p (particle speed) relationship or *Mie − Grüneisen* [4.1] EOS. The most complicated formula of the EOS will be the one proposed by Osborne [4.2]. This model includes the second order polynomial function of density and internal energy.

In the calculation of the burned product pressure of a high explosive, the JWL EOS is quite suitable. The JWL EOS for high explosive is described in Section 4.9.

4.2 The *Mie − Grüneisen* EOS and the Simple u_s, u_p Model

The *Mie − Grüneisen* [4.1] EOS is

$$P = P_H + \rho\gamma(\epsilon - \epsilon_H). \tag{4.1}$$

In Eq. (4.1), P is the pressure, ρ the density, γ the *Grüneisen* parameter, ϵ the internal energy, P_H and ϵ_H are the pressure and internal energy on the principal Hugoniot curve. The pressure P and the internal energy ϵ are function of density and temperature, while P_H and ϵ_H are function of density only. The *Grüneisen* parameter is defined as

$$\gamma = \frac{1}{\rho}(\frac{\partial P}{\partial \epsilon})_\rho. \tag{4.2}$$

A good approximation for γ is

$$\rho\gamma = \rho_0\gamma_0, \tag{4.3}$$

Table 4.1 Parameters for the $Mie - Grüneisen$ EOS.

Material	$\rho_0(g/cm^3)$	$C_0(Km/s)$	s	γ_0
Stainless Steel 304	7.9	4.57	1.49	2.17
Copper OFHC 1/2 Hard	8.93	3.94	1.49	1.96
Water	0.9979	2.393	1.333	0.50
Granite	2.672	3.712	1.086	0.90
Bronze	8.733	3.814	1.452	2.12

or

$$\gamma = \frac{\rho_0 \gamma_0}{\rho}, \quad (4.4)$$

where ρ_0 and γ_0 are at zero Kelvin states. Based on the experimental data, the shock wave velocity u_s and the particle velocity u_P can be related to by

$$u_s = C_0 + s u_P, \quad (4.5)$$

where C_0 and s are parameters dependent on material. For example, some of the material used for oil well perforator problems as described in Chapter 5 are given in Table 4.1.

Using Eq. (4.5) and the Hugoniot equation (4.1) one can derive the Hugoniot pressure

$$P_H = \frac{\rho_0 C_0 \eta}{(1 - s\eta)^2}, \quad (4.6)$$

where

$$\eta = 1 - \frac{\rho_0}{\rho}. \quad (4.7)$$

Also, from the Hugoniot relationship the internal energy ϵ_H can be written as

$$\epsilon_H = \frac{1}{2} \frac{P_H \mu}{\rho}, \quad (4.8)$$

Table 4.2 Osborne coefficients.

	Tungsten	Aluminum	Steel	Uranium	Copper
A_1	21.67419	1.1867466	4.9578323	2.4562457	4.9578323
A_2	14.93338	0.762995	3.6883726	4.6163216	3.6883726
B_0	10.195827	3.4447654	7.4727361	4.3432909	7.4727361
B_1	12.263234	1.5452573	11.519148	0.76214541	11.519148
B_2	9.3051515	0.96429632	5.5251138	6.4410793	5.5251138
C_0	0.48248861	0.43381656	0.39492613	0.31988993	0.39492613
C_1	0.48248861	0.54873462	0.52883412	0.46744784	0.52883412
D_0	7.0	1.5	3.6	2.2	0.6000001
ρ_0	19.17	2.806	7.9	18.983	8.899

where

$$\mu = 1 - \frac{\rho_0}{\rho}. \tag{4.9}$$

During the hydrodynamic code calculation, for example, at step 14 in Section 6.5, the density ρ and internal energy ϵ are available from the code computation, one will use Eq. (4.6) for computing P_H, Eq. (4.8) for ϵ_H, and Eq. (4.1) for P. The material dependent constants ρ_0, C_0, s, γ_0 are obtained from Table 4.1. More Hugoniot data are provided in Appendix C.

4.3 The Osborne Model

The Osborne model [4.2] is a quadratic form polynomial that calculates the pressure as a function of density ratio and internal energy. The expression is

$$P(Mbar) = \frac{A_1\mu + A_2\mu|\mu| + (B_0 + B_1\mu + B_2\mu^2)\epsilon + (C_0 + C_1\mu)\epsilon^2}{\epsilon + D_0}, \tag{4.10}$$

where

$$\mu = \frac{\rho}{\rho_0} - 1. \tag{4.11}$$

In Eq. (4.10), ρ is the density (g/cm^3), ρ_0 is the initial density (g/cm^3), ϵ is the specific internal energy $(Mbar - cm^3/g)$, $A_1, A_2, B_0, B_1, B_2, C_0, C_1$ and D_0 are coefficients that are material dependent. The coefficients for some material are given in Table 4.2.

4.4 The Tillotson Equation of State

The Tillotson [4.3] equation of state is

$$P = P_c = \left[a + \frac{b}{\frac{I}{I_0\eta^2}+1}\right] I\rho + A_1\mu + B_1\mu^2, \qquad \text{for } I < I_s, \tag{4.12}$$

$$P = P_E = aI\rho + \left[\frac{bI\rho}{\frac{I}{I_0\eta^2}+1} + A_1\mu e^{-\beta(\frac{\rho_0}{\rho}-1)}\right] e^{-\alpha(\frac{\rho_0}{\rho}-1)^2}, \qquad \text{for } I > I_s', \tag{4.13}$$

and

$$P = \frac{(I - I_s)P_E + (I_s' - I)P_c}{I_s' - I_s}, \qquad \text{for } I_s < I < I_s'. \tag{4.14}$$

For example, the parameters for the oil shale are: $a = 0.5$; $b = 1.0$; $A_1 = 0.28$; $B_1 = 0.11$; $\alpha = 5.0$; $\beta = 5.0$; $I_0 = 0.11$; $I_s = 0.032$; $I_s' = 0.16$; and $\rho_0 = 2.3(g/cm^3)$. The oil shale data are used in Chapter 5 for the oil well perforator problems.

4.5 Introduction to the Constitutive Relationship

Various expressions for the shear modulus, G, and the yield strength, Y, are available. Here we simply present options made available for use in either Eulerian or Lagrangian code.
(a) Hydrodynamic flow

$$G = Y = S^{ij} = 0. \tag{4.15}$$

(b) Elastic perfect plastic flow

$$G = G_0 = \text{constant}, \quad (4.16)$$

$$Y = Y_0 = \text{constant}. \quad (4.17)$$

(c) Strain-hardened elastic-plastic model

(1) Quadratic EOS.
(2) Steinberg–Guinan model.
(3) Combination of (i) and (ii) for multi-material problems.

4.6 Quadratic Model

In this model, the dynamic yield strength, Y, and the melting energy, E_m, are expressed in terms of pressure and internal energy as below

$$Y = (Y_0 + \alpha P)\left(1 - \frac{E}{E_m}\right), \quad (4.18)$$

and

$$E_m = E_{m0} + E_{m1}(1-v) + E_{m2}(1-v)^2, \quad (4.19)$$

where Y_0 = yield strength; α = coefficient for pressure dependence; E_{m0}, E_{m1}, E_{m2} = coefficients for melting function; and v = normalized specific volume, i.e., $v = \frac{\rho_0}{\rho}$.

4.7 Steinberg–Guinan Model

This model is adopted from Steinberg and Guinan [4.4], who used the following constitutive relations for high-strain rate.

$$G = G_0\left\{1 + b\frac{P}{\eta^{1/3}} + h\left[\frac{E - E_0(x)}{3R'} - 300\right]\right\}exp\left[-\frac{fE}{E_m(x) - E}\right], \quad (4.20)$$

$$Y = Y_0(1+\beta e)^n\left\{1 + qb\frac{P}{\eta^{1/3}} + h\left[\frac{E-E_0(x)}{3R'} - 300\right]\right\}$$

$$\times exp\left[-\frac{gE}{E_m(x)-E}\right], \qquad (4.21)$$

$$(1+\beta e)^n \leq Y_{max}, \qquad (4.22)$$

$$E_m(x) = E_0(x) + 3R'T_m(x), \qquad (4.23)$$

$$T_m(x) = \frac{T_{m0}[exp(2ax)]}{(1-x)^\alpha}, \qquad (4.24)$$

$$\alpha = 2\left(\gamma_0 - a - \frac{1}{3}\right), \qquad (4.25)$$

$$E_0(x) = \int_0^x P(x)dx - 3R'TAD, \qquad (4.26)$$

and

$$TAD = \frac{300[exp(ax)]}{(1-x)^{(\gamma_0-a)}}, \qquad (4.27)$$

where P = pressure ($Mbar$),
G_0 = shear modulus ($Mbar$) at reference state($T = 300°K, P = 0, e = 0$),

$$b = \frac{1}{G_0}\frac{\partial G}{\partial T} = \frac{G_P}{G_0}, \qquad (4.28)$$

$$h = \frac{1}{G_0}\frac{\partial G}{\partial P} = \frac{G_T}{G_0}, \qquad (4.29)$$

ρ = density(g/cc),

$$\eta = \frac{\rho}{\rho_0}, \qquad (4.30)$$

E = energy($Mbar - cm^3/cm_0^3$),
A = atomic weight($g/mole$),
R = gas constant = $8.314 \times 10^{-5} Mbar - cm^3/(mole - °K)$,

$$R' = \frac{R\rho_0}{A}, \qquad (4.31)$$

f, g = shaping parameters in the melting region,
β, n = work hardening parameters,
q = ratio of pressure dependence of yield strength to that of the shear modulus,
e = equivalent plastic strain,
Y_{max} = maximum value of yield strength at $T = 300°K$ and $P = 0$,
T_{m0} = melting temperature at $\rho = \rho_0(°K)$,
T_m = melting temperature at $P = 0(°K)$,

$$x = 1 - \frac{1}{\eta} = 1 - \frac{v}{v_0}, \qquad (4.32)$$

$$\gamma = \text{thermodynamic gamma} = \frac{C_P}{C_v}, \qquad (4.33)$$

a = correction of first-order volume correction to γ.

For the Eulerian code, the integral expression in Eq. (4.26) is replaced by a polynomial Form:

$$\int_0^x P(x)dx = y^2 \frac{T_1}{T_2}, \qquad (4.34)$$

where

$$y = \eta - 1 = \frac{\rho}{\rho_0}, \qquad (4.35)$$

$$T_1 = \alpha_0 + [\alpha_1 + (\alpha_2 + \alpha_3 y)y]y, \qquad (4.36)$$

and

$$T_2 = \beta_0 + [\beta_1 + (\beta_2 + \beta_3 y)y]y. \qquad (4.37)$$

For example, for copper, the coefficients are: $\alpha_0 = 0.68462$; $\alpha_1 = -0.00868$; $\alpha_2 = 0.26429$; $\alpha_3 = 0.0068119$; $\beta_0 = 1.0$; $\beta_1 = 0.00903$; $\beta_2 = 0.170902$; and $\beta_3 = 0.0417657$. Eqs. (4.20) and (4.21) are in the same form except for the work-hardening term in Eq. (4.21). The temperature dependence is characterized by

$$G_0 h \left[\frac{E - E_0(x)}{3R'} - 300 \right] = G_T(T - 300). \qquad (4.38)$$

The coefficients of the Steinberg–Guinan model of some material are given in Table 4.3.

4.8 Steinberg's New Model

In 1991, Steinberg [4.5] reports the material properties for 37 elemental metals and alloys, 34 of which have a rate-independent constitutive model and 7 of which have a rate-dependent model. In addition, there are 9 plastic materials plus graphite, LiF and WC. Most of the material in this section is adopted from Steinberg's report [4.5].

The rate-independent model is described in Ref. [4.6], and other applications of the model are given in Refs. [4.7] and [4.8]. The principal equations are as follows:

$$Y = Y_0 f(\epsilon_p) G(P,T)/G_0, \qquad (4.39)$$

where

$$Y_0 f(\epsilon_p) = Y_0 [1 + \beta(\epsilon_p + \epsilon_i)]^n \leq Y_{max}, \qquad (4.40)$$

and

$$G(P,T) = G_0 \left[1 + \frac{AP}{\eta^{1/3}} - B(T-300)\right], \qquad (4.41)$$

where ϵ_p is the equivalent plastic strain and ϵ_i the initial plastic strain.

Here

$$A = \frac{1}{G_0}\frac{dG}{dP}, \qquad (4.42)$$

and

$$B = \frac{1}{G_0}\frac{dG}{dT}. \qquad (4.43)$$

Melting is based on a modified Lindemann law and is described in Ref. [4.6]. The melting temperature T_m is given by

$$T_m = T_{m0} exp\left[2a\left(1-\frac{1}{\eta}\right)\right]\eta^{2(\gamma_0 - a - \frac{1}{3})}. \qquad (4.44)$$

When T equals or exceeds T_m, Y and G are set to zero. Because the EOS is energy-based, it is easier to test for melt on the basis of energy. The

Table 4.3 The Coefficients of the Steinberg–Guinan model for Tungsten, Aluminum, Copper, and Uranium.

Material	Tungsten	Aluminum	Copper	Uranium
G_0	1.6	0.276	0.477	0.844
Y_0	0.022	0.0029	0.0012	0.0012
β	7.7	125.0	36	16000.
n	0.13	0.1	0.45	0.26
Y_{max}	0.04	0.0068	0.0064	0.0168
b	1.375	7.971	3.1446541	4.739
h	-0.0001375	-0.0067159	-0.000377358	-0.0008056
q	1.0	1.0	1.0	1.0
f	0.001	0.001	0.001	0.001
g	0.001	0.001	0.001	0.001
$R'(Mbar/°K)$	0.000008671	0.000008326	0.00001164	0.00000663
T_{m0}	4520	1220.0	1790	1710
$\gamma_0 - a$	0.27	0.49	0.52	0.92
a	1.4	1.7	1.5	1.5

melt energy E_m is defined as

$$E_m = E_c + c_p T_m, \qquad (4.45)$$

where E_c is given

$$E_c = E(\eta) - 300 c_p exp\left[a\left(1 - \frac{1}{\eta}\right)\right] \eta^{\gamma_0 - a}, \qquad (4.46)$$

and $E(\eta)$ is the integral of PdV on the zero Kelvin isotherm.
A simplified model is

$$Y = Y_0(1 + \beta\epsilon_p)^n \left[1 + \frac{CP}{\eta^{1/3}} - m(T - 300)\right], \qquad (4.47)$$

Table 4.4 Coefficients used in Equations (4.40) and (4.41) for some materials.

Material	$G_0(GPa)$	$Y_0(GPa)$	β	n	C	m	$T_{melt}(ev)$
Stainless Steel 304	0.77	0.0034	43.	0.35	2.26	0.000455	
Copper OFHC Hard	0.477	0.0012	36.	0.45	2.83	0.000377	
Bronze	0.409	0.00123	39.	0.27	3.03	0.000411	
Aluminum 6061-T6	0.33	0.324	125.	0.10	6.52	0.000616	0.08
Gold	0.425	0.02	49.	0.39	3.75	0.000311	0.092

and

$$G = G_0 \left[1 + \frac{CP}{\eta^{1/3}} - m(T - 300)\right]. \quad (4.48)$$

The parameters for some material are given in Table 4.4. A Fortran 77 code 'EOSGY' which will calculate the yield strength, Y, and the shear modulus, G, is given as supplementary materials (see Appendix D) with all of the data described in Ref. [4.5]. The computer code 'EOSGY' will be described in next section.

The formula of the constitutive model with the data for metals subjected to large strain, high strain rate, and high temperature are given by Johnson and Cook [4.9]. The basic model is well suited for computations because it uses the information that is readily available in most of the computer codes. The von Mises flow stress is given by

$$\sigma_{eq} = Y_0(1 + \beta \epsilon_p^n) \left[1 + C \ln\left(\frac{\dot{\epsilon}_p}{\dot{\epsilon}_0}\right)\right](1 - \theta^m), \quad (4.49)$$

where

$$\dot{\epsilon}_0 = 1.0/s,$$

$$\theta = \frac{T - T_0}{T_m - T_0}, \quad (4.50)$$

Table 4.5 The Coefficients of Eq. (4.49) (Johnson–Cook Model) for some material.

Material	θ	$Y_0(GPa)$	β	n	C	m	$T_{melt}(ev)$
Copper		0.0897	3.26	0.31	0.025	1.09	0.119
Steel 4340		0.793	0.643	0.26	0.014	1.03	0.157
Tungsten		1.507	0.117	0.12	0.016	1.0	0.151
Granite Westerly	300°	0.048	1.0	1.0	0.0	0.5	

and
$$T_0 = 300°K.$$
The coefficients of Eq. (4.42) for some material are given in Table 4.5. In Eq. (4.49), the definitions of the variables are

σ_{eq} : von Mises flow stress
ϵ_p : equivalent plastic strain
$\dot{\epsilon}_p$: equivalent plastic strain rate
θ : reduced temperature
T_m : melt temperature.

4.8.1 Program EOSGY

Program "EOSGY", written in FORTRAN 77, computes the shear modulus G and yield strength Y using the Steinberg's material models [4.4, 4.5]. Steinberg's 1973 model uses 32 parameters to calculate G and Y while the 1991 model uses only 18 parameters. We suggest the reader to use the 1991 model since the data file "INPUT8" contains the complete parameters of 56 materials as provided by the 1991 model and the code "EOSGY" will read the file "INPUT8". Also, the 1991 model is more accurate than the 1973 model. If one is interested in using the 1973 model, one has to prepare the 32 parameters for the 1973 model by himself because currently the "EOSGY" code does not have the data file for the 1973 model. In order to run the 1973 model, one has to set $GY1973 = 1$ and $GY1991 = 0$ at the beginning of the code. In the program "EOSGY", there are two subroutines, one for computing G and Y using the 1973 model (subroutine GANDY), the other one for 1991 model (subroutine GANDYN).

The input data file for "EOSGY" code is called "input1" which contains the total number of material in the first line, and, then, the material numbers. For example, in the file "input1", the first line is 20 that means we are going to compute 20 different materials. The second line is the real material number and they are 141, 142, 143, 144, 145, 146, 147, 148, 149, 150, 151, 152, 153, 154, 155, 156, 137, 138, 139 and 140. In the second line, 141 is the material number for Zinc, 142 is for Zirconium, 156 is for Tungsten Carbide and 140 is for Vanadium. The input file "input1" is the only input data necessary to run the code.

Suppose one is interested in calculating the G and the Y for only one material, say, copper (OFHC, 1/2 Hard), one will set $GY1991 = 1$, $IB = 1, IE = 1, JB = 1$, and $JE = 1$ inside the code "EOSGY". The input data file "input1" should have
1
107.

After running the code, one should have the values of $G1(1,1,1)$ and $Y1(1,1,1)$ provided by the output file "output1".

The first 21 lines of the "output1" file contains the same information as the "input1" file. The 22nd line is
Material ID=101, coefficients=0.271, 0.0004, 400., 0.270, 0.0048, 6.52, 0.000455, 0.0000884, 2.71.
The meaning of this line is
101 Aluminum (1100-0)
$G(0) = 0.271$, shear modulus at reference state
$Y(0) = 0.0004$, yield strength at reference state
$\beta = 400.$
$n=0.27$
$Y(max) = 0.0048$, maximum yield strength at reference state
$c=6.52$
$m=0.000455$
$C_p=0.0000884$
$\rho = 2.71$, density.

4.9 High Explosive

4.9.1 Introduction

A programmed burn model of high explosive (HE) is adequate for the Lagrangian and Eulerian codes calculations. Huygens' principle and the

Chapman–Jouguet theory are used in defining the detonation velocity and location where the high-explosive energy is released. Precalculated burn information is implemented into the code before the hydrodynamic actions take place.

In HE programmed burn, the basic assumption is that the detonation wave front travels in all directions at the Chapman–Jouguet detonation velocity. Information concerning the energy released from HE such as burn time (BT) and burn interval (BI) is precalculated and stored in the code at time $t = 0$. During the run, when the problem time T^n at cycle n becomes greater than the BT value of a HE but less than (BT + BI), a fraction of the specific energy for the particular HE is deposited in the cell. This fraction is given by $(T^n - \text{BT})/\text{BI}$. On the next cycle, at T^{n+1}, if T^{n+1} is still less than (BT + BI), another fraction $(T^{n+1} - T^n)/\text{BI}$ of specific energy is deposited into the cell. This continues until all the energy is deposited into the cell and $T^n > (\text{BT} + \text{BI})$. Then the cell is completely burned.

4.9.2 JWL Equation of State

A number of different equations of state have been developed to describe the pressure due to the HE burn. The Jones-Wilkins-Lee (JWL) EOS [4.10] is accepted as one that accurately describes the expansion process for nearly ideal explosives:

$$P = A\left[1 - \left(\frac{\omega}{R_1 V}\right)\right] e^{-R_1 V} + B\left[1 - \left(\frac{\omega}{R_2 V}\right)\right] e^{-R_2 V} + \frac{\omega E}{V}, \quad (4.51)$$

and

$$P_s = A e^{-R_1 V} + B e^{-R_2 V} + C V^{-(\omega+1)}, \quad (4.52)$$

where
P = pressure,
P_s = pressure along the expansion isentrope, i.e., pressure as a function of volume at constant entropy,
A, B, C = linear coefficient in units of P and P_s,
R_1, R_2, ω = nonlinear, unitless coefficients,
$V = v/v_0$ = detonation product volume (v)/initial high-explosive volume (v_0),
E = detonation energy per unit volume.

The coefficients are obtained by comparing EOS calculations with experimental expansion data. In the Appendix C, Tables C.6 and C.7 list the JWL parameters for a number of explosives, along with the C-J detonation parameters used in obtaining this information.

4.9.3 Small Variation of JWL EOS

In the Eulerian code, we use a slightly different form of JWL EOS which is

$$P(Mbar) = A\left[1 - \left(\frac{\omega}{R_1 V}\right)\right] e^{-R_1 V} + B\left[1 - \left(\frac{\omega}{R_2 V}\right)\right] e^{-R_2 V} + \frac{\omega(E - E_1)}{V}, \quad (4.53)$$

For Octol 75/25 the coefficients are: $A = 7.486$ Mbar; $B = 0.1338$ Mbar; $R_1 = 4.5$; $R_2 = 1.2$; $\omega = 0.38$; $E_1 = 0.272\ Mbar-cm^3/cm^3$; V is the normalized specific volume (see Section 4.9.2); and E the detonation energy ($Mbar - cm^3/cm^3$). For the Chapman–Jouguet parameters: $\rho_0 = 1.821\ g/cm^3$, detonation velocity; $D = 0.849 cm/\mu s$; $E_0 = 0.098\ Mbar-cm^3/cm^3$; and $E_{chemical} = E_0 + E_1 = 0.37\ Mbar - cm^3/cm^3$.

In Eq. (4.53), the $E_1 = 0.272\ Mbar - cm^3/cm^3$ is calculated by setting $P = 0$, $E = 0$, and $V = 1$ in Eq. (4.53), and using other parameters from the JWL EOS Table 4.6. In the code calculation the high explosive energy released for the hydrodynamic energy equation is the term $E_{chemical}$. In calculating E_1, it is very convenient to use the following unit conversion factors for pressure and energy: 1 Mbar = 100 GPa; $1\ Mbar - cm^3/cm^3 = 10^{11} J/m^3$.

4.9.4 Computer Code for Two-Dimensional Programmed Burn

In the programmed burn model the time, at which the HE in each mesh cell detonates, is calculated a priori by calculating the arrival times of waves emanating from the prescribed detonation point or points. The energy released from the HE burn is then add to the energy equation as the energy sources.

As shown in Fig. 4.1, z is at the cell center where p_1 and p_2 are at the cell corners. The triangle $\Delta p_1 p_2 z$ is called side. The Lund [4.11] algorithm requires that the burn times for all but one of the vertices of the triangle be known, and the unknown time is computed from these known values. The algorithm iteratively propagates the solution away from the detonators until the entire burn time field has been calculated.

EOS, Constitutive Relationship and High Explosive   147

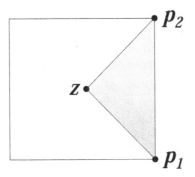

Fig. 4.1 For two-dimensional HE burn, the triangle $\triangle p_1 p_2 z$ is called a side where z is at cell center and p_1 and p_2 are at the grid points.

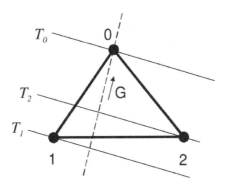

Fig. 4.2 The planar burn time assumes constant gradient $G_x = \frac{\delta t}{\delta x}$ (planar detonation at constant detonation velocity) within the triangle and computes G_x, G_y and t_0 in terms of known burn time values t_1 and t_2 and the detonation velocity D of the HE. The three lines represent the planar burn front times at each of the three vertices.

The first part of the program, HEDET2, uses the line of sight (LOS) approximation to compute the burn time of each grid point from the detonators. Therefore, the burn time is equal to the distance divided by the detonation velocity, D, assuming that there is no obstacle on the path of the line of sight (LOS). If there are two detonators in the problem, we calculate the burn time of the first detonator for every grid points and save the values of the burn time. Then, we calculate the burn time for each grid points again using the second detonator. Finally, we choose the smaller value of the burn time from those two detonators. These values of burn time will be used in the second-part calculations.

In the second part of the code, HEDET2, the HE burn time at one vertex of a triangle, t_0 as shown in Fig. 4.2, can be calculated if the times at the other two vertices, t_1 and t_2, are known from calculations of other triangles, or from the first-part calculations. The calculated time is a trial solution which must satisfy a number of criteria, discussed below, before it can be accepted as the vertex HE lighting time.

Therefore, in two-dimension, three equations are needed in order to determine the burn time, t_0, at one vertex of the triangle, by the HEDET2 algorithm. In addition to t_0, the unknowns in the finite-difference solution are the two derivatives of the time with respect to the coordinates x and y. We solve the following equations

$$\left(\frac{\delta t}{\delta x}\right)^2 + \left(\frac{\delta t}{\delta y}\right)^2 = \frac{1}{D^2}, \qquad (4.54)$$

$$t_1 = t_0 + \frac{\delta t}{\delta x}(x_1 - x_0) + \frac{\delta t}{\delta y}(y_1 - y_0), \qquad (4.55)$$

and

$$t_2 = t_0 + \frac{\delta t}{\delta x}(x_2 - x_0) + \frac{\delta t}{\delta y}(y_2 - y_0). \qquad (4.56)$$

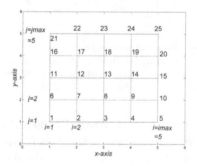

Fig. 4.3 A two-dimensional problem with $imax = 5$ in the x-axis and $jmax = 5$ in the y-axis. There are 25 grid points with $KKDL = 1$ and the detonator is located at $i = 1$ and $j = 1$ (or grid point 1).

Equations (4.55) and (4.56) are found from taking a Taylor series about the vertex for which we are calculating the time. The derivatives are evaluated at the t_0 vertex. Equations (4.54)-(4.56) are solved for the three unknowns t_0, $\frac{\delta t}{\delta x}$, and $\frac{\delta t}{\delta y}$. It should be noted that Eq. (4.54) is a nonlinear equation. A quadratic equation results, this giving two solutions for t_0. The larger value is chosen.

The burn times computed by HEDET2 are quite good with maximum error of less than 3.95% for typical coarse grid and approaches zero with fine mesh. Especially, for two-detonators problem, the code is very robust and easy to use. For example, a two-dimensional and one point detonation problem with $i = 1, 2, \ldots, imax$ and $j = 1, 2, \ldots, jmax$, we have to input the data of $imax$, $jmax$ and set $KKDL = 1$ with the location of the detonator. Figure 4.3 shows a two-dimensional problem with $imax = 5$, $jmax = 5$ and $KKDL = 1$. At the beginning of the subroutine HEDET2D, the dimension of some arrays must be reset with some numbers consistent to the input data $imax$ and $jmax$.

For two-detonators problem, we just set $KKDL = 2$ and define the location of the second detonator. Therefore, for a new problem, we have to do the following at the beginning of subroutine HEDET2D.

1. Define the maximum grid numbers $imax$ and $jmax$.
2. Define the grid sizes dx and dy.
3. Set $KKDL = 1$ for one detonator problem and $KKDL = 2$ for two detonators problem.
4. Define the detonation velocity D.
5. Reset the dimension of some arrays.
6. Set $KKPLL = (imax - 1) * (jmax - 1) * 4$.
7. Set $KKZLL = imax * jmax + (imax - 1) * (jmax - 1)$.
8. Set $KKSLL = KKPLL$.
9. Set $KKPHE = imax * jmax$.

The output file from the code calculation is called OUTPUT1 which provides the burn times of every grid points using the variable PHET. The burn time at the cell center is called ZHET. Users can use PHET or ZHET for their hydrodynamic calculations with burn interval (BI) as described in Section 4.9.1. Sample calculations and more detail descriptions of the Lund model can be obtained from Ref. [4.11]. A complete FORTRAN code of HEDET2 is provided in the supplementary materials (see Appendix D) with file named HEDET2.

150   Computational Solid Mechanics for Oil Well Perforator Design

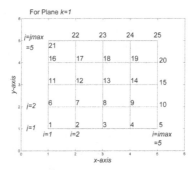

Fig. 4.4   A three-dimensional problem with $imax = 5$ in the x-axis, $jmax = 5$ in the y-axis and $kmax = 3$ in the z-axis. There are 75 grid points with $KKDL = 1$ and the detonator is located at $i = 1$, $j = 1$ and $k = 1$ (or grid point 1). The first 25 grid points on $k = 1$ plane are shown.

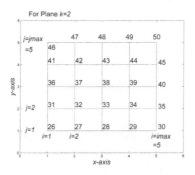

Fig. 4.5   A three-dimensional problem with $imax = 5$ in the x-axis, $jmax = 5$ in the y-axis and $kmax = 3$ in the z-axis. There are 75 grid points with $KKDL = 1$ and the detonator is located at $i = 1$, $j = 1$ and $k = 1$ (or grid point 1). The second 25 grid points on $k = 2$ plane are shown.

4.9.5   Computer Code HEDET3 for Three-Dimensional Programmed Burn

Currently, the three-dimensional code HEDET3 uses only the line of sight approximation to compute the burn time for one or two detonators. The more sophisticated model using Taylor series expansion to obtain the $\frac{\delta t}{\delta x}$, $\frac{\delta t}{\delta y}$, and $\frac{\delta t}{\delta z}$ is still under developing. For three-dimensional model, at the beginning of the subroutine HEDET3D, we have to provide the following information.
1. Define the maximum grid numbers $imax$, $jmax$ and $kmax$.
2. Define the grid sizes dx, dy and dz.
3. Set $KKDL = 1$ for one detonator problem and $KKDL = 2$ for two detonators problem.

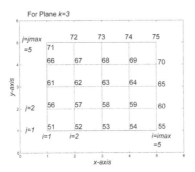

Fig. 4.6 A three-dimensional problem with $imax = 5$ in the x-axis, $jmax = 5$ in the y-axis and $kmax = 3$ in the z-axis. There are 75 grid points with $KKDL = 1$ and the detonator is located at $i = 1$, $j = 1$ and $k = 1$ (or grid point 1). The third 25 grid points on $k = 3$ plane are shown.

4. Define the detonation velocity D.
5. Reset the dimension of some arrays.
6. Set $KK4LL = 4$.
7. Set $KKDLL = 2$.
8. Set $KK3LL = 3$.
9. Set $KKPLL = (imax-1)*(jmax-1)*(kmax-1)*7 + imax*jmax*kmax$.
10. Set $KKZLL = KKPLL$.
11. Set $KKSLL = (imax-1)*(jmax-1)*(kmax-1)*24$.
12. Set $KKSL = KKSLL$.

The output file OUPUT1 provides the burn time at each grid point with the variable PHET and that at the cell center is ZHET. The FORTRAN code HEDET3 is included in the supplementary materials (see Appendix D). Figure 4.4 shows a three-dimensional problem with $imax = 5$, $jmax = 5$, $kmax = 3$ and $KKDL = 1$ with 25 grid points for $k = 1$ plane. Figure 4.5 shows the next 25 grid points for $k = 2$ plane. The last 25 grid points are shown in Fig. 4.6 for $k = 3$ plane. Some sample calculations can be found in the Ref. [4.11].

4.10 Derivation of the Hugoniot Relations

4.10.1 Introduction

In this section the Hugoniot relations are derived from the conservation equations for mass, momentum and energy. The pressure is used for the

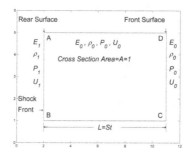

Fig. 4.7 A cylindrical pipe of diameter A and length L filled with un-shocked material at initial conditions E_0, ρ_0, P_0 and U_0. The shock with velocity S coming from the left side which is the rear surface.

governing equations because we are dealing with material in gas or liquid phases.

4.10.2 *Conservation of Mass*

As shown in Fig. 4.7, the shock front is coming from the left side of the cylindrical pipe with the following definitions.

E_0: internal energy of the un-shocked material.
ρ_0: density of the un-shocked material.
P_0: pressure of the un-shocked material.
U_0: velocity of the un-shocked material.
A: cross section area.
L: length of the cylindrical pipe.
S: shock speed.
t: total time for the shock to go through the pipe from $\overline{AB}$ to $\overline{CD}$.
U_1: velocity of the shocked material.
ρ_1: density of the shocked material.
P_1: pressure of the shocked material.
E_1: internal energy of the shocked material.

Also, let assume $U_0 = 1 cm/sec$, $U_1 = 7.75 cm/sec$, $L = 9 cm$ and $t = 1 sec$. Then, Fig. 4.8 shows that with shock velocity of $S = 9 cm/sec$ it will

EOS, Constitutive Relationship and High Explosive

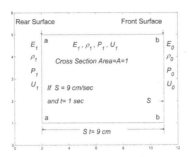

Fig. 4.8 A cylindrical pipe of diameter $A=1$ and length $L=9cm$ filled with shocked material with conditions E_1, ρ_1, P_1 and U_1. The shock velocity is $S=9cm/sec$. It takes $1sec$ for the shock to go from the rear surface to the front surface.

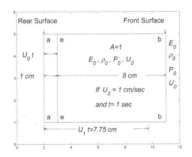

Fig. 4.9 A cylindrical pipe of diameter $A=1$ and length $L=9cm$ filled with un-shocked material with conditions E_0, ρ_0, P_0 and U_0. The un-shocked material has a velocity of $U_0 = 1cm/sec$. In $1sec$ the $U_0 = 1cm/sec$ will move the rear surface from $\overline{aa}$ to $\overline{ee}$ which is $1cm$. In $1sec$, the shocked velocity U_1 will travel a distance of $7.75cm$. Therefore, the rear surface will be moved by the shock with a distance of $7.75cm$.

take 1 sec for the shock to move the left side from $\overline{aa}$ to $\overline{bb}$. or the distance

$$St = (9) \times (1) = 9cm. \quad (4.57)$$

Figure 4.9 shows that the un-shocked material has its own velocity of $U_0 = 1cm/sec$, therefore, in 1 sec, the left boundary will move from $\overline{aa}$ to $\overline{ee}$ or

$$U_0 t = (1) \times (1) = 1cm. \quad (4.58)$$

From Fig. 4.9, it is obviously that the material contained in the pipe $\overline{eebb}$ is the only material that the shock will go through in 1 sec. The

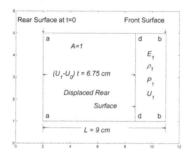

Fig. 4.10 A cylindrical pipe of diameter $A = 1$ and length $L = 9cm$ filled with un-shocked material with initial conditions E_0, ρ_0, P_0 and U_0. The shock wave will move the rear surface at $\overline{aa}$ to $\overline{dd}$ in 1 sec. Since $U_0 = 1cm/sec$ and $U_1 = 7.75cm/sec$. In 1sec the $(U_1 - U_0)t = (7.75 - 1)(1) = 6.75cm$ is the displaced distance for the rear surface.

distance $\overline{eb}$ is $9 - 1 = 8cm$ as shown in Fig. 4.9. Therefore, in 1 sec with $U_0 = 1cm/sec$ and $U_1 = 7.75cm/sec$ the shock will move the left boundary $\overline{aa}$ to $\overline{dd}$ as shown in Fig. 4.10. That is

$$(U_1 - U_0)t = (7.75 - 1) \times (1) = 6.75cm. \quad (4.59)$$

As shown in Fig. 4.10, we know that the un-shocked material in the volume contained by $\overline{aadd}$ will be shocked into the volume contained by $\overline{ddbb}$. From Fig. 4.7, the un-shocked mass is

$$(mass)_{before\ shocked} = StA\rho_0, \quad (4.60)$$

$$(mass)_{after\ shocked} = [St - (U_1 - U_0)t]A\rho_1, \quad (4.61)$$

where the relative velocity is $U_1 - U_0$. From Eqs. (4.60) and (4.61) we have

$$StA\rho_0 = [St - (U_1 - U_0)t]A\rho_1, \quad (4.62)$$

or

$$S\rho_0 = S\rho_1 - (U_1 - U_0)\rho_1, \quad (4.63)$$

or

$$S(\rho_0 - \rho_1) = -(U_1 - U_0)\rho_1, \quad (4.64)$$

or

$$S = (U_1 - U_0)\frac{\rho_1}{\rho_1 - \rho_0} = (U_1 - U_0)\frac{\frac{1}{V_1}}{\frac{1}{V_1} - \frac{1}{V_0}}, \quad (4.65)$$

or

$$S = (U_1 - U_0)\frac{V_0}{V_0 - V_1},\tag{4.66}$$

where

$$V_1 = \frac{\rho_0}{\rho_1},\tag{4.67}$$

and

$$V_0 = \frac{\rho_0}{\rho_0} = 1.\tag{4.68}$$

Equation (4.66) is the Hugoniot relation in conservation of mass. Here the volume are referred to the density ρ_0, making the relative volume V_0 equal to 1. V_0 is carried through the equations, even though it is 1, in order to describe the general case where the volume are referred to a reference density ρ_{ref} that is not the density ρ_0, ahead of the shock. In this case, the result is $V_0 = \rho_{ref}/\rho_0$ and $V_1 = \rho_{ref}/\rho_1$.

4.10.3 Conservation of Momentum

Since momentum is equal to *Force* × *time*, with the notations shown in Fig. 4.7, the momentum for the length L is

$$Momentum = force \times time = (P_1 A - P_0 A) \times t = (P_1 A t - P_0 A t).\tag{4.69}$$

Also,

$$Momentum = force \times time$$
$$= mass \times acceleration \times time = mass \times velocity.\tag{4.70}$$

Therefore, Eq. (4.69) should be equal to the change in momentum

$$(P_1 A t - P_0 A t) = mass \times U_1 - mass \times U_0,\tag{4.71}$$

or

$$(P_1 A t - P_0 A t) = \rho_0 A L U_1 - \rho_0 A L U_0,\tag{4.72}$$

or

$$(P_1 - P_0)t = \rho_0 S t U_1 - \rho_0 S t U_0,\tag{4.73}$$

or
$$P_1 - P_0 = \rho_0 S(U_1 - U_0). \tag{4.74}$$

Equation (4.66) can be written as
$$(U_1 - U_0) = \frac{S(V_0 - V_1)}{V_0}. \tag{4.75}$$

Substituting Eq. (4.75) into Eq. (4.74) results
$$(P_1 - P_0) = \rho_0 S \frac{S(V_0 - V_1)}{V_0}, \tag{4.76}$$

or
$$S^2 = \frac{V_0}{\rho_0} \frac{P_1 - P_0}{(V_0 - V_1)}. \tag{4.77}$$

Equation (4.74) can be written as
$$S = \frac{P_1 - P_0}{\rho_0(U_1 - U_0)}. \tag{4.78}$$

Substituting Eq. (4.66) into Eq. (4.78) results
$$\frac{P_1 - P_0}{\rho_0(V_0 - V_1)} = (U_1 - U_0)\frac{V_0}{V_0 - V_1}, \tag{4.79}$$

then
$$(P_1 - P_0)(V_0 - V_1) = \rho_0 U_0 (U_1 - U_0)^2, \tag{4.80}$$

the above equation is the Hugoniot relation in the conservation of momentum.

4.10.4 Conservation of Energy

Conservation of energy requires that the net work on the mass be equal to the change in total energy (kinetic and internal energy). Since work = force X distance. Therefore, for the rear end moving from surface $\overline{AB}$ to $\overline{CD}$ as shown in Fig. 4.7, the work done to the system by the shock wave is
$$W = P_1 A \times L, \tag{4.81}$$

or
$$W = P_1 A \times St. \tag{4.82}$$

For the shocked material moving, the input work to the system is

$$W = (P_1 A) \times (U_1 t). \tag{4.83}$$

For the rear end moving from surface $\overline{aa}$ to $\overline{ee}$ as shown in Fig. 4.9, the work is done by the system, therefore, it is a negative work to the system. The negative work is

$$-W = -(P_0 A) \times (U_0 t). \tag{4.84}$$

Therefore, the total work is

$$(P_1 A)(U_1 t) - (P_0 A)(U_0 t). \tag{4.85}$$

The change of kinetic energy and internal energy is

$$L\rho_0 A [\frac{1}{2}(U_1^2 - U_0^2) + E_1 - E_0], \tag{4.86}$$

where E_1 and E_0 are internal energy per unit mass and $L = ST$. From Eqs. (4.85) and (4.86), one gets

$$P_1 U_1 - P_0 U_0 = S\rho_0 [\frac{1}{2}(U_1^2 - U_0^2)] + S\rho_0 (E_1 - E_0), \tag{4.87}$$

or

$$P_1 U_1 - P_0 U_0 = \frac{1}{2} S\rho_0 (U_1 - U_0)(U_1 + U_0) + S\rho_0 (E_1 - E_0). \tag{4.88}$$

Equation (4.78) can be written as

$$S\rho_0 = \frac{P_1 - P_0}{(U_1 - U_0)}. \tag{4.89}$$

Substituting $S\rho_0$ from Eq. (4.89) into Eq. (4.88), the result is

$$P_1 U_1 - P_0 U_0 = \frac{P_1 - P_0}{U_1 - U_0}(\frac{1}{2})(U_1 - U_0)(U_1 + U_0) + S\rho_0 (E_1 - E_0). \tag{4.90}$$

Substituting Eq. (4.66) into Eq. (4.90), we have

$$P_1 U_1 - P_0 U_0 = \frac{P_1 - P_0}{U_1 - U_0}(\frac{1}{2})(U_1 - U_0)(U_1 + U_0)$$
$$+ \rho_0 V_0 (E_1 - E_0) \frac{U_1 - U_0}{V_0 - V_1}. \tag{4.91}$$

Equation (4.91) can be rewritten as

$$P_1 U_1 - P_0 U_0 = \frac{1}{2}(P_1 - P_0)(U_1 + U_0) + \rho_0 V_0 (E_1 - E_0) \frac{U_1 - U_0}{V_0 - V_1}. \tag{4.92}$$

Since

$$P_1U_1 - P_0U_0 - \tfrac{1}{2}(P_1 - P_0)(U_1 + U_0)$$

$$= P_1U_1 - P_0U_0 - \frac{1}{2}[P_1U_1 + P_1U_0 - P_0U_1 - P_0U_0], \quad (4.93)$$

$$= \frac{1}{2}[P_1(U_1 - U_0) + P_0(U_1 - U_0)], \quad (4.94)$$

$$= \frac{1}{2}(U_1 - U_0)(P_1 + P_0). \quad (4.95)$$

Substituting Eq. (4.95) into Eq. (4.92), one gets

$$\frac{1}{2}(U_1 - U_0)(P_1 + P_0) = \rho_0 V_0 (E_1 - E_0) \frac{U_1 - U_0}{V_0 - V_1}. \quad (4.96)$$

Equation (4.96) can be simplified as

$$\rho_0 V_0 (E_1 - E_0) = \frac{1}{2}(P_1 + P_0)(V_0 - V_1). \quad (4.97)$$

Let

$$\epsilon = \rho_0 V_0 E. \quad (4.98)$$

Equation (4.97) becomes

$$\epsilon_1 - \epsilon_0 = \frac{1}{2}(P_1 + P_0)(V_0 - V_1). \quad (4.99)$$

The unit of ϵ is

$$\epsilon = \rho_0 V_0 E = (\frac{gm}{cm^3})(1)(\frac{joule}{gm}) = \frac{joule}{cm^3}. \quad (4.100)$$

Equations (4.90), (4.91), (4.92), (4.96), (4.97) and (4.99) are the conservation of energy for the Hugoniot relation.

4.11 The Shock-Change Equations

4.11.1 *Introduction*

The shock-change equations [4.12, 4.13] describes the time (or distance) rate of change of the leading wave in rate-dependent elastic-plastic solids, shock-induced phase change, and shock initiation of solid explosives. There are numerous ways to express the rate of change of variables along the shock path and there are several approximations to exact relationships: both in Eulerian (spatial) and Lagrangian (material) coordinates. We give here a general derivation of the shock-change equation in one dimension and list its many forms.

4.11.2 The Shock-Change Equation

The shock-change equation is the relationship between derivatives of quantities in terms of x and t (or X and t) and derivatives of variables following the shock front, which moves with speed U into undisturbed material at rest. The planar shock front is assumed to be propagating in the x (Eulerian spatial coordinate) or X (Lagrangian spatial coordinate) direction, $\rho dx = \rho_0 dX$. In this section, we use

U: shock speed.
u_1: velocity of shocked material.
ρ_0: density of un-shocked material.
ρ: density of shocked material.
σ: longitudinal stress.

In Eulerian coordinates x and t, the mass and momentum conservation laws and material constitutive equation are given by ($u = u_1 =$ particle velocity, $\sigma = \sigma_{11} =$ longitudinal stress and $\rho =$ material density)

$$\dot{\rho} + \rho \frac{\partial u}{\partial x} = 0, \tag{4.101}$$

$$\rho \dot{u} + \frac{\partial \sigma}{\partial x} = 0, \tag{4.102}$$

$$\dot{\sigma} + \frac{\rho}{\rho_0} C_\ell^2 \dot{\rho} = F, \tag{4.103}$$

where C_ℓ is the longitudinal sound speed, a dot over a variable signifies differentiation along a particle path. (Lagrangian differentiation). Micromechanical effects are contained in the function F. If we define D_t to be the time-derivative operator following the shock path $dx/dt = U(t)$, then

$$D_t = \frac{\partial}{\partial t} + U \frac{\partial}{\partial x}, \tag{4.104}$$

$$(\dot{\ }) = D_t - (U - u)\frac{\partial}{\partial x}, \tag{4.105}$$

where U is the shock velocity, u is the shocked particle velocity.

The jump conditions across a discontinuity moving into undisturbed material at rest from momentum and continuity equations are

$$\sigma = \rho_0 U u, \tag{4.106}$$

$$\rho(U - u) = \rho_0 U. \tag{4.107}$$

We define two quantities specifying the elastic shock response (the subscript H refers to conditions along the elastic Hugoniot curve)

$$Z_H = (\frac{d\sigma}{du})_H, \qquad (4.108)$$

$$C_H^2 = (\frac{d\sigma}{d\rho})_H. \qquad (4.109)$$

Equation (4.106) can be derived from Eq. (4.74) which is

$$P_1 - P_0 = \rho_0 S(U_1 - U_0), \qquad (4.110)$$

and with $P_0 = 0, P_1 = \sigma, S = U, U_1 = u$, and $U_0 = 0$, we have

$$\sigma = \rho_0 U u. \qquad (4.111)$$

Equation (4.107) can be derived from Eq. (4.65) which is

$$S = (U_1 - U_0)\frac{\rho_1}{\rho_1 - \rho_0}, \qquad (4.112)$$

with $S = U, U_1 = u$ and $U_0 = 0$, it follows

$$U = u\frac{\rho_1}{\rho_1 - \rho_0}, \qquad (4.113)$$

or

$$U(\rho_1 - \rho_0) = u\rho_1, \qquad (4.114)$$

or

$$U\rho_1 - u\rho_1 = U\rho_0, \qquad (4.115)$$

or

$$\rho_1(U - u) = U\rho_0. \qquad (4.116)$$

Equation (4.116) is the same as Eq. (4.107). Equation (4.116) can be written as

$$\frac{\rho_0}{\rho} = \frac{U - u}{U}, \qquad (4.117)$$

or

$$1 - \frac{\rho_0}{\rho} = 1 - \frac{U - u}{U}. \qquad (4.118)$$

Let

$$1 - \frac{\rho_0}{\rho} = \epsilon, \qquad (4.119)$$

then
$$1 - \frac{U-u}{U} = \epsilon, \tag{4.120}$$

or
$$u = U\epsilon. \tag{4.121}$$

On Hugoniot cure with dividing Eq. (4.121) by Eq. (4.106) results
$$\epsilon\sigma = \rho_0 u^2. \tag{4.122}$$

Equation (4.122) can be written as
$$\sigma\frac{d\epsilon}{du} + \epsilon\frac{d\sigma}{du} = 2\rho_0 u. \tag{4.123}$$

Since
$$\frac{d\epsilon}{du} = \frac{d\epsilon}{d\sigma}\frac{d\sigma}{du}. \tag{4.124}$$

Substituting Eq. (4.124) into Eq. (4.123), one gets
$$(\frac{d\epsilon}{d\sigma}\sigma + \epsilon)\frac{d\sigma}{du} = 2\rho_0 u. \tag{4.125}$$

Also, from Eq. (4.119) we have
$$\frac{d\sigma}{d\epsilon} = \frac{d\sigma}{d\rho}\frac{d\rho}{d\epsilon} = \frac{\rho^2}{\rho_0}\frac{d\sigma}{d\rho} = \frac{\rho^2}{\rho_0}C_H^2. \tag{4.126}$$

Substituting Eq. (4.126) into Eq. (4.125), one gets
$$(\frac{\rho_0\sigma}{\rho C_H^2} + \epsilon)\frac{d\sigma}{du} = 2\rho_0 u, \tag{4.127}$$

or
$$(\frac{\rho_0\sigma}{\rho\epsilon C_H^2} + 1)\frac{d\sigma}{du} = 2\rho_0\frac{u}{\epsilon} = 2\rho_0 U, \tag{4.128}$$

or
$$\frac{d\sigma}{du} = \frac{2\rho_0 U}{1 + \frac{\rho_0^2 U^2}{\rho^2 C_H^2}}, \tag{4.129}$$

or
$$Z_H = \frac{2\rho_0 U}{1 + (\frac{\rho_0 U}{\rho C_H})^2}. \tag{4.130}$$

This allows us to write
$$D_t\sigma = Z_H D_t u, \tag{4.131}$$

or

$$D_t\sigma = C_H^2 D_t\rho. \tag{4.132}$$

The quantities Z_H and C_H define the differential properties of the elastic Hugoniot. At low-stress amplitudes $Z_H \approx \rho_0 C_{\ell 0}$ and $C_H \approx C_{\ell 0}$, where $C_{\ell 0}$ is the adiabatic longitudinal elastic sound speed at $\rho = \rho_0$. Equation (4.101) can be written as

$$\rho_0 U \dot{\rho} = \rho_0 U(-\rho\frac{\partial u}{\partial x}) = -\rho_0 \rho U \frac{\partial u}{\partial x}, \tag{4.133}$$

and

$$\rho^2(D_t u - \dot{u}) = \rho^2[\frac{\partial u}{\partial t} + U\frac{\partial u}{\partial x} - (\frac{\partial u}{\partial t}) + U\frac{\partial u}{\partial x} + (U-u)\frac{\partial u}{\partial t}], \tag{4.134}$$

or

$$\rho^2(D_t u - \dot{u}) = \rho^2(U-u)\frac{\partial u}{\partial x}. \tag{4.135}$$

Adding Eqs. (4.133) and (4.135), the result is

$$\rho_0 U \dot{\rho} + \rho^2(D_t u - \dot{u}) = -\rho_0 \rho U\frac{\partial u}{\partial x} + \rho^2(U-u)\frac{\partial u}{\partial x}. \tag{4.136}$$

Substituting Eq. (4.107) into the right hand side of Eq. (4.136), one gets

$$\rho_0 U \dot{\rho} + \rho^2(D_t u - \dot{u}) = -\rho\rho(U-u)\frac{\partial u}{\partial x} + \rho^2(U-u)\frac{\partial u}{\partial x}, \tag{4.137}$$

or

$$\rho_0 U \dot{\rho} + \rho^2(D_t u - \dot{u}) = 0. \tag{4.138}$$

From Eq. (4.105) we have

$$D_t\sigma - \dot{\sigma} = D_t\sigma - D_t\sigma + (U-u)\frac{\partial \sigma}{\partial x}, \tag{4.139}$$

or

$$D_t\sigma - \dot{\sigma} = (U-u)\frac{\partial \sigma}{\partial x}. \tag{4.140}$$

Substituting Eq. (4.102) into Eq. (4.140), one gets

$$D_t\sigma - \dot{\sigma} = (U-u)(-\rho\dot{u}). \tag{4.141}$$

Substituting Eq. (4.107) into Eq. (4.141), the result is

$$D_t\sigma - \dot{\sigma} = -\rho_0 U \dot{u}, \tag{4.142}$$

therefore
$$D_t\sigma - \dot{\sigma} + \rho_0 U \dot{u} = 0. \quad (4.143)$$

Equations (4.138) and (4.143) are valid only at the shock front.
Using Eq. (4.103) one can write
$$F + [(\frac{\rho C_\ell}{\rho_0 U})^2 - 1]\rho_0 U u = \dot{\sigma} - \frac{\rho}{\rho_0}C_\ell^2\dot{\rho} + [(\frac{\rho C_\ell}{\rho_0 U})^2 - 1]\rho_0 U u. \quad (4.144)$$

Equation (4.143) can be written as
$$\dot{\sigma} = D_t\sigma + \rho_0 U \dot{u}. \quad (4.145)$$

Substituting Eq. (4.145) into the right hand side of Eq. (4.144) the result is
$$\text{RHS of Eq. (4.144)} = D_t\sigma + \rho_0 U \dot{u} - \frac{\rho}{\rho_0}C_\ell^2\dot{\rho} + \left[\left(\frac{\rho C_\ell}{\rho_0 U}\right)^2 - 1\right]\rho_0 U u, \quad (4.146)$$

or
$$\text{RHS of Eq. (4.144)} = D_t\sigma - \frac{\rho}{\rho_0}C_\ell^2\dot{\rho} + \frac{(\rho C_\ell)^2}{\rho_0 U}\dot{u}. \quad (4.147)$$

Since
$$[1 + (\frac{\rho C_\ell}{\rho_0 U})^2(\frac{\rho_0 U}{Z_H})]D_t\sigma = D_t\sigma + \frac{(\rho C_\ell)^2}{\rho_0 U Z_H}D_t\sigma. \quad (4.148)$$

Equation (4.131) can be written as
$$\frac{D_t\sigma}{Z_H} = D_t u. \quad (4.149)$$

Substituting Eq. (4.149) into the right hand side of Eq. (4.148) one gets
$$[1 + (\frac{\rho C_\ell}{\rho_0 U})^2(\frac{\rho_0 U}{Z_H})]D_t\sigma = D_t\sigma + \frac{(\rho C_\ell)^2}{\rho_0 U}D_t u, \quad (4.150)$$

or
$$[1 + (\frac{\rho C_\ell}{\rho_0 U})^2(\frac{\rho_0 U}{Z_H})]D_t\sigma = D_t\sigma + \frac{(\rho C_\ell)^2}{\rho_0 U}(\frac{\partial u}{\partial t} + U\frac{\partial u}{\partial x}), \quad (4.151)$$

or
$$[1 + (\frac{\rho C_\ell}{\rho_0 U})^2(\frac{\rho_0 U}{Z_H})]D_t\sigma = D_t\sigma + \frac{(\rho C_\ell)^2}{\rho_0 U}\dot{u} + \frac{(\rho C_\ell)^2}{\rho_0}\frac{\partial u}{\partial x}. \quad (4.152)$$

Substituting Eq. (4.101) into the right hand side of Eq. (4.152) we have
$$[1 + (\frac{\rho C_\ell}{\rho_0 U})^2(\frac{\rho_0 U}{Z_H})]D_t\sigma = D_t\sigma + \frac{(\rho C_\ell)^2}{\rho_0 U}\dot{u} + \frac{\rho(C_\ell)^2}{\rho_0}\dot{\rho}. \quad (4.153)$$

Since the right hand side of Eq. (4.153) is the same as the right hand side of Eq. (4.147) we conclude that the left hand side of Eq. (4.150) is equal to the left hand side of Eq. (4.144), therefore

$$[1 + (\frac{\rho C_\ell}{\rho_0 U})^2(\frac{\rho_0 U}{Z_H})]D_t\sigma = F + [(\frac{\rho C_\ell}{\rho_0 U})^2 - 1]\rho_0 U u. \qquad (4.154)$$

The above equation can be rewritten as

$$D_t\sigma = \frac{F + [(\frac{\rho C_\ell}{\rho_0 U})^2 - 1]\rho_0 U \dot{u}}{1 + (\frac{\rho C_\ell}{\rho_0 U})^2(\frac{\rho_0 U}{Z_H})}. \qquad (4.155)$$

This is expressed in terms of the particle acceleration immediately behind the shock front. Equation (4.155) can be expressed in terms of the Lagrangian stress gradient $(\partial\sigma/\partial X)$, and the Lagrangian longitudinal sound speed $C_L = \rho C_\ell/\rho_0$. Substituting $C_\ell = \rho_0 C_L/\rho$ and Eq. (4.102) into the numerator of the right hand side of Eq. (4.155) one gets

$$NU = F + [(\frac{C_L}{U})^2 - 1](\rho_0 U)\frac{1}{\rho}(-\frac{\partial\sigma}{\partial x}). \qquad (4.156)$$

Substituting $\rho dx = \rho_0 dX$ into Eq. (4.156), the result is

$$NU = F - [(\frac{C_L}{U})^2 - 1](\rho_0 U)(\frac{\partial\sigma}{\partial\rho_0 X}), \qquad (4.157)$$

or

$$NU = F - [(\frac{C_L}{U})^2 - 1](U)(\frac{\partial\sigma}{\partial X}). \qquad (4.158)$$

The denominator of Eq. (4.155) is

$$DE = 1 + (\frac{C_L}{U})^2(\frac{\rho_0 U}{Z_H}). \qquad (4.159)$$

Substituting Eqs. (4.158) and (4.159) into Eq. (4.155), the result is

$$D_t\sigma = \frac{F - [(\frac{C_L}{U})^2 - 1](U)(\frac{\partial\sigma}{\partial X})}{1 + (\frac{C_L}{U})^2(\frac{\rho_0 U}{Z_H})}. \qquad (4.160)$$

If we accept the assumption that the elastic wave can be treated to good approximation as a mathematical discontinuity, then the stress decay at the elastic front is given by Eqs. (4.155) and (4.160) in terms of the material-dependent and amplitude-dependent wave speed: C_ℓ (the isentropic longitudinal elastic sound speed), U (the finite-amplitude elastic shock velocity), and C_H as defined in Eq. (4.109).

For elastic compression the entropy increase is small and can come only from effects of thermal conduction. Thus, to a very good approximation at low compression in most material

$$C_H \approx C_\ell. \qquad (4.161)$$

Equation (4.130) can be written as

$$\frac{\rho_0 U}{Z_H} = \frac{\rho_0 U[1 + (\frac{\rho_0 U}{\rho C_H})^2]}{2\rho_0 U}. \tag{4.162}$$

Substituting Eq. (4.161) into Eq. (4.162) results

$$\frac{\rho_0 U}{Z_H} = \frac{1}{2}[1 + (\frac{\rho_0 U}{\rho C_\ell})^2]. \tag{4.163}$$

Substituting Eq. (4.163) into the denominator on the right hand side of Eq. (4.155), the result is

$$1 + (\frac{\rho C_\ell}{\rho_0 U})^2 (\frac{\rho_0 U}{Z_H}) = \frac{3}{2} + \frac{1}{2}(\frac{\rho C_\ell}{\rho_0 U})^2. \tag{4.164}$$

Substituting Eq. (4.164) into Eq. (4.155), one gets

$$D_t \sigma \approx \frac{F + [(\frac{\rho C_\ell}{\rho_0 U})^2 - 1]\rho_0 U \dot{u}}{\frac{3}{2} + \frac{1}{2}(\frac{\rho C_\ell}{\rho_0 U})^2}. \tag{4.165}$$

An additional approximation can be used in the case when $\rho C_\ell \approx \rho_0 U$ and the acceleration behind the elastic shock front is small, so that

$$F >> [(\frac{\rho C_\ell}{\rho_0 U})^2 - 1]\rho_0 U \dot{u}. \tag{4.166}$$

In this case

$$D_t \sigma \approx \frac{F}{2}, \tag{4.167}$$

which can be very useful approximation in estimating decay rates at low-stress amplitudes or for nearly linear material behavior such that

$$\rho C_\ell = \rho_0 U. \tag{4.168}$$

Equation (4.168) holds, for example, in material which obey isentropic elastic relationship

$$\sigma = C_{11}(1 - \frac{\rho_0}{\rho}), \tag{4.169}$$

where C_{11} is the constant isentropic longitudinal modulus.

Equation (4.155) was derived with rate of decay dependent upon F and $\dot{u}$, we could just as well have derived expression for $D_t \sigma$ in terms of F and $\dot{\rho}$ or F and $\dot{\sigma}$ or other combinations. For nearly constant U and from Eq. (4.106) we have

$$\dot{\sigma} = \rho_0 U \dot{u}. \tag{4.170}$$

Substituting Eq. (4.170) into Eq. (4.155) one gets

$$D_t\sigma = \frac{F + [(\frac{\rho C_\ell}{\rho_0 U})^2 - 1]\dot{\sigma}}{1 + (\frac{\rho C_\ell}{\rho_0 U})^2(1 + \frac{\rho_0 U}{Z_H})}. \tag{4.171}$$

When micro-mechanical effects are neglected, i.e., $F \approx 0$, Equation (4.103) becomes

$$\dot{\sigma} = \frac{\rho}{\rho_0} C_\ell^2 \dot{\rho}. \tag{4.172}$$

Also, we can assume

$$\rho C_\ell \approx \rho_0 U, \tag{4.173}$$

or

$$(C_\ell)^2 \approx (\frac{\rho_0 U}{\rho})^2. \tag{4.174}$$

Substituting Eq. (4.174) into Eq. (4.172) one gets

$$\dot{\sigma} = \frac{\rho}{\rho_0}(\frac{\rho_0 U}{\rho})^2 \dot{\rho} = \frac{\rho}{\rho_0} U^2 \dot{\rho}. \tag{4.175}$$

Substituting Eqs. (4.175) and (4.173) into Eq. (4.171) one gets

$$D_t\sigma = \frac{F + [(\frac{\rho C_\ell}{\rho_0 U})^2 - 1]\frac{\rho_0}{\rho} U^2 \dot{\rho}}{1 + (\frac{\rho_0 U}{Z_H})}. \tag{4.176}$$

Equation (4.131) can be written as

$$D_t u = \frac{D_t \sigma}{Z_H}. \tag{4.177}$$

Applying D_t to Eq. (4.106) one gets

$$D_t\sigma = \rho_0 u D_t U + U D_t(\rho_0 u), \tag{4.178}$$

or

$$\rho_0 u D_t U = D_t\sigma - U D_t(\rho_0 u), \tag{4.179}$$

or

$$\rho_0 u D_t U = D_t\sigma - \rho_0 U D_t(u). \tag{4.180}$$

Substituting Eq. (4.177) into Eq. (4.180) we have

$$\rho_0 u D_t U = (1 - \frac{\rho_0 U}{Z_H}) D_t\sigma. \tag{4.181}$$

Applying D_t to Eq. (4.107) one gets

$$(U - u)D_t\rho + \rho D_t(U - u) = \rho_0 D_t U . \tag{4.182}$$

By applying Eq. (4.177) to Eq. (4.182) it follows

$$(U - u)D_t\rho + \rho D_t U - \frac{\rho}{Z_H}D_t\sigma = \rho_0 D_t U . \tag{4.183}$$

Equation (4.181) can be written as

$$D_t U = \frac{1}{\rho_0 u}(1 - \frac{\rho_0 U}{Z_H})D_t\sigma . \tag{4.184}$$

Equation (4.183) can be written as

$$(U - u)D_t\rho = \frac{\rho}{Z_H}D_t\sigma + (\rho_0 - \rho)D_t U . \tag{4.185}$$

Substituting Eq. (4.184) into Eq. (4.185) one gets

$$(U - u)D_t\rho = \frac{\rho}{Z_H}D_t\sigma + (\rho_0 - \rho)(\frac{1}{\rho_0 u})(1 - \frac{\rho_0 U}{Z_H})D_t\sigma , \tag{4.186}$$

or

$$(U - u)D_t\rho = [\frac{\rho}{Z_H} + (\frac{\rho_0 - \rho}{\rho_0 u})(1 - \frac{\rho_0 U}{Z_H})]D_t\sigma , \tag{4.187}$$

or

$$D_t\rho = (\frac{\rho}{\rho_0 U})[\frac{\rho}{Z_H} + (\frac{\rho_0 - \rho}{\rho_0 u})(1 - \frac{\rho_0 U}{Z_H})]D_t\sigma , \tag{4.188}$$

or

$$D_t\rho = (\frac{\rho}{(\rho_0 U)^2})\{\frac{\rho_0 U}{Z_H}[\rho - \frac{(\rho_0 - \rho)U}{u}] + \frac{(\rho_0 - \rho)U}{u}\}D_t\sigma . \tag{4.189}$$

Equation (4.107) can be written as

$$(\rho_0 - \rho)\frac{U}{u} = -\rho . \tag{4.190}$$

Substituting Eq. (4.190) into Eq. (4.189) one gets

$$D_t\rho = (\frac{\rho}{(\rho_0 U)^2})[\frac{\rho_0 U}{Z_H}(\rho + \rho) + (-\rho)]D_t\sigma , \tag{4.191}$$

or

$$D_t\rho = (\frac{\rho}{(\rho_0 U)^2})(\frac{2\rho\rho_0 U}{Z_H} - \rho)D_t\sigma , \tag{4.192}$$

or

$$D_t\rho = (\frac{\rho}{\rho_0 U})^2(\frac{2\rho_0 U}{Z_H} - 1)D_t\sigma . \tag{4.193}$$

4.11.3 *Summary of the Shock-Change Equation*

By defining $A = (\rho C_\ell / \rho_0 U)^2$ the shock-change equations with exact relationships are given below

$$D_t \sigma = \frac{F + (A-1)\rho_0 U \dot{u}}{1 + A(\frac{\rho_0 U}{Z_H})} = \frac{F + (A-1)\rho_0 U^2 (\partial u / \partial X)}{1 + A(\frac{\rho_0 U}{Z_H})}, \qquad (4.194)$$

$$D_t \sigma = \frac{F + [A-1](\frac{\rho_0 U}{\rho})^2 \dot{\rho}}{1 + (\frac{\rho_0 U}{Z_H})} = \frac{F + [A-1](\frac{\rho_0 U}{\rho})^2 U \frac{\partial \rho}{\partial X}}{\frac{3\rho_0 U}{Z_H} + A(1 - \frac{2\rho_0 U}{Z_H})}, \qquad (4.195)$$

$$D_t \sigma = \frac{F + [A-1]\dot{\sigma}}{1 + A(1 + \frac{\rho_0 U}{Z_H})} = \frac{F + [A-1](U)\frac{\partial \sigma}{\partial X}}{1 + A(1 + \frac{\rho_0 U}{Z_H})}. \qquad (4.196)$$

In the above equations $D_t \sigma$ is the decay rate of stress. When the shock goes through the material the temperature of the material will rise and the stress will decrease. If the shock is strong enough, then, the material may become plastic regime and eventually become liquid state. At liquid state there is only pressure instead of stress. In the material constitutive relationship the melting of the material is defined by the melting temperature and melting energy as given in Eqs. (4.44) and (4.45).

4.12 The Theoretical Spall Strength

The theoretical spall strength is expected to place an upper bound on spall strength calculated from more realistic theories. In certain materials, under the very rapid tensile loading rates achieved in impact situation, measured spall strength not too far removed from theoretical spall strength are actually achieved. The calculations of theoretical strength are based on energy concepts.

Three-parameter models of the intermolecular potential have been developed which provide reasonable descriptions of the thermodynamic behavior of solids. Examples include the Morse potential, the exponential-six potential, and, more recently, a form proposed by Rose et al. [4.14] for metals.

To derive an explicit expression for the theoretical spall strength, an analytic representation of the cold compression-tension curve is used based on the Morse potential of the form

$$U(v) = U_{coh} \left\{ exp\left[\frac{-2(v - v_0)}{a}\right] - 2exp\left[\frac{-(v - v_0)}{a}\right] \right\}. \qquad (4.197)$$

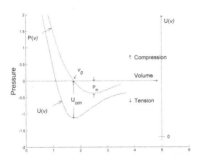

Fig. 4.11 The cold compression-tension behavior of condensed matter. The volume dependence of energy and pressure or tension are illustrated. The cohesive energy and maximum tension (theoretical spall strength) are properties of the material.

In Eq. (4.197), U_{coh} is the specific cohesive energy, $v = 1/\rho$ is the specific volume, ρ is the density and v_0 is the specific volume at zero pressure as shown in Fig. 4.11. The final parameter a is constrained by the relation for the bulk modulus

$$v\frac{d^2U}{dv^2} = B_0 . \qquad (4.198)$$

Applying d/dv to Eq. (4.197), it follows

$$\frac{dU}{dv} = -\frac{2U_{coh}}{a} exp\left[\frac{-2(v-v_0)}{a}\right] + \frac{2U_{coh}}{a} exp\left[\frac{-(v-v_0)}{a}\right], \qquad (4.199)$$

and

$$\frac{d^2U}{dv^2} = \frac{4U_{coh}}{a^2} exp\left[\frac{-2(v-v_0)}{a}\right] - \frac{2U_{coh}}{a^2} exp\left[\frac{-(v-v_0)}{a}\right] . \qquad (4.200)$$

When $v = v_0$, Eq. (4.200) becomes

$$\frac{d^2U}{dv^2} = \frac{2U_{coh}}{a^2} . \qquad (4.201)$$

Substituting Eq. (4.201) into Eq. (4.198) and let $v = v_0$, it follows

$$v_0\frac{2U_{coh}}{a^2} = B_0 , \qquad (4.202)$$

or

$$a = \sqrt{\frac{2v_0 U_{coh}}{B_0}}. \quad (4.203)$$

The cold pressure follows directly from Eq. (4.197)

$$P(v) = -\frac{dU}{dv} = \frac{2U_{coh}}{a}\left\{exp\left[\frac{-2(v-v_0)}{a}\right] - exp\left[\frac{-(v-v_0)}{a}\right]\right\}. \quad (4.204)$$

Within the present analysis, the theoretical spall strength P_{th} is determined from the minimum of Eq. (4.204)

$$\frac{dP}{dv} = \frac{2U_{coh}}{a^2}\left\{2exp\left[\frac{-2(v-v_0)}{a}\right] - exp\left[\frac{-(v-v_0)}{a}\right]\right\} = 0, \quad (4.205)$$

or

$$exp\left[\frac{-(v-v_0)}{a}\right] = \frac{1}{2}. \quad (4.206)$$

Substituting Eq. (4.206) into Eq. (4.204), it follows

$$P(v) = \frac{2U_{coh}}{a}(-\frac{1}{4}), \quad (4.207)$$

Table 4.6 Theoretical spall strength of selected materials.

Metal	$B_0(GP_a)$	$\rho(kg/m^3)$	$U_{coh}(MJ/kg)$	$P_{th}(GP_a)$
Aluminum	72.2	2710	11.9	17.1
Beryllium	100	1820	35.7	28.5
Copper	137	8930	5.32	28.5
Titanium	105	4510	9.78	24.1
Iron	168	7870	7.41	35.0
Tantalum	200	16660	4.31	42.4
Tungsten	323	19250	4.55	59.5
Uranium	98.7	19050	2.19	22.7
Mercury	38.2	13530	0.33	4.6
Tin	111	7300	2.53	16.0

or

$$P^2(v) = \frac{4U_{coh}^2}{a^2}\left(\frac{1}{16}\right). \tag{4.208}$$

Substituting Eq. (4.203) into Eq. (4.208), one gets

$$P_{th} = \sqrt{\frac{U_{coh}B_0}{8v_0}}. \tag{4.209}$$

Note that the theoretical spall strength now depends upon the cohesive energy as well as the bulk modulus. Representative values for selected metals are shown in Table 4.6.

Bibliography

4.1 McQueen, R., Marsh, S., Taylor, J., Fritz, J. and Carter, W. (1970). The equation of state of solids from shock wave measurements, *High-Velocity Impact*, edited by Kinslow R. Academic Press, New York.
4.2 Jacobson, J. (1993). *Private communication, Los Alamos National Lab, Los Alamos, NM*.
4.3 Tillotson, J. (1962). *Metallic equation of state for hypervelocity impact*, General Atomic Report GA-3216.
4.4 Steinberg, DJ and Guinan, MW. (1973). *Constitutive relations for the Kospall code*, Lawrence National Laboratory report UCID-16326.
4.5 Steinberg, DJ. (1991). *Equation of state and strength properties of selected materials*, UCRL-MA-106439, Lawrence Livermore National Laboratory.
4.6 Steinberg, DJ, Cochran, SG and Guinan, MW. (1980). A constitutive model for metals applicable at high-stain rate, *J. Appl. Physics*, 51, 1498.
4.7 Steinberg, DJ. (1987). Constitutive model used in computer simulation of time-resolved, shock-wave data, *Int. J. Impact Eng.*, 5, 603.
4.8 Steinberg, DJ and Lund, CM. (1989). A constitutive model for strain rates from 10^{-4} to $10^6 s^{-1}$, *J. Appl. Physics*, 65, 1528.
4.9 Johnson, GR and Cook, WH. (1983). *A constitutive model and data for metals subjected to large strains, high strain rates and high temperature*, Proceedings of 7th Int'l Symposium on Ballistics, The Hague, The Netherland, April 1983.
4.10 Lee, E, Horning, H and Kury, J. (1968). *Adiabatic expansion of high explosive detonation products*, Lawrence National Laboratory report UCRL-50422.
4.11 Mandell, D, Burton, D and Lund, CM. (1998). *High explosive programmed burn in the FLAG code*, Los Alamos National Laboratory Report LA-13406, UC-741.
4.12 Ahrens, TJ and Duvall, GE. (1966). Stress relation behind elastic shock waves in rocks, *J. Geophys. Res.* 71, 4349-4360 (1966).

4.13 Clifton, RJ. (1971). On the analysis of elastic/visco-plastic wave of finite uniaxial strain, in Shock Waves and the Mechanical Properties of Solid, edited by Burke JJ and Weiss V. Syracuse University Press (1971).

4.14 Rose, JH, Smith, JR, Guinea, F and Ferrante, J. (1984). Universal features of the equation of state of metals, *Phys. Rev. B* 29, 2963.

Chapter 5

The Exact Dimensions of the Perforators

Notations

M_a, M_b, M_c, M_d markers
P pressure ($Mbar$)
R, r radial coordinate (cm)
s distance (cm)
t time (μsec)
u_R displacement in the R direction (cm)
u_Z displacement in the Z direction (cm)
Z, z axial coordinate (cm)

Greek letters

δ_{ij} Kronecker delta (no unit)
ϵ specific internal energy per mass ($\frac{Mbar-cm^3}{g}$)
γ ratio of the specific heat, i.e., $\gamma = C_P/C_V$ (no unit)
η normalized density $= \frac{\rho}{\rho_0}$ (no unit)
θ angular coordinate
λ half of the grid size in R direction, i.e., $\lambda = \Delta R/2$ (cm)
μ normalized density minus one ($=\eta - 1 = \frac{\rho}{\rho_0} - 1$)(no unit)
ρ density ($\frac{g}{cm^3}$)

Subscripts

0 initial value
R derivative with respect to R coordinate
t derivative with respect to time
Z derivative with respect to Z coordinate

Superscript

n time at n time-step, i.e., $t^n = t_0 + n \cdot \Delta t$

5.1 Introduction

In this chapter the exact dimensions of seven perforators will be discussed and the coordinates of the liners and the cases will be provided in seven tables. These perforators, i.e., Perforators B, E, G, L, M, N and P, use stainless steel as cases, RDX as high explosive and copper or bronze as liners. Perforators B, E and G are also described in Ref. [5.1.]

The liners are constructed using circular arcs and straight lines while most of the steel case is using straight lines. The liners are usually made by pressing the bronze or copper powder into the desired shape. Same process is used to produce the high explosive part except that the powder of the HE should have very fine grain and uniform density once the HE is pressed.

There are two computer codes used to produce the information described in this chapter. The first one is called "Program Curve" which is written in FORTRAN 77 language and it will run under the Compaq FORTRAN compiler. This program provides the coefficients used by Steinberg's material model as described in Chapter 4. It also calculates the total mass of the liner as well as that of the high explosive. The second program is called PlotPB which will draw the actual dimensions of the Perforator B with coordinates and labels as shown in Fig. 5.1. The program PlotPB is written in MATLAB language to produce the plot shown in Fig. 5.1, therefore, if one would like to run the program PlotPB, one should have MATLAB software.

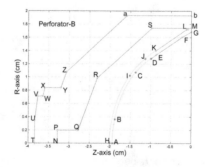

Fig. 5.1 Perforator B of diameter 3.45 cm with copper as liner and RDX with 2% wax as the high explosive.

Table 5.1 The coordinates of the labelling points for Perforator B.

Label	Z (cm)	R (cm)	Label	Z (cm)	R (cm)
A	-1.9363	0.0000	N	-3.2838	0.0000
B	-1.8794	0.3617	Q	-2.7798	0.1981
C	-1.3714	1.0691	R	-2.3067	0.9855
D	-0.9904	1.2806	S	-0.9739	1.7399
E	-0.8633	1.3373	T	-3.8390	0.0000
F	-0.2283	1.6037	U	-3.8390	0.3469
G	0.0000	1.6839	V	-3.7628	0.7112
H	-2.0086	0.0000	W	-3.6538	0.7112
I	-1.5138	1.0210	X	-3.5896	0.8382
J	-1.1330	1.2788	Y	-3.2015	0.8382
K	-0.8790	1.3968	Z	-3.0652	1.0744
L	-0.1169	1.7221	a	-1.5835	1.9304
M	-0.0761	1.7246	b	0.0000	1.9304

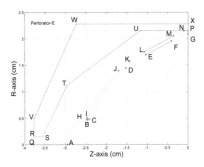

Fig. 5.2 Perforator E with diameter 4.3 cm uses bronze as the liner, RDX with 2% wax as high explosive and iron as the case.

Besides program PlotPB, there are six other programs, namely, PlotFig1, PlotFig6, PlotFig11, PlotFigwhl, PlotPE and PlotPG that will produce Figs. 5.2, 5.3, 5.4, 5.5, 5.6, and 5.7.

5.2 Perforator B

This is a simple perforator of diameter 3.45 cm with copper as liner and RDX with 2% wax as the high explosive. The coordinates of the liner as well as the stainless-steel-304 case are listed in Table 5.1. The circular arc $\widehat{AB}$ has the center located at $R=-0.03026$ cm and $Z=-0.5658$ cm as shown in Fig. 5.1. Since the scale in Z-axis is different from that of R-axis in Fig. 5.1, one has to be very cautious about using the data from the Table 5.1 with the plot shown in Fig. 5.1. The center of the arc $\widehat{BC}$ is located at R=-0.0693 cm and Z=-0.5188 cm. For arc $\widehat{CD}$, the center is at R=-1.4231 cm and Z=-0.2778 cm. For arc $\widehat{HI}$, the center is at R=-0.06806 cm and Z=-0.5675 cm. For arc $\widehat{IJ}$, the center is at R=-0.31063 cm and Z=-0.3361 cm. For arc $\widehat{JK}$, the center is at R=-1.42315 cm and Z=-0.2778 cm. The rest points shown in Fig. 5.1 are all of straight lines and the data are provided in Table 5.1.

5.3 Perforator E

This perforator is different from other perforators at the apex section which is occupied by high explosive only. In Fig. 5.2, the circular arc $\widehat{AB}$ is the surface of the high explosive, there is no liner material in this section. Perforator E with diameter 4.3 cm uses bronze as the liner, RDX with 2% wax as high explosive and iron as the case. The HE is detonated at point Q. The circular arc $\widehat{AB}$ has a center located at R=0.0 cm and Z=-0.2778 cm. For arc $\widehat{DE}$, the center is at R=-1.2778 cm and Z=-0.1765 cm. For arc $\widehat{FG}$, the center is at R=0.11506 cm and Z=0.4058 cm. For arc $\widehat{IJ}$, the center is at R=-0.56464 cm and Z=-0.3233 cm. For arc $\widehat{JK}$, the center is at R=-1.73939 cm and Z=0.4876 cm. For arc $\widehat{KL}$, it is the same as arc

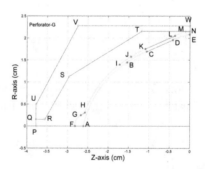

Fig. 5.3 The diameter of Perforator G is 4.3 cm and it has the RDX with 2% wax as the HE and iron as the case.

$\widehat{JK}$. For arc $\widehat{MN}$, the center is at R=-0.88773 cm and Z=0.8867 cm. The coordinates of the rest points are provided in Table 5.2.

Table 5.2 The coordinates of the labelling points for Perforator E.

Label	Z (cm)	R (cm)	Label	Z (cm)	R (cm)
A	-2.9540	0.0000	M	-0.3560	2.0541
B	-2.4714	0.4826	N	-0.0990	2.1498
C	-2.4221	0.4826	P	-0.0090	2.1499
D	-1.5267	1.4441	Q	-3.7744	0.0000
E	-1.0395	1.6940	R	-3.7744	0.1524
F	-0.4066	1.9578	S	-3.5585	0.1524
G	-0.0000	2.0877	T	-2.9647	1.1214
H	-2.4714	0.4726	U	-1.1745	2.1549
I	-2.4713	0.5054	V	-3.7744	0.4902
J	-1.7093	1.3944	W	-2.7414	2.2796
K	-1.4316	1.5718	X	0.0000	2.2796
L	-1.0752	1.7542			

Table 5.3 The coordinates of the labelling points for Perforator G.

Label	Z (cm)	R (cm)	Label	Z (cm)	R (cm)
A	-2.6080	0.0000	L	-0.3649	2.0541
B	-1.5356	1.4440	M	-0.1080	2.1498
C	-1.0484	1.6940	N	-0.0089	2.1499
D	-0.4155	1.9578	P	-3.7744	0.0000
E	-0.0089	2.0877	Q	-3.7744	0.1524
F	-2.8140	0.0000	R	-3.5585	0.1524
G	-2.6692	0.2508	S	-2.9647	1.1214
H	-2.5671	0.3097	T	-1.1745	2.1549
I	-1.7183	1.3944	U	-3.7744	0.4902
J	-1.4406	1.5718	V	-2.7414	2.2796
K	-1.0841	1.7542	W	0.0000	2.2796

5.4 Perforator G

Figure 5.3 shows the plot of Perforator G which uses copper as the liner with convex-shaped hemi-spherical solid liner near the apex region. The diameter of Perforator G is 4.3 cm and it has the RDX with 2% wax as the HE and iron as the case.

Table 5.3 gives the coordinates of each points with labels as shown in Fig. 5.3. The circular arc $\widehat{AB}$ has a center located at $R=-0.5580$ cm and $Z=-0.3495$ cm. The arc $\widehat{BC}$ is centered at $R=-1.27787$ cm and $Z=0.1675$ cm. The arc $\widehat{DE}$ has a center located at $R=0.11506$ cm and $Z=0.3970$ cm. The back side of the liner starts with the arc $\widehat{FG}$ centered at $R=0.0$ cm and

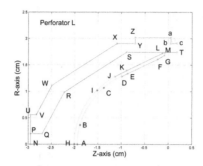

Fig. 5.4 The diameter of Perforator L is 3.5 cm and it has the RDX with 2% wax as the HE and stainless-steel-304 as the case.

$Z=-2.5245$ cm. The arc $\widehat{HI}$ is centered at $R=-0.56464$ cm and $Z=-0.3322$ cm. The arc $\widehat{IJ}$ is centered at $R=-1.73939$ cm and $Z=0.4788$ cm. The arc $\widehat{JK}$ has the same center as arc $\widehat{IJ}$. The arc $\widehat{LM}$ has a center located at $R=-0.88773$ cm and $Z=0.8778$ cm. The coordinates of the rest points are given in Table 5.3.

5.5 Perforator L

This perforator is very popular in the oil industries for well perforation. Its diameter is 3.5 cm with spherical liner near the apex section and conical base using copper as liner, RDX with 2% wax as HE, and stainless-steel-304 as the case.

The coordinates of each labelling points are given in Table 5.4. In Fig. 5.4, the circular arc $\widehat{AB}$ has a center located at $R=-0.03026$ cm and $Z=-0.5658$ cm. The center of arc $\widehat{BC}$ is at $R=-0.18$ cm and $Z=-0.38$ cm. For arc $\widehat{CD}$, it is at $R=-0.03$ cm and $Z=-0.53$ cm.

The center for arc $\widehat{HI}$ is at $R=-0.06806$ cm and $Z=-0.5675$ cm. For arc $\widehat{IJ}$, it is at $R=-0.31063$ cm and $Z=-0.3361$ cm. The coordinates of all other labelling points are provided in Table 5.4. It is noticed that the structure of the Stainless-steel-304 case is very simple.

5.6 Perforator M

This is a 3.5 cm charge diameter perforator that uses copper as the liner, RDX with 2% wax as the HE, and SS-304 as the case. As shown in Fig. 5.5, the apex section has a small convex liner with much thicker liner near the

The Exact Dimensions of the Perforators

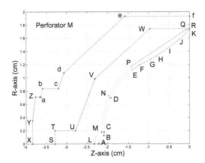

Fig. 5.5 The diameter of Perforator M is 3.5 cm and it has the RDX with 2% wax as the HE, copper as the liner and stainless-steel-304 as the case.

Table 5.4 The coordinates of the labelling points for Perforator L.

Label	Z (cm)	R (cm)	Label	Z (cm)	R (cm)
A	-1.9363	0.0000	P	-2.9471	0.1981
B	-1.8794	0.3617	Q	-2.6956	0.1981
C	-1.3714	1.0690	R	-2.2234	0.9855
D	-0.9904	1.2806	S	-0.8897	1.7399
E	-0.8633	1.3373	T	0.1822	1.7399
F	-0.2283	1.6040	U	-2.9471	0.5563
G	-0.0761	1.6839	V	-2.8214	0.5563
H	-2.0086	0.0000	W	-2.4864	1.1137
I	-1.5138	1.0210	X	-1.0878	1.9050
J	-1.1330	1.2788	Y	-0.7068	1.9050
K	-0.8790	1.3968	Z	-0.7068	2.0193
L	-0.1169	1.7221	a	0.0552	2.0193
M	-0.1169	1.7221	b	0.0552	1.9050
N	-2.9471	0.0000	c	0.1822	1.9050

base section. The case is thicker near the Z-axis and is thinner when R becomes larger. This perforator makes a large hole on the casing wall, therefore, it is a very good shaped charge. The arc $\widehat{AB}$ has a center located at R=-0.0 cm and Z=-2.0897 cm. The center of arc $\widehat{CD}$ is at R=0.1778 cm and Z=-1.18561 cm. The center of arc $\widehat{DE}$ is at R=0.0 cm and Z=-0.93168 cm. For the back side of the liner, arc $\widehat{LM}$ is with center at R=0.0 cm and Z=-2.14562 cm. The arc $\widehat{MN}$ has a center located at R=0.1778 cm and Z=-1.18561 cm. The arc $\widehat{NP}$ has a center located at R=0.0 cm and Z=-0.93168 cm. The coordinates of the rest labelling points are given in Table 5.5.

180   Computational Solid Mechanics for Oil Well Perforator Design

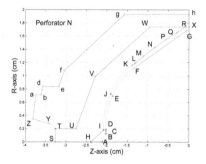

Fig. 5.6 The diameter of Perforator N is 3.5 cm and it has the RDX with 2% wax as the HE and stainless-steel-304 as the case.

Table 5.5 The coordinates of the labelling points for Perforator M.

Label	Z (cm)	R (cm)	Label	Z (cm)	R (cm)
A	-2.2167	0.0000	R	0.0000	1.7399
B	-2.0897	0.1270	S	-3.2838	0.0000
C	-2.0897	0.1778	T	-3.2838	0.198
D	-1.9262	0.6964	U	-2.7798	0.1981
E	-1.4061	1.1176	V	-2.3067	0.9855
F	-1.2387	1.1870	W	-0.9739	1.7399
G	-0.9437	1.2795	X	-3.8390	0.0000
H	-0.8160	1.3270	Y	-3.8390	0.3469
I	-0.5560	1.4377	Z	-3.7628	0.7112
J	-0.2490	1.5766	a	-3.6538	0.7112
K	0.0000	1.7272	b	-3.5896	0.8382
L	-2.3234	0.0000	c	-3.2015	0.8382
M	-2.1456	0.1778	d	-3.0652	1.0744
N	-1.9720	0.7284	e	-1.5835	1.9304
P	-1.4279	1.1690	f	0.0000	1.9304
Q	-0.0828	1.7272			

5.7 Perforator N

This perforator also has a diameter of 3.5 cm with much thicker liner near the base section. At the apex section, there is a conical shaped liner at the high explosive side and a spherical arc shape at the front side as shown in Fig. 5.6. The liner is made of copper, HE is RDX with 2% wax, and the case is made from SS-304. The arc $\widehat{AB}$ has a center located at $R=0.0$ cm

Table 5.6 The coordinates of the labelling points for Perforator N.

Label	Z (cm)	R (cm)	Label	Z (cm)	R (cm)
A	-2.1042	0.0000	R	-0.1016	1.7399
B	-2.0162	0.1270	S	-3.2617	0.0000
C	-2.0162	0.1821	T	-3.2617	0.1981
D	-2.0162	0.1821	U	-2.7577	0.1981
E	-1.8541	0.6964	V	-2.2846	0.9855
F	-1.3534	1.1092	W	-0.9517	1.7399
G	-0.0508	1.6891	X	0.0000	1.7399
H	-2.3876	0.0000	Y	-3.4102	0.2752
I	-2.0721	0.1821	Z	-3.8169	0.3469
J	-1.8999	0.7284	a	-3.7407	0.7112
K	-1.3761	1.1602	b	-3.6137	0.7112
L	-1.3065	1.1912	d	-3.5674	0.8382
M	-1.1082	1.2961	e	-3.1793	0.8382
N	-0.8275	1.4295	f	-3.0430	1.0744
P	-0.5669	1.5455	g	-1.5613	1.9304
Q	-0.2905	1.6574	h	0.0000	1.9304

and Z=-1.92826 cm. $\overline{BC}$ and $\overline{CD}$ are straight lines. The arc $\overarc{DE}$ has a center located at R=0.1821 cm and Z=-1.11963 cm. The arc $\overarc{EF}$ has a center located at R=0.0 cm and Z=-0.85955 cm. Again, $\overline{HI}$ is the conical surface of the liner. The arc $\overarc{IJ}$ has a center located at R=0.1821 cm and Z=-1.11963 cm. The arc $\overarc{JK}$ has a center located at R=0.0 cm and Z=-0.85955 cm. The arc $\overarc{YT}$ has a center located at R=0.0 cm and Z=3.4587 cm. The coordinates of the rest labelling points on Fig. 5.6 are given in Table 5.6. It is noticed that the case are much thicker when R is small. The HE is detonated at point S.

5.8 Perforator P

Both of the perforators P and N use the same material for the liner, the high explosive and the case. Also, both of them have thicker liner near the base section and thinner liner near the apex section. Figure 5.7 shows that some sections of the SS-304 case are constructed using circular arcs instead of straight line. Both of the front side and the back side of the liner near the apex section use the circular arcs, i.e., $\overarc{AB}$ and $\overarc{HI}$. This perforator makes excellent penetration on the well pipes. The arc $\overarc{AB}$ in Fig. 5.7

182 Computational Solid Mechanics for Oil Well Perforator Design

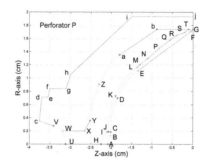

Fig. 5.7 The diameter of Perforator P is 3.5 cm and it has the RDX with 2% wax as the HE and stainless-steel-304 as the case.

has a center located at $R=0.1760$ cm and $Z=-1.9283$ cm. That of arc $\widehat{CD}$ is at $R=0.1821$ cm and $Z=-1.1196$ cm. For arc $\widehat{DE}$ it is at $R=0.0$ cm and $Z=-0.8595$ cm. For arc $\widehat{HI}$ it is at $R=0.1760$ cm and $Z=-1.9283$ cm. That of arc $\widehat{JK}$ is at $R=0.1821$ cm and $Z=-1.1196$ cm. That of arc $\widehat{KL}$ is at $R=0.0$ cm and $Z=-0.8595$ cm. For arc $\widehat{HI}$ it is at $R=0.1760$ cm and $Z=-1.9283$ cm. For arc $\widehat{VU}$ it is at $R=0.0$ cm and $Z=-3.4587$ cm. That of arc $\widehat{YZ}$ is at $R=0.1821$ cm and $Z=-1.1196$ cm. That of arc $\widehat{Za}$ is the same as arc $\widehat{YZ}$. The coordinates of the rest labelling points in Fig. 5.7 are giving in Table 5.7.

5.9 The Penetrating Characteristic of the Penetrators

5.9.1 *Introduction*

In this section, seven typical perforators will be described with the design characteristics of the liner shape, the steel casing dimensions and the material used for the liner. The thickness and the curvature of the liner are very sensitive to the perforated sizes of the gun and oil well walls. The shapes of the casing also have some effects on the shock wave reflection from the casing inner surface which is in contact with the high explosive. The reflected shock wave will then compress the liner to form the jet.

In the cylindrical coordinate system, the surfaces of the liner and the casing of the perforator can be constructed using straight lines and circular arcs. The outside casing of the perforator is made of stainless steel and the liner is made by pressing the copper or bronze powder with small amount of wax. High explosive is usually made from PBX-9501 or RDX which can

Table 5.7 The coordinates of the labelling points for Perforator P.

Label	Z (cm)	R (cm)	Label	Z (cm)	R (cm)
A	-2.0842	0.0000	T	-0.1016	1.7399
B	-2.0162	0.1270	V	-3.4102	0.2752
C	-2.0162	0.1821	U	-3.1217	0.0000
D	-1.8541	0.6964	W	-3.1897	0.1981
E	-1.3534	1.1092	X	-2.6577	0.1981
F	-0.0508	1.7091	Y	-2.5246	0.3555
G	-0.0010	1.7399	Z	-2.3041	0.9064
H	-2.2542	0.0000	a	-1.8034	1.3492
I	-2.1162	0.1821	b	-0.8817	1.7399
J	-2.0721	0.1881	c	-3.7569	0.3469
K	-1.8999	0.7284	d	-3.7007	0.7112
L	-1.3761	1.1602	e	-3.5437	0.7112
M	-1.3065	1.2012	f	-3.5074	0.8382
N	-1.1082	1.2961	g	-3.1093	0.8382
P	-0.8275	1.4495	h	-3.0030	1.0544
Q	-0.5669	1.5755	i	-1.5013	1.9304
R	-0.4005	1.6474	j	0.0000	1.9304
S	-0.2005	1.6954			

sustain high temperature environment such as a few thousand feet below the earth surface.

In the supplementary materials (see Appendix D), there is a file called 'EULE2D-Fig' which can be opened by using Adobe Reader. The file EULE2D-Fig contains forty figures with eight different perforators. Each perforator is illustrated by five figures that include the configuration of the perforator and the penetrations of the gun and the oil-well walls by the liner jet.

The water gaps, i.e., the distance between the gun and casing walls, are 0.0 cm, 1.27 cm, 2.54 cm and 3.81 cm. For each water gap, there are three plots for times equal to 10.0 μsec, 30 μsec and 50 μsec respectively.

5.9.2 Copper Liner of a Diameter 3.5 cm

This is the Perforator L as shown in Fig. 5.8. The high explosive RDX with 2% wax is in the yellow color, the copper liner in the red, the SS-304 in the green, the gun and casing walls in dark blue, water in light blue and the Westerly granite in the green (for $Z > 2.3$ cm).

The HE is detonated at R=0.0 cm and Z=-2.9 cm with only one point detonation. The red saw teeth shape plotted in the front perimeter of the liner, i.e., the interface between the liner and the air, comes from the deficiency of the graphical program. It is not a part of the liner.

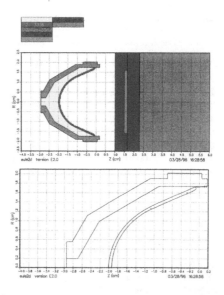

Fig. 5.8 Copper liner consisting of spherical apex and conical base of a diameter 3.5 cm. The HE is RDX with 2% wax and stainless-steel-304 as the case.

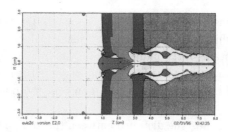

Fig. 5.9 The penetration of the copper jet at time 50 μsec, there is a small portion of the gun tube extruded to $Z \approx 1.8$ cm where $R=0.5$ cm.

Figure 5.9 shows the penetration of the copper jet at time 50 μsec, there is a small portion of the gun tube extruded to $Z \approx 1.8$ cm where $R=0.5$ cm. In order to reduce the extrusion of the gun tube after the shaped charge is fired, we make a small ditch between the gun and well casing walls as shown with blue color in the upper plot of Fig. 5.8.

There are four markers with blue color located at $Z=-1.9$ cm (M_a), -1.95 cm (M_b), -1.91 cm (M_c) and -0.28 cm (M_d) as shown in the lower plot of Fig. 5.8. These markers are fixed on the liner material and will follow the movement of the liner material when the liner is deformed. From

The Exact Dimensions of the Perforators 185

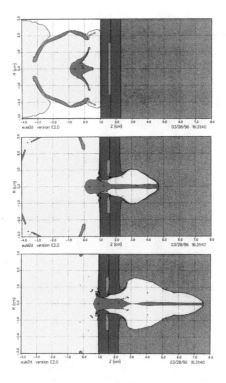

Fig. 5.10 The penetration of the copper jet at time 10 μsec (top), 30 μsec (middle), and 50 μsec (bottom) for the Perforator L. The water gap between the gun and the casing walls is 0.0 cm.

the positions of these markers one can determine which part of the liner will form the jet front or the slug in the back side. These markers are very useful to understand the jet formation and jet particulation. In the top plot of Fig. 5.10, the marker M_a is located at R=0.0 cm and Z=-0.7 cm, M_b at R=0.1 cm and Z=-0.1 cm, M_c at R=0.2 cm and Z=-0.15 cm and M_d at R=0.4 cm and Z=-0.3 cm.

In the central plot of Fig. 5.10, i.e., $t = 30$ μsec, the coordinate of marker M_a is at R=0.0 cm and Z=0.3 cm, M_b is at R=0.0 cm and Z= 4.6 cm, M_c is at R=0.1 cm and Z= 3.3 cm, and M_d is at R=0.1 cm and Z= 1.25 cm. At time $t = 50$ μsec, the bottom plot of Fig. 5.10 shows that M_a is at R=0.0 cm and Z=0.99 cm, M_b is at R=0.0 cm and Z= 7.6 cm, M_c is at R=0.05 cm and Z= 6.9 cm, and M_d is at R=0.07 cm and Z= 2.25 cm. The penetration depth of the liner jet into the Westerly granite is about

5.4 cm. Figure 5.9 shows that M_b is located at $R=0.0$ cm and $Z=8.0$ cm, which means that the penetration depth of the liner jet into the granite is about 4.5 cm that is smaller than the one shown in the bottom plot of Fig. 5.10, i.e., 5.4 cm.

Therefore, if the water gap between the gun and casing walls is larger, the liner jet will be cooled down by the water and its kinetic energy will go down hence the penetration depth into the granite is smaller.

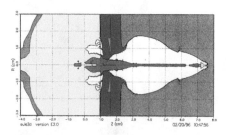

Fig. 5.11 The penetration of the copper jet at time 50 μsec for the Perforator M. The water gap between the gun and the casing walls is 0.0 cm.

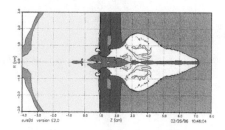

Fig. 5.12 The penetration of the copper jet at time 50 μsec for the Perforator N. The water gap between the gun and the casing walls is 0.0 cm.

5.9.3 *Perforators Described in Figs. 6 and 11 of the File EULE2D-Fig*

At time $t = 50$ μsec, Figs. 5.11, 5.12 and the bottom plot of Fig. 5.10 show that the openings of the casing walls are 1.166 cm (Fig. 5.10, i.e., Perforator L), 1.332 cm (Fig. 5.11, i.e., Perforator M) and 1.58 cm (Fig. 5.12, i.e., Perforator N). Therefore, the best perforator among these three perforators is the one shown in Fig. 5.13 since it produces the largest opening on the casing pipe. However, the sharp conical shape at the apex section present

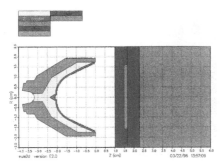

Fig. 5.13 A 3.5-cm charge diameter with a conical shape liner near the apex region. This is Perforator N.

some difficulty in manufacturing. That is why Perforator N is not very practical.

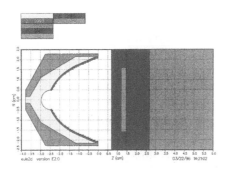

Fig. 5.14 Bronze liner with 1-cm diameter hole at the apex region and the charge diameter is 4.3 cm known as Perforator E.

5.9.4 Perforators Described in Figs. 16, 26 and 31 of the File EULE2D-Fig

Perforator E (Fig. 5.14), Perforator Q (Fig. 5.15) and Perforator R (Fig. 5.16) have the same dimensions on the iron case, the contained high explosives and the liners. These perforators are of 4.3 cm charge diameter with a 1.0 cm diameter hole at the apex section. The liner materials are bronze and copper with RDX and 2% wax as high explosive. The gun and the outside casing tube are steel and the surrounding rock is Westerly granite.

At time $t = 50$ μsec, the penetration depth of the liner into the granite is deepest by the shaped charge shown in Fig. 5.15, next by Fig. 5.16, and

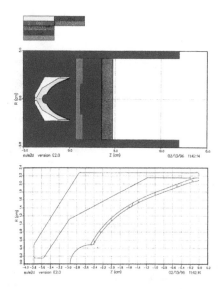

Fig. 5.15 A charge diameter of 4.3 cm with bronze liner near the apex section and copper near the base region known as Perforator Q.

shallowest by the one shown in Fig. 5.14. This means using bronze and copper of almost equal amount can produce the deepest penetration into the outside rock.

At time $t = 30$ μsec, the upper plot of Fig. 5.17 shows that the liner rod comprises the bronze in the core section and the copper formed as an outside layer surrounding the core.

At time $t = 50$ μsec, the lower plot of Fig. 5.17 shows that the bronze jet is broken and some long thin sections of the bronze liner are formed. Perforator E produces the largest casing tube opening but the shortest penetration depth, while Perforator Q results in smallest opening with deepest penetration for the case when the water gap between the gun and well walls is zero.

Since Perforators E, Q and R have the same dimensions in the liner and the case, one can use the coordinates of Perforator E described in Fig. 5.2 to plot Perforators Q and R. For Perforator Q, the material interface between the bronze and the copper is indicated in the lower plot of Fig. 5.15 at $R=1.45$ cm and $Z = -1.52$ cm. For Perforator R, the material interfacial line is located at $R=1.61$ cm and $Z=-1.24$ cm as shown in the lower plot of Fig. 5.16. From Fig. 5.18, it is clear that the bronze jet is the primary material for penetrating the casing pipe.

The Exact Dimensions of the Perforators 189

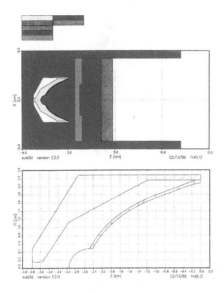

Fig. 5.16 This is the Perforator R which has the same dimension as described in Fig. 5.15 except that the length of the bronze liner is longer and that of the copper liner is shorter.

At time $t = 50\,\mu sec$, the bottom plot of Fig. 5.18 shows that the copper material is responsible for the penetration of the outside granite. The gun pipe opening at time $50\,\mu sec$ with water gap of 3.81 cm are almost the same for Perforators E, Q and R.

5.9.5 *Special Design of 4.3 cm Charge Diameter Shaped Charge*

Figure 5.19 shows a special design of 4.3 cm charge diameter shaped charge (Perforator S) that creates a large hole in the first wall and a small hole in the farther wall if there is a gap between these two walls. The liner is made of two materials, the one near the conical base section is the copper and that near the apex section is the teflon. The teflon liner does not contribute any penetration to both walls.

From the plots of Figs. 5.20, 5.21 and 5.22, it is obvious that the hole size of the gun wall is much bigger then that of the casing wall. Therefore, if one is interested in creating a large hole in the first wall, Perforator S is a good choice. It is noticed that the hole size in the first wall is almost twice

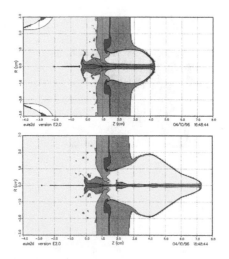

Fig. 5.17 The jet penetration into the gun, casing, and the granite is shown at time $t = 30$ μsec and $t = 50$ μsec for the Perforator Q.

as big as that of the second wall when the water gap is 3.81 cm as shown in Fig. 5.22.

Since the liner near the apex section does not contribute any penetration to the walls, it is suggested that a low density material like teflon is a good candidate.

5.10 Program Curve

The program "CURVE" is written in FORTRAN 77 language which requires the Compaq FORTRAN software to compile and to execute. The subroutine "CURVEG" will generate the data for plotting the curve using the C-Sharp plotting routines or using MATLAB software. Since it is very cumbersome to use C-Sharp routines, therefore, we recommend to the users to use "MATLAB" for plotting the shaped charge components. All of the perforators discussed in this section use the "MATLAB" software for plotting purpose.

For Perforator E, the input data for the program "CURVE" is
P-E
4
6
8

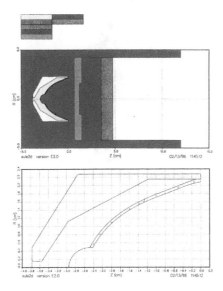

Fig. 5.16 This is the Perforator R which has the same dimension as described in Fig. 5.15 except that the length of the bronze liner is longer and that of the copper liner is shorter.

At time $t = 50\,\mu sec$, the bottom plot of Fig. 5.18 shows that the copper material is responsible for the penetration of the outside granite. The gun pipe opening at time $50\,\mu sec$ with water gap of 3.81 cm are almost the same for Perforators E, Q and R.

5.9.5 Special Design of 4.3 cm Charge Diameter Shaped Charge

Figure 5.19 shows a special design of 4.3 cm charge diameter shaped charge (Perforator S) that creates a large hole in the first wall and a small hole in the farther wall if there is a gap between these two walls. The liner is made of two materials, the one near the conical base section is the copper and that near the apex section is the teflon. The teflon liner does not contribute any penetration to both walls.

From the plots of Figs. 5.20, 5.21 and 5.22, it is obvious that the hole size of the gun wall is much bigger then that of the casing wall. Therefore, if one is interested in creating a large hole in the first wall, Perforator S is a good choice. It is noticed that the hole size in the first wall is almost twice

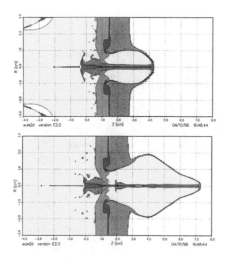

Fig. 5.17 The jet penetration into the gun, casing, and the granite is shown at time $t = 30$ μsec and $t = 50$ μsec for the Perforator Q.

as big as that of the second wall when the water gap is 3.81 cm as shown in Fig. 5.22.

Since the liner near the apex section does not contribute any penetration to the walls, it is suggested that a low density material like teflon is a good candidate.

5.10 Program Curve

The program "CURVE" is written in FORTRAN 77 language which requires the Compaq FORTRAN software to compile and to execute. The subroutine "CURVEG" will generate the data for plotting the curve using the C-Sharp plotting routines or using MATLAB software. Since it is very cumbersome to use C-Sharp routines, therefore, we recommend to the users to use "MATLAB" for plotting the shaped charge components. All of the perforators discussed in this section use the "MATLAB" software for plotting purpose.

For Perforator E, the input data for the program "CURVE" is
P-E
4
6
8

The Exact Dimensions of the Perforators

5
4
CIRC -2.9540 0.0 -2.4714 0.4826 3.0 -2.4714 -0.0
LINE -2.4714 0.4826 -2.4221 0.4826
CIRC -2.4221 0.4824 -1.5267 1.4441 3.0 -0.3406 -0.5580
CIRC -1.5267 1.4441 -1.0395 1.6940 3.0 0.1765 -1.2778
LINE -1.0395 1.6940 -0.4066 1.9578
CIRC -0.4066 1.9578 -0.0000 2.08769 3.0 0.4058 0.11506

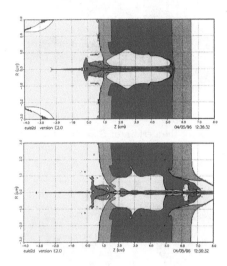

Fig. 5.18 The jet penetration into the gun and the casing is shown at time $t = 30$ μsec and $t = 50$ μsec for the Perforator R.

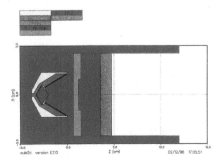

Fig. 5.19 This is the Perforator S of 4.3-cm diameter shaped charge that creates a large hole in the first wall and a small hole in the farther wall if there is a gap between these two walls.

CIRC -2.9540 0.0 -2.4714 0.4726 3.0 -2.4714 0.0
LINE -2.4714 0.4826 -2.4713 0.5054
CIRC -2.4713 0.5054 -1.7093 1.394415 3.0 -0.3233 -0.56464
CIRC -1.7093 1.394415 -1.4316 1.571824 3.0 0.4876 -1.73939
CIRC -1.4316 1.571824 -1.0752 1.754219 3.0 0.4876 -1.73939
LINE -1.0752 1.754219 -0.3560 2.05409
CIRC -0.3560 2.05409 -0.0990 2.149850 3.0 0.8867 -0.88773

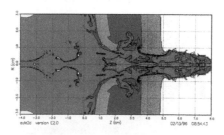

Fig. 5.20 The penetration of the Perforator S at time $t = 50$ μsec with the water gap of 2.08 cm.

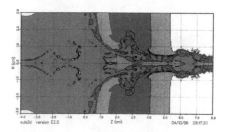

Fig. 5.21 The penetration of the Perforator S at time $t = 50$ μsec with the water gap of 2.54 cm.

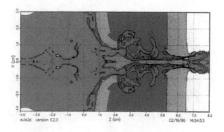

Fig. 5.22 The penetration of the Perforator S at time $t = 50$ μsec with the water gap of 3.81 cm. This Perforator creates a large hole in the first wall and a small hole in the farther wall.

LINE -0.0090 2.149856 -0.0000 2.149856
LINE -3.7744 0.0 -3.7744 0.1524
LINE -3.7744 0.1524 -3.5585 0.1524
LINE -3.5585 0.1524 -2.96468 1.12141
LINE -2.96468 1.12141 -1.17449 2.15493
LINE -1.17449 2.15493 0.0 2.15493
LINE -3.7744 0.0 -3.77444 0.1524
LINE -3.7744 0.15240 -3.77442 0.4902
LINE -3.7744 0.4902 -2.7414 2.27965
LINE -2.7414 2.2796 0.0 2.2796

The meanings of the above data are:
P-E: Perforator E.
4: There are four groups of data which will describe the surface of the perforator.
6: The first group of the data has 6 input data lines that are CIRC, LINE, CIRC, CIRC, LINE and CIRC. They are belonged to the front surface of the liner.
8: The second group of the data has 8 input data lines that are CIRC, LINE, CIRC, CIRC,CIRC, LINE, CIRC and LINE. They are belonged to the back surface of the liner.
5: The third group of the data has 5 input data lines that are LINE, LINE, LINE, LINE and LINE. They are belonged to the inside surface of the case.
4: The last group of the data has 4 input data lines that are LINE, LINE, LINE and LINE. They are belonged to the outside surface of the case.

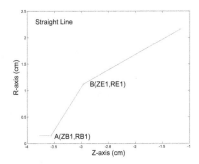

Fig. 5.23 The straight line starting at point $A(ZB1, RB1)$ and ending at point $B(ZE1, RE1)$.

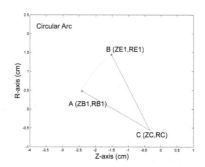

Fig. 5.24 The circular arc starting at point $A(ZB1, RB1)$ and ending at point $B(ZE1, RE1)$ where $C(ZC, RC)$ is the center of the circle.

The CIRC is for the circular arc and its data structure is:
ZB1=-2.9540, Z-coordinate of the starting point.
RB1= 0.0, R-coordinate of the starting point.
ZE1=-2.4714, Z-coordinate of the ending point.
RE1= 0.4826, R-coordinate of the ending point.
RAD= 3.0, this input data has never been used inside the code.
ZC=-2.4714, the Z-coordinate of the center of the circle.
RC= 0.0, the R-coordinate of the center of the circle.

The LINE is for the straight line and its data structure is:
ZB1=-2.4714, Z-coordinate of the starting point.
RB1= 0.4826, R-coordinate of the starting point.
ZE1=-2.4221, Z-coordinate of the ending point.
RE1= 0.4826, R-coordinate of the ending point.

Figure 5.23 shows the starting point $A(ZB1, RB1)$ and the ending point $B(ZE1, RE1)$ for the straight line $\overline{AB}$.

In Fig. 5.24, we have a circular arc $\widehat{AB}$ which starts at point $A(ZB1, RB1)$ and ends at point $B(ZE1, RE1)$ with the circular center located at point $C(ZC, RC)$. Since the scalings of the graphical window are different in the R and the Z directions, the lengths of the radius $\overline{AC}$ and $\overline{BC}$ look different even their lengths is the same.

For ellipse, Fig. 5.25 shows the 90° elliptic arc starting at point $A(ZB1, RB1)$ and ends at point $B(ZE1, RE1)$ with the elliptic center

The Exact Dimensions of the Perforators   195

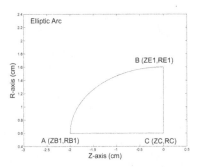

Fig. 5.25 The elliptic arc starting at point $A(ZB1, RB1)$ and ending at point $B(ZE1, RE1)$ where $C(ZC, RC)$ is the center of the ellipse. The major axis for the ellipse is $\overline{AC}$ where $\overline{BC}$ is the minor one.

located at point $C(ZC, RC)$. The major axis for the ellipse is $\overline{AC}$ where $\overline{BC}$ is the minor one.

The program "CURVE" will also calculate the volume and the mass of the liner, high explosive and the case. The results of the calculations can be found in the output file called "output1".

5.11 Plotting Programs Using MATLAB

The perforator is plotted by using the MATLAB software. For example, in order to plot the configuration of the perforator E, we use the program

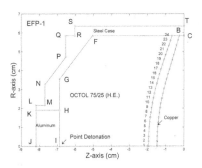

Fig. 5.26 The explosive formed projectile uses copper as liner, OCTOL 75/25 as HE, steel as the case with aluminum block next to the HE near the Z-axis.

"plotPE.m" to get the actual dimensions and the labels of each point for the liner and the case. If one is interested in doing his own design of the

Table 5.8 The coordinates of the labelling points for the copper liner.

Label	1	2	3	4	5	6	7
$R\,(cm)$	0.0	0.2540	0.5080	0.7620	1.0160	1.2700	1.5240
Left $Z\,(cm)$	-1.9562	-1.9534	-1.9455	-1.9320	-1.9135	-1.8896	-1.8601
Right $Z\,(cm)$	-1.4856	-1.4839	-1.4762	-1.4625	-1.4427	-1.4183	-1.3889
Label	8	9	10	11	12	13	14
$R\,(cm)$	1.7780	2.0320	2.2860	2.5400	2.7940	3.0480	3.3020
Left $Z\,(cm)$	-1.8251	-1.7850	-1.7390	-1.6877	-1.6308	-1.5686	-1.5010
Right $Z\,(cm)$	-1.3541	-1.3139	-1.2682	-1.2164	-1.1590	-1.0960	-1.0272
Label	15	16	17	18	19	20	21
$R\,(cm)$	3.5560	3.8100	4.0640	4.3180	4.5720	4.8260	5.0800
Left $Z\,(cm)$	-1.4283	-1.3516	-1.2704	-1.1850	-1.0971	-1.0072	-0.9165
Right $Z\,(cm)$	-0.9543	-0.8786	-0.8004	-0.7196	-0.6391	-0.5593	-0.4808
Label	22	23	24				
$R\,(cm)$	5.3340	5.5880	5.8420				
Left $Z\,(cm)$	-0.8261	-0.7377	-0.6534				
Right $Z\,(cm)$	-0.4041	-0.3289	-0.2540				

perforator, one can modify the program "plotPE.m" and run the modified program using the MATLAB software. The coordinates of each label-point can be found in the system MATLAB file or in the input data file "inp-P-E". The plotting programs for Perforators L, M, N, P, E and G are plotFig1, plotFig6, plotFig11, plotwhl, plotPE and plotPG respectively.

5.12 The Exact Dimensions of Explosive Formed Projectile and Shaped Charge

5.12.1 *Copper Explosive Formed Projectile*

As shown in Fig. 5.26, this EFP uses copper as liner, OCTOL 75/25 as high explosive, steel as the case, and aluminum block next to the explosive near the Z-axis. The detonation initiates at point I. The coordinates of the liner are shown in Table 5.8 and that of the case are shown in Table 5.9. The front part of the liner is occupied by air, however, we use vacuum instead of air in the simulation calculation. This EFP is very similar to the one shown on page 219 in Ref. [5.1].

Table 5.9 The coordinates of the labelling points for the case and the aluminum block.

Label	B	C	F	G	H	I	J
$R\,(cm)$	5.842	5.842	5.842	3.5415	1.905	0.0	0.0
$Z\,(cm)$	-0.254	0.0	-5.0368	-6.8834	-6.8834	-6.8834	-8.1788
Label	K	L	M	N	P	Q	R
$R\,(cm)$	1.905	2.159	2.159	3.2639	4.707	5.842	5.842
$Z\,(cm)$	-8.1788	-8.1788	-7.6911	-7.6911	-6.5278	-6.5278	-6.0223
Label	S	T					
$R\,(cm)$	6.35	6.35					
$Z\,(cm)$	-6.0223	0.0					

5.12.2 Bi-conical Copper Liner Shaped Charge

This shaped charge has a bi-conical copper liner with hemi-spherical apex as shown in Fig. 5.27. The lead block, $\overline{abcd}$, is the wave shaper which is embedded inside the high explosive. The steel case, labelled with points

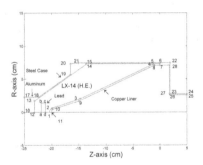

Fig. 5.27 This shaped charge has a bi-conical copper liner with hemi-spherical apex, steel case and lead wave shaper. The high explosive is LX-14.

18, 19,..., 28, holds the copper liner and the $LX - 14$ high explosive. The copper liner is labelled by points 1, 2,..., 10 and 11 with the hemi-spherical apex labelled by points 1, 2, 10 and 11; the first conical section by points 2, 3, 9 and 10; the second conical section by points 3, 4, 8 and 9. The $LX - 14$ high explosive is detonated at point 12. The coordinates of the liner are given in Table 5.10 while that of the steel case and the lead wave shaper are given in Table 5.11.

Table 5.10 The coordinates of the labelling points for the copper liner.

Label	1	2	3	4	5	6	7
$R\,(cm)$	0.0	0.6329	2.0169	7.2313	7.3075	7.3075	7.0637
$Z\,(cm)$	-20.4266	-19.9409	-14.7761	-1.1919	-1.1919	0.0	0.0
Label	8	9	10	11			
$R\,(cm)$	7.0637	1.6906	0.4662	0.0			
$Z\,(cm)$	-0.9907	-14.9882	-19.8877	-20.2514			

Table 5.11 The coordinates of the labelling points for the steel case and the lead wave shaper.

Label	12	13	14	15	16	17	18
$R\,(cm)$	0.0	1.7145	7.2364	7.3075	0.0	2.3012	2.3012
$Z\,(cm)$	-23.1444	-23.1444	-13.5808	-13.5808	-25.	-25.	-22.3824
Label	19	20	21	22	23	24	25
$R\,(cm)$	5.6937	7.3075	7.4345	7.4345	2.7514	2.7514	2.6014
$Z\,(cm)$	-16.5069	-16.5069	-15.0439	1.9	1.9	5.0	5.0
Label	26	27	28	29	30	31	32
$R\,(cm)$	2.6014	2.7514	7.0637	0.0	1.4185	1.4185	0.0
$Z\,(cm)$	1.9	1.65	1.65	-21.9430	-21.9430	-21.1632	-21.1632

5.12.3 Small Charge Diameter Conical Shaped Charge

The small charge diameter conical shaped charge is shown in Fig. 5.28 with the charge diameter of 6.35 cm. The conical copper liner has a hemispherical apex. The high explosive $LX-14$ is detonated at point 1. The booster is confined by points 1, 2, 3 and 4. There is no case on the outside surface of the high explosive. On Fig. 5.28, the scale on the R-axis is different from that of the Z-axis. Figure 5.29 is the same plot as Fig. 5.28 with even scale in both R and Z axes.

The purpose of this small shaped charge is to defeat the reactive armor. It is attached to the regular shaped charge and detonated before the regular shaped charge detonates. Therefore, the high explosive box of the reactive armor will be defeated by this small shaped charge. Table 5.12 lists the coordinates of the liner, the booster and the outside surface of the $LX-14$ explosive.

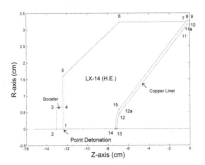

Fig. 5.28 This small shaped charge has a conical copper liner with hemi-spherical apex and booster. The high explosive is LX-14.

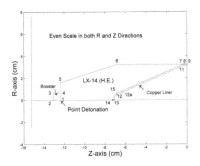

Fig. 5.29 This small shaped charge is the same as the one shown in Fig. 5.28 except this plot has even scale in both R and Z directions.

5.12.4 Small Charge Diameter Shaped Charge with Wave Shaper

This conical shaped charge with small charge diameter has a wave shaper as shown in Fig. 5.30. The purpose of this shaped charge is to defeat the reactive armor just like the one described in the previous section. The conical copper liner has a charge diameter of 6.24 cm (or 2.457 $inch$) with semi-spherical apex. The main charge is $LX - 14$ which is detonated at point 1. There is no case on the outside surface of the HE. The liner is labelled by points 8, 9, 10 and 11. The HE is labelled by points 1, 2, 3, 4, 5, 6, 7, 8, 11, 12 and 15, while wave shaper is by points 12, 13, 14 and 15. The coordinates of the copper liner are given in Table 5.13. The coordinates of the labelled points as shown in Fig. 5.30 are given in Table 5.14.

Table 5.12   The coordinates of the labelling points for the small shaped charge shown in Fig. 5.29.

Label	1	2	3	4	5	6	7
$R\,(cm)$	0.0	0.0	0.260	0.260	0.6275	1.271	1.271
$Z\,(cm)$	-4.88	-5.14	-5.14	-4.88	-4.88	-2.6855	-0.1401
Label	8	9	10	11	11a	12	12a
$R\,(cm)$	1.283	1.283	1.244	1.157	0.1244	0.1919	0.27
$Z\,(cm)$	-0.0975	-0.0	-0.0	-0.297	-0.0817	-2.6855	-2.492
Label	13	14	15				
$R\,(cm)$	0.0	0.0	0.2355				
$Z\,(cm)$	-2.815	-2.862	-2.7032				

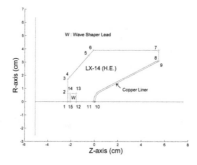

Fig. 5.30   This small shaped charge has a lead wave shaper with copper as liner, $LX-14$ as high explosive.

5.12.5   Non-axisymmetric Tantalum EFP Warhead

The dimensions of the liner, OCTOL explosive, booster, detonator and the steel case of the non-axisymmetric tantalum EFP warhead as described on pages 226–231 in Ref. [5.1] is presented in Fig. 5.31. The dimension of the tantalum liner is $80 \times 120 mm$ (or $3.15 \times 4.72 inch$) as shown in Fig. 5.32 and nominally $2mm$ (or $0.075 inch$) thick as shown in Fig. 5.31. In Fig. 5.31, the locations of the liner are defined by points 1, 2, 3 and 4, that of OCTOL explosive are by points 1, 2, 5 and 6, that of steel case are by points 3, 2, 5, 6, 7, 8, 9, 10 and 11. In Fig. 5.31, the dimension of the detonator is: $Y_{min} = 0.409, Y_{max} = 0.575, Z_{min} = -2.8145$, and $Z_{max} = -2.3145$. The dimension of the booster is: $Y_{min} = 0.2619, Y_{max} = 0.7261, Z_{min} = -2.3145$, and $Z_{max} = -2.1645$. There are 10 detonators,

Table 5.13 The coordinates of the copper liner for the small shaped charge with wave as shaper shown in Fig. 5.30.

Point	1	2	3	4	5	6	7
$Z\,(cm)$	0.0	0.0793	0.1586	0.2379	0.3172	0.3965	0.4756
Upper $R\,(cm)$	0.1851	0.2071	0.3155	0.3525	0.3894	0.4264	0.4634
Lower $R\,(cm)$	0.0	0.2090	0.2734	0.3116	0.3486	0.3856	0.4226
Point	8	9	10	11	12	13	14
$Z\,(cm)$	0.5551	0.6344	0.7137	0.7930	0.8723	0.9516	1.0309
Upper $R\,(cm)$	0.5004	0.5373	0.5743	0.6113	0.6483	0.6853	0.7222
Lower $R\,(cm)$	0.4595	0.4965	0.5335	0.5705	0.6075	0.6444	0.6914
Label	15	16	17	18	19	20	21
$Z\,(cm)$	1.1102	1.1895	1.2686	1.3491	1.4274	1.5067	1.5950
Upper $R\,(cm)$	0.7592	0.7962	0.8332	0.8701	0.9071	0.9441	0.9811
Lower $R\,(cm)$	0.7184	0.7554	0.7923	0.8293	0.8663	0.9035	0.9403
Label	22	23	24	25	26	27	28
$Z\,(cm)$	1.6653	1.7445	1.8230	1.9032	1.9825	2.0618	2.1411
Upper $R\,(cm)$	1.0181	1.0550	1.0920	1.1290	1.1660	1.2029	1.2204
Lower $R\,(cm)$	0.9772	1.0142	1.0512	1.0982	1.1251	1.1621	1.1991

Table 5.14 The coordinates of the labelling points as shown in Fig. 5.30 for the small shaped charge with wave shaper.

Point	1	2	3	4	5	6	7
$Z\,(cm)$	-0.9018	-0.9018	-0.9018	-0.8198	-0.1998	-0.0408	2.1553
$R\,(cm)$	0.0	0.2435	0.6674	0.75	1.375	1.5355	1.5355
Point	8	9	10	11	12	13	14
$Z\,(cm)$	2.1553	2.1553	0.0	-0.0408	-0.6118	-0.6118	-0.8198
$R\,(cm)$	1.2285	1.2058	0.0	0.0	0.0	0.2435	0.2435
Label	15						
$Z\,(cm)$	-0.8198						
$R\,(cm)$	0.0						

i.e. circles $E, F, G, H, K, L, M, N, P$ and Q, as given in Fig. 5.32. The diameter of the detonator is 0.166 $inch$. The center of K is at $X_K = 1.456, Y_K = 0.494$. The distance between H and K is 0.728. The center of Q is at $X_Q = 1.456, Y_Q = -0.494$. Also, in Fig. 5.32, the steel case is indicated by points A, B, C, D, S and R, and the liner by points T, U, V and W. The coordinates of the tantalum liner are given in Table 5.15. The

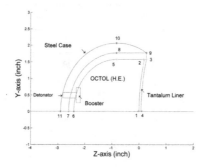

Fig. 5.31 The $Y - Z$ cross section (at $X = 0$) view of the non-axisymmetric tantalum EFP warhead.

coordinates of the labelled points 1, 2, 3,..., 11 as shown in Fig. 5.31 are given in Table 5.16.

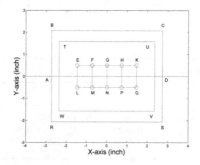

Fig. 5.32 In the $Y - X$ coordinates, the tantalum liner is described by points T, U, V and W, while the ten detonators are indicated by circles $E, F, G, H, K, L, M, N, P$ and Q, the outside case is by points A, B, C, D, S and R.

5.12.6 Copper Hemi-spherical Liner with Energetic Explosive

As described on pages 194, 198-201 in Ref. [5.1], this hemi-spherical shaped charge uses copper as liner, and energetic $PBX-W-113$ as explosive which is detonated from the whole surface. There are 44 detonators to make a uniform surface detonation as shown in Fig. 5.33. For each detonator, there is a small circle of diameter 1.125 $inch$ which is drilled through the aluminum alloy $7075 - T6$ plate of 0.25 $inch$ thick. The large circle is only

Table 5.15 The coordinates of the tantalum liner as shown in Fig. 5.31.

Label	1	2	3	4	5	6	7
Y (inch)	0.0	0.0787	0.1575	0.2362	0.3150	0.3937	0.4724
left Z (inch)	0.0	0.0005	0.0019	0.0043	0.0077	0.0121	0.0174
right Z (inch)	0.0754	0.0761	0.0777	0.0803	0.0839	0.0885	0.0940
Label	8	9	10	11	12	13	14
Y (inch)	0.5512	0.6299	0.7057	0.7874	0.8651	0.9449	1.0236
left Z (inch)	0.0237	0.0310	0.0392	0.0485	0.0587	0.0699	0.0821
right Z (inch)	0.1005	0.1080	0.1164	0.1259	0.1362	0.1478	0.1600
Label	15	16	17	18	19	20	21
Y (inch)	1.1021	1.1811	1.2595	1.3385	1.4173	1.4951	1.5748
left Z (inch)	0.0944	0.1096	0.1248	0.1411	0.1584	0.1768	0.1962
right Z (inch)	0.1740	0.1879	0.2033	0.2198	0.2373	0.2559	0.2755

Table 5.16 The coordinates of the labelling points as shown in Fig. 5.31.

Point	1	2	3	4	5	6	7
Z (inch)	0.0	0.1962	0.2755	0.0754	-0.8305	-2.4053	-2.6025
Y (inch)	0.0	1.5748	1.5748	0.0	1.5748	0.0	0.0
Point	8	9	10	11			
Z (inch)	-0.8305	0.2735	-0.8305	-2.8975			
Y (inch)	1.772	1.772	2.067	0.0			

drilled to 0.338 $inch$ deep with 0.193 $inch$ diameter. The channel is 0.047 $inch$ wide and 0.040 $inch$ deep. A circular bore with 0.062 $inch$ diameter and 0.047 $inch$ deep is drilled at the corner of the groove. There are 54 circular bore holes for the whole detonator plate. At the center of the aluminum plate, there is a small circular groove with radius of 0.08 $inch$ and depth of 0.102 $inch$. This circular groove is connected to a channel which is 0.07 $inch$ wide and 0.05 $inch$ deep. This channel makes $-45°$ from the Z-axis and connects to a large circular donut which has an inner radius of 0.25 $inch$, an outer radius of 0.383 $inch$ and a depth of 0.07 $inch$. Also, this donut is connected to four main channels which are 0.047 $inch$ wide and 0.04 $inch$ deep.

Figure 5.34 shows the structure of the copper hemi-spherical shaped charge with the energetic explosive $PBX - W - 113$. The copper liner has an outside diameter of 8.89 cm (center at $R = 0.0$ cm and $Z = 0.0$ cm)

and a thickness of 0.47498 cm. The aluminum case is 0.635 cm in thickness with an outside diameter of 21.59 cm. The Z coordinates of the case are from Z=-23.787 cm to Z=0.0 cm. The coordinates of the polyethylene are given in Table 5.17. The radius of the detonator is 10.16 cm and the coordinates of the detonator are from Z= -23.787 cm to Z= -21.635 cm. For the booster (Detasheet), they are from Z= -21.635 cm to Z= -21.0 cm.

For $PBX-W-113$ the Chapman–Jouguet parameters are: ρ_0= 1.672 g/cm^3, D= 0.8311 $cm/\mu s$, E_0= 0.087 $Mbar-cm^3/cm^3$, and $E_{chemical} = E_0+E_1 = 0.742086 \, Mbar-cm^3/cm^3$. The JWL EOS (as given in Eq. 4.53) coefficients are: A= 9.50448 $Mbar$, B= 0.10915 $Mbar$, $R_1 = 5.0$, $R_2 = 1.4$ $\omega = 0.40$, E_1= undetonated energy, E = the detonation energy.

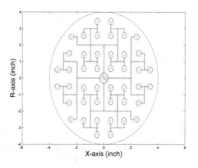

Fig. 5.33 There are 44 detonators to make a uniform surface detonation for the boosters. The grooves are 0.047 inch in width and 0.04 inch in depth.

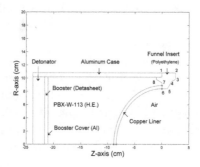

Fig. 5.34 The structure of the hemi-spherical shaped charge with the energetic explosive $PBX-W-113$.

Table 5.17 The coordinates of the labelling points as shown in Fig. 5.34 for the polyethylene.

Point	1	2	3	4	5	6	7
$R\,(cm)$	10.795	10.795	9.8	9.53	8.415	8.415	9.3
$Z\,(cm)$	0.0	2.53	2.53	1.8	1.0	0.0	0.0
Point	8	9					
$R\,(cm)$	9.7	10.16					
$Z\,(cm)$	-1.0	-1.0					

5.13 Computer Programs

Available for download as supplementary materials are some computer programs. Instructions for download are found in Appendix D. Most of these files are discussed in this chapter but some of them are described in Chapter 4. Their names are as follows:

1. MATLAB program for plotting Perforator B: PlotPB.

2. There are six other programs, namely, PlotFig1, PlotFig6, PlotFig11, PlotFigwhl, PlotPE and PlotPG that will produce Fig. 5.4, Fig. 5.5, Fig. 5.6, Fig. 5.7, Fig. 5.2 and Fig. 5.3. These programs also use MATLAB software.

3. The FORTRAN 77 program that will calculate the shear modulus G and yield strength Y: EOSGY.

4. Program CURVEG will calculate the mass and volume of the liner, the high explosive and the casing of a shaped charge: CURVEG.

5. The FORTRAN 77 program HEDET2 will calculate the burn time for two-dimensional problems as described in Section 4.9.4.

6. The FORTRAN 77 program HEDET3 will calculate the burn time for three-dimensional problems as described in Section 4.9.5.

7. The file EULE2D-Fig contains forty figures with eight different perforators. Each perforator is illustrated by five figures that include the configuration of the perforator and the penetration of the gun and the oil-well walls by the liner.

The author uses DELL compatible personal computer and a Compaq FORTRAN Compiler to create an executable code for EOSGY, CURVEG, HEDET2 and HEDET3. The author thinks that the codes may also run on some standard FORTRAN compiler. All of the plotting programs require a MATLAB software.

Bibliography

5.1. Lee, WH. (2006). *Computer simulation of shaped charge problems*, World Scientific Publishing Co., Singapore.

Chapter 6

Two-Dimensional Lagrangian Method for Radiation Diffusion

Notations

c speed of light $(3 \times 10^{10} cm/s)$
E radiation energy density $(jerks/cm^3)$
E^m material internal energy per unit volume $(jerks/cm^3)$
$\dot{e}^{ij}$ strain rate deviator tensor $(1/shake)$
$\vec{F}$ radiation flux vector $(jerks/cm^2/shake)$
G shear modulus of elasticity $(jerks/cm^3)$
$I(\vec{r},\nu,\vec{\Omega},t)$ specific intensity of the radiation field defined as the rate of energy flow per unit frequency and solid angle across a unit area oriented normal to the direction of propagation at point $\vec{r}$, frequency ν, in the direction Ω, at time t $(jerks/cm^2)$
J volume Jacobian (cm^3)
j Jacobian of the transformation between (R, Z) and (k, ℓ) (cm^2)
K Lagrangian coordinate used in figures and difference equation
k Lagrangian coordinate
L Lagrangian coordinate used in figures and difference equation
ℓ Lagrangian coordinate
M mass constant (g)
P^m material pressure $(jerks/cm^3)$
$(P^r)^{ij}$ radiation energy tensor $(jerks/cm^3)$
Q artificial viscosity $(jerks/cm^3)$
Q_A artificial viscosity, either $_1Q_A$ or $_3Q_A$ as defined by Eqs. (6.69) and (6.70)

Q_B artificial viscosity, either $_2Q_B$ or $_4Q_B$ defined by Eqs. (6.71) and (6.72)
$\widetilde{R}$ normal vector to $\vec{r}$, i.e., $\widetilde{R} = (Z, -R)$
$\hat{R} = R$ for cylindrical coordinate and $\hat{R} = 1$ for Cartesian coordinate
$\vec{r}$ position vector (R, Z)
S^{ij} stress deviator tensor $(jerks/cm^3)$
$\dot{S}^{ij}$ stress rate deviator tensor $(jerks/cm^3/shake)$
$S^{RR}, S^{ZZ}, S^{RZ}, S^{\theta\theta}$ stress deviator components $(jerks/cm^3)$
t time $(shake)$
u absolute value of the velocity vector, i.e., $u = |\vec{u}|$ $(cm/shake)$
$\vec{u}$ velocity vector (U, V) $(cm/shake)$
W energy source $(jerks/g/shake)$
$\dot{W}$ rate of energy source $(jerks/g/shake^2)$
Y^0 yield stress in simple tension $(jerks/cm^3)$

Greek letters

γ ratio of the specific heat, i.e., $\gamma = C_P/C_V$ (no unit)
δ^{RR}, δ^{ZZ} correction terms for the rigid body rotation $(jerks/cm^3/shake)$
ϵ specific internal energy per mass $(jerks/g)$
η weighting function defined in Ref. [6.1]
ξ weighting function defined in Ref. [6.1]
ρ density $(\frac{g}{cm^3})$
τ specific volume $(\frac{cm^3}{g})$
ϕ azimuthal angle (radian)
$\vec{\Omega}$ unit vector in the direction of the photon transport

Subscripts

0 initial value
R derivative with respect to R coordinate
t derivative with respect to time
Z derivative with respect to Z coordinate

Superscript

n time at n time-step, i.e., $t^n = t_0 + n \cdot \Delta t$

6.1 Introduction

Lagrangian method is most suitable for small deformation problems. However, it can solve large deformation problems if the code has the rezone and the sliding line capabilities. The governing equations of the mass, the momentum, and the energy may be discretized and solved by finite difference or finite element methods. In this book, only finite difference Lagrangian method will be discussed.

Two-dimensional Lagrangian methods are still used extensively in solving multi-material problems with strong shock. In this chapter, we use Schulz's two-dimensional radiation hydrodynamic code as a framework [6.1]. Three new features are implemented into the basic code, namely, material strength model, artificial viscosity in the direction of local acceleration, and radiation diffusion calculation. Finite difference equations for deviatoric stresses and strains, shear modulus and yield strengths are written compatibly with Schulz's special formulations. Material properties (i.e. shear modulus and yield strength) are dependent on pressure, compression and equivalent plastic strain. Modifications to the artificial viscosity are made so that the acceleration from the artificial viscosity will project onto the unit vector in the direction of local acceleration. This projection eliminates 'false heating' and helps maintain stable computational grids. Also, a very sharp shock front location will be obtained through this method.

In many two-dimensional Lagrangian radiation hydrodynamic calculations, finite difference approximation and Monte Carlo method are very popular in solving the radiation transport problems. We use implicit scheme to solve the radiation-hydrodynamic partial differential equation.

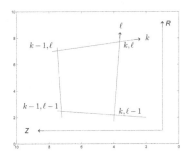

Fig. 6.1 A typical computational zone for cell (k, ℓ).

6.2 Definition of Variable and Notation

In this section, we describe the definitions of variable and notation which will be used for the governing equations such as the conservations of mass, momentum, and energy.

A typical zone of the present computation is shown in Fig. 6.1. On this mesh, we then have two types of variables, zone variables, defined at zone centers and point variables, defined at mesh points. The zone variables are defined so that the mass of zone k, ℓ appears as $M_{k-\frac{1}{2}, \ell-\frac{1}{2}}$ and similarly for the other zone-centered quantities. We will occasionally, for brevity, use k, ℓ instead of $k - \frac{1}{2}, \ell - \frac{1}{2}$, and $k-1, \ell$ instead of $k - \frac{3}{2}, \ell$, etc., in Fig. 6.1, the following definitions hold:

$R(k, \ell, t)$ = Eulerian coordinate (Cartesian or cylindrical) in cm.
$Z(k, \ell, t)$ = Eulerian coordinate (always Cartesian) in cm.
$\vec{R}$ = Vector (R, Z).
k = Lagrangian coordinate. It is a logical coordinate without length unit.
ℓ = Lagrangian coordinate. It is a logical coordinate without length unit.
j = Jacobian of the transformation between (R, Z) and (k, ℓ) in cm^2.

Letting $R_k = \partial R/\partial k$, $R_\ell = \partial R/\partial \ell$ etc., then

$$j = R_k Z_\ell - R_\ell Z_k. \tag{6.1}$$

Define $\hat{R} = R$ for cylindrical coordinates and $\hat{R} = 1$ for Cartesian coordinate, then a volume Jacobian may be defined as

$$J = \hat{R} j. \tag{6.2}$$

We now want to obtain the relations which take us from Eulerian space derivatives to their corresponding Lagrangian counterparts:

$$\frac{\partial}{\partial R} = \frac{\partial k}{\partial R}\frac{\partial}{\partial k} + \frac{\partial \ell}{\partial R}\frac{\partial}{\partial \ell}, \tag{6.3}$$

$$\frac{\partial}{\partial Z} = \frac{\partial k}{\partial Z}\frac{\partial}{\partial k} + \frac{\partial \ell}{\partial Z}\frac{\partial}{\partial \ell}. \tag{6.4}$$

Expressions are required which relate $\frac{\partial k}{\partial R}$ ... $\frac{\partial \ell}{\partial Z}$ to R_k ... Z_ℓ. For arbitrary g, we have

$$\frac{\partial g}{\partial k} = R_k \frac{\partial g}{\partial R} + Z_k \frac{\partial g}{\partial Z}, \tag{6.5}$$

$$\frac{\partial g}{\partial \ell} = R_\ell \frac{\partial g}{\partial R} + Z_\ell \frac{\partial g}{\partial Z}. \tag{6.6}$$

Letting $g = k$, we can solve for $\frac{\partial k}{\partial R}$ and $\frac{\partial k}{\partial Z}$, and letting $g = \ell$ provides $\frac{\partial \ell}{\partial R}$ and $\frac{\partial \ell}{\partial Z}$. The result is

$$\frac{\partial k}{\partial R} = \frac{Z_\ell}{j}, \tag{6.7}$$

$$\frac{\partial \ell}{\partial R} = -\frac{Z_k}{j}, \tag{6.8}$$

$$\frac{\partial k}{\partial Z} = -\frac{R_\ell}{j}, \tag{6.9}$$

and

$$\frac{\partial \ell}{\partial Z} = \frac{R_k}{j}, \tag{6.10}$$

which gives

$$\frac{\partial}{\partial R} = \frac{Z_\ell}{j}\frac{\partial}{\partial k} - \frac{Z_k}{j}\frac{\partial}{\partial \ell}, \tag{6.11}$$

and

$$\frac{\partial}{\partial Z} = -\frac{R_\ell}{j}\frac{\partial}{\partial k} + \frac{R_k}{j}\frac{\partial}{\partial \ell}. \tag{6.12}$$

Now define a vector $\overleftrightarrow{R}$ which lags $\vec{R}$ by 90° as the normal vector to $\vec{R}$, thus

$$\overleftrightarrow{R} = (Z, -R). \tag{6.13}$$

We now define the gradient operator in Lagrange space as $\vec{\nabla} \to \vec{D}$ where

$$\vec{D} = \frac{1}{j}[\overleftrightarrow{R}_\ell \frac{\partial}{\partial k} - \overleftrightarrow{R}_k \frac{\partial}{\partial \ell}], \tag{6.14}$$

or

$$\vec{D} = \frac{1}{j}[\frac{\partial}{\partial k}(\overleftrightarrow{R}_\ell)... - \frac{\partial}{\partial \ell}(\overleftrightarrow{R}_k)...]. \tag{6.15}$$

Hence, for arbitrary function f and vector $\vec{f}$, we have

$$\vec{\nabla} f = \vec{D} f, \tag{6.16}$$

and

$$\vec{\nabla} \cdot \vec{f} = \frac{1}{R}\vec{D} \cdot (\hat{R}\vec{f}). \tag{6.17}$$

Lagrange time derivatives, i.e., partial derivatives with respect to time with k and ℓ fixed, are written as

$u(k, \ell, t) = \frac{\partial R}{\partial t} = \dot{R} = R_t = R$ velocity in cm/shake,

and

$v(k, \ell, t) = \frac{\partial Z}{\partial t} = \dot{Z} = Z_t = Z$ velocity in cm/shake,

with $\vec{u}$ = the vector (u, v) and 1 shake = 10^{-8} seconds.

In addition to the variables already defined, i.e., R, Z, u, v, which are point variables and j, a zone variable, we have the following definitions where (Z) implies a zone variable, (P) a point variable, and (I) an interface variable:

M = mass $(g)(Z)$
$\tau = 1/\rho$ = specific volume $(cm^3/g)(Z)$
P^m = material pressure $(jerks/cm^3)(Z)$
where 1 jerk = 10^{16} ergs
E^m = material energy per unit volume $(jerks/cm^3)(Z)$
ϵ^m = material energy per unit volume $(jerks/cm^3)(Z)$
C_v = material specific heat at constant specific volume $(jerks/g/keV)(Z)$
c = speed of light = 300 $(cm/shake)(Z)$
$I(\vec{r}, \nu, \vec{\Omega}, t)$ = specific intensity of the radiation field defined as the rate of energy flow per unit frequency and solid angle across a unit area oriented normal to the direction of propagation at point $\vec{r}$, frequency ν, in direction $\vec{\Omega}$, at time t. $(jerks/cm^2)(Z)$

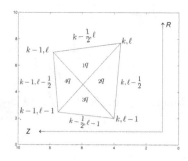

Fig. 6.2 A typical quadrilateral zone in cell (k, ℓ) space.

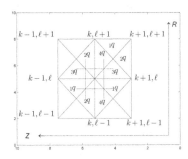

Fig. 6.3 Centering of q, p, E with triangular subzone, only $_1q, _2q, _3q$ and $_4q$ are shown.

$E = \frac{1}{c} \int_0^\infty d\nu \int_{4\pi} I(\nu, \vec{\Omega}) d\vec{\Omega}$

$\quad$ = radiation energy density $(jerks/cm^3)(Z)$, $\hfill (6.18)$

$\vec{F} = \int_0^\infty d\nu \int_{4\pi} \vec{\Omega} I(\nu, \vec{\Omega}) d\vec{\Omega}$

$\quad$ = radiative flux $(jerks/cm^2/shakes)(I)$, $\hfill (6.19)$

$P^r = \frac{1}{c} \int_0^\infty d\nu \int_{4\pi} \vec{\Omega}\vec{\Omega} I(\nu, \vec{\Omega}) d\vec{\Omega}$

$\quad$ = radiation energy tensor $(jerks/cm^3)(Z)$, $\hfill (6.20)$

$\mu_a(\nu)$ = absorption coefficient at frequency ν $(cm^{-1})(Z)$
$\mu'_a(\nu) = \mu_a(\nu)(1 - e^{-h\nu/kT})$
$\quad$ = absorption coefficient corrected for induced emission$(cm^{-1})(Z)$ $\hfill (6.21)$

$\overline{\mu}^P$ = Planck absorption coefficient $(cm^{-1})(Z)$
$K^P = \overline{\mu}^P \tau$ = Planck opacity $(cm^2/g)(Z)$
$\overline{\mu}^R$ = Rosseland absorption coefficient $(cm^{-1})(Z, I)$
$K^R = \overline{\mu}^R \tau$ = Rosseland opacity $(cm^2/g)(Z, I)$
T = material temperature $(keV)(Z)$
$\varphi = aT^4$ = radiative source function $(jerks/cm^3)(Z)$
a = radiation constant $(0.0137 jerks/cm^3/keV^4)(Z)$

Difference over the k variable will be represented by Δ, i.e.,

$\Delta = \frac{\partial}{\partial k}$ and

$$\Delta \vec{R}^n_{k+\frac{1}{2},\ell} = \vec{R}^n_{k+1,\ell} - \vec{R}^n_{k,\ell}. \hfill (6.22)$$

Similarly, we use δ for $\frac{\partial}{\partial \ell}$ and

$$\delta \vec{R}^n_{k,\ell+\frac{1}{2}} = \vec{R}^n_{k,\ell+1} - \vec{R}^n_{k,\ell}. \tag{6.23}$$

A typical zone in k,ℓ space is shown in Fig. 6.2 with pressure defined at cell center. There are four artificial viscosities $_1q$, $_2q$, $_3q$, and $_4q$ defined along the four sides. We further break up each zone into four triangles as shown in Fig. 6.3. Each of these triangles has an associated material internal energy, E_1, E_2, E_3, E_4, and likewise for the pressure and artificial viscosity.

There are two weighted functions which must be defined for use in the momentum equation. They are obtained as follows. Define

$$\omega_{k+\frac{1}{2},\ell} = [\frac{1}{4}(\Delta \vec{R}^n_{k+\frac{1}{2},\ell+\frac{1}{2}} + \Delta \vec{R}^n_{k+\frac{1}{2},\ell-\frac{1}{2}})^2]^{\frac{1}{2}}, \tag{6.24}$$

$$\omega_{k,\ell+\frac{1}{2}} = [\frac{1}{4}(\delta \vec{R}^n_{k+\frac{1}{2},\ell+\frac{1}{2}} + \delta \vec{R}^n_{k-\frac{1}{2},\ell+\frac{1}{2}})^2]^{\frac{1}{2}}, \tag{6.25}$$

$$\xi'_{k,\ell} = \frac{2\omega_{k-\frac{1}{2},\ell}}{\omega_{k+\frac{1}{2},\ell} + \omega_{k-\frac{1}{2},\ell}}, \tag{6.26}$$

and

$$\eta'_{k,\ell} = \frac{2\omega_{k,\ell-\frac{1}{2}}}{\omega_{k,\ell+\frac{1}{2}} + \omega_{k,\ell-\frac{1}{2}}}. \tag{6.27}$$

The weighted functions are then given by

$$\xi_{k,\ell} = \max[0.6, \min(\xi'_{k,\ell}, 1.4)], \tag{6.28}$$

and

$$\eta_{k,\ell} = \max[0.6, \min(\eta'_{k,\ell}, 1.4)]. \tag{6.29}$$

Essentially these weighted functions are used to weigh the various pressure difference and artificial viscosity difference terms in the momentum equation based on the proximity of the point of their formation to the point k,ℓ. Actually, it's more of an inverse weight.

6.3 The Governing Equation

The basic Lagrangian equation for the radiation hydrodynamics are given by Pomraning [6.3]. With the addition of the stress deviatoric tensors we have
the position equation

$$\frac{\partial \vec{r}}{\partial t} = \vec{u}, \qquad (6.30)$$

the continuity equation

$$\rho J = M, \qquad (6.31)$$

the momentum equation

$$\rho \frac{D}{Dt}(\vec{u} + \frac{\vec{F}}{\rho c^2}) + \nabla(P^m + Q) + \nabla \cdot [(P^r)^{ij} - \frac{\vec{u}\vec{F}}{c^2}] - S^{ij}_{,j} = 0, \qquad (6.32)$$

and the energy equation

$$\rho \frac{D}{Dt}[\frac{u^2}{2} + \frac{1}{\rho}(E^m + E)] + \nabla \cdot [\vec{F} + (P^m + Q - E)\vec{u}] - (S^{ij}\vec{u})_{,j} = W. \qquad (6.33)$$

Note that $(P^r)^{ij}$ in Eq. (6.32) and S^{ij} in both Eqs. (6.32) and (6.33) are tensors with, j representing the derivative.

The governing Eqs. (6.30)–(6.33) are split into two major blocks and solved by finite difference method. The first block solves the hydrodynamics and the elastic-plastic flow to obtain new values of pressure, energy, density, and velocity as described in Section 6.5. The radiation transport equation with the energy source is solved by radiation diffusion method in the second block that is described in Section 6.6. At the end of the radiation calculations, we update the pressure and the energy due to their changes in the radiation process. Therefore, in the first block, the conservation equations (6.30)–(6.33) in cylindrical coordinates without the external work become

$$\frac{\partial \vec{r}}{\partial t} = \vec{u}, \qquad (6.34)$$

$$\rho J = M, \qquad (6.35)$$

$$\rho \frac{D\vec{u}}{Dt} + \nabla(P^m + Q) - S^{ij}_{,j} = 0, \qquad (6.36)$$

and

$$\rho \frac{D}{Dt}(\frac{u^2}{2} + \frac{E^m}{\rho}) + \nabla \cdot (P^m + Q)\vec{u} - (S^{ij}\vec{u})_{,j} = 0. \qquad (6.37)$$

6.4 Equation of State

A stress-supporting medium deforming under a wide range of stress exhibits a variety of different physical characteristics. Depending on its retention of elastic character, the flow may be elastic or plastic. When melting, it behaves like a fluid; therefore, in each regime of flow, we need appropriate models or equations of state and criteria defining the transition from one regime to another; for example, when material flows elastically, Hook's law for the deviator stresses can be written as

$$S_t^{RR} = 2G(U_R + \frac{\rho_t}{3\rho}) + \delta^{RR}, \tag{6.38}$$

$$S_t^{ZZ} = 2G(V_Z + \frac{\rho_t}{3\rho}) + \delta^{ZZ}, \tag{6.39}$$

$$S_t^{RZ} = G(U_Z + V_R), \tag{6.40}$$

and

$$S_t^{\theta\theta} + S_t^{RR} + S_t^{RZ} = 0. \tag{6.41}$$

For plastic flows, Prandtl and Reuss considered both plastic and elastic strain simultaneously and arrived at the following flow equation

$$\dot{S}^{ij} = 2G\dot{e}^{ij} - \frac{G\dot{W}}{\frac{1}{3}(Y^0)^2}S^{ij}, \tag{6.42}$$

where $\dot{W}(\sum_{ij} S^{ij}\dot{e}^{ij})$ is the plastic work/unit volume, G and Y are obtained from Steinberg–Guinan model which is described in Chapter 4 and Ref. [4.4].

6.5 Calculation Procedures and Finite Differences

Before we express Eqs. (6.34)–(6.37) in finite difference form, it is useful for the reader to get familiar with the notations and some derivatives as given in Section 6.2. The scalar form of the conservation equations (6.34)–(6.37) can be written as the position equations

$$\frac{\partial R}{\partial t} = U, \tag{6.43}$$

$$\frac{\partial Z}{\partial t} = V, \tag{6.44}$$

the continuity equation

$$\rho(\frac{\partial R}{\partial k}\frac{\partial Z}{\partial \ell} - \frac{\partial R}{\partial \ell}\frac{\partial Z}{\partial k}) = M, \tag{6.45}$$

the momentum equations

$$\frac{\partial U}{\partial t} = -\frac{1}{\rho j}\left(\frac{\partial Z}{\partial \ell}\frac{\partial P}{\partial k} - \frac{\partial Z}{\partial k}\frac{\partial P}{\partial \ell} + \frac{\partial Z}{\partial k}\frac{\partial S^{RR}}{\partial \ell} - \frac{\partial Z}{\partial \ell}\frac{\partial S^{RR}}{\partial k}\right)$$

$$+\frac{1}{\rho j}\left(\frac{\partial R}{\partial k}\frac{\partial S^{RZ}}{\partial \ell} - \frac{\partial R}{\partial \ell}\frac{\partial S^{RZ}}{\partial k}\right)$$

$$-\frac{1}{\rho R}(S^{RR} - S^{\theta\theta}) - \frac{1}{M}\frac{\partial}{\partial k}\left[RQ_A \frac{\frac{\partial Z}{\partial \ell}\frac{\partial U}{\partial k} - \frac{\partial R}{\partial \ell}\frac{\partial V}{\partial k}}{(\frac{\partial U}{\partial k})^2 + (\frac{\partial V}{\partial k})^2}\left(\frac{\partial U}{\partial k}\right)\right]$$

$$+\frac{1}{M}\frac{\partial}{\partial \ell}\left[RQ_B \frac{\frac{\partial Z}{\partial k}\frac{\partial U}{\partial \ell} - \frac{\partial R}{\partial k}\frac{\partial V}{\partial \ell}}{(\frac{\partial U}{\partial \ell})^2 + (\frac{\partial V}{\partial \ell})^2}\left(\frac{\partial U}{\partial \ell}\right)\right], \qquad (6.46)$$

$$\frac{\partial V}{\partial t} = -\frac{1}{\rho j}\left(-\frac{\partial R}{\partial \ell}\frac{\partial P}{\partial k} + \frac{\partial R}{\partial k}\frac{\partial P}{\partial \ell} + \frac{\partial R}{\partial k}\frac{\partial S^{ZZ}}{\partial \ell} - \frac{\partial R}{\partial \ell}\frac{\partial S^{ZZ}}{\partial k}\right)$$

$$+\frac{1}{\rho j}\left(\frac{\partial Z}{\partial \ell}\frac{\partial S^{RZ}}{\partial k} - \frac{\partial Z}{\partial k}\frac{\partial S^{RZ}}{\partial \ell}\right) + \frac{S^{RZ}}{\rho R}$$

$$+\frac{1}{M}\frac{\partial}{\partial k}\left[RQ_A \frac{\frac{\partial Z}{\partial \ell}\frac{\partial U}{\partial k} - \frac{\partial R}{\partial \ell}\frac{\partial V}{\partial k}}{(\frac{\partial U}{\partial k})^2 + (\frac{\partial V}{\partial k})^2}\left(\frac{\partial V}{\partial k}\right)\right]$$

$$-\frac{1}{M}\frac{\partial}{\partial \ell}\left[RQ_B \frac{\frac{\partial Z}{\partial k}\frac{\partial U}{\partial \ell} - \frac{\partial R}{\partial k}\frac{\partial V}{\partial \ell}}{(\frac{\partial U}{\partial \ell})^2 + (\frac{\partial V}{\partial \ell})^2}\left(\frac{\partial V}{\partial \ell}\right)\right], \qquad (6.47)$$

and the energy equation

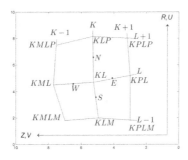

Fig. 6.4 The logical meshes and index as plotted on the physical plane (i.e., R, Z coordinates).

$$\rho\frac{\partial \epsilon}{\partial t} = -P\nabla\cdot\vec{u} - \frac{Q_A}{j}\left(\frac{\partial Z}{\partial \ell}\frac{\partial U}{\partial k} - \frac{\partial R}{\partial \ell}\frac{\partial V}{\partial k}\right) - \frac{Q_B}{j}\left(-\frac{\partial Z}{\partial k}\frac{\partial U}{\partial \ell} + \frac{\partial R}{\partial k}\frac{\partial V}{\partial \ell}\right)$$

$$+\frac{S^{RR}}{j}\left(\frac{\partial Z}{\partial \ell}\frac{\partial U}{\partial k} - \frac{\partial Z}{\partial k}\frac{\partial U}{\partial \ell}\right) + S^{\theta\theta}\left(\frac{U}{R}\right) + \frac{S^{ZZ}}{j}\left(\frac{\partial R}{\partial k}\frac{\partial V}{\partial \ell} - \frac{\partial R}{\partial \ell}\frac{\partial V}{\partial k}\right)$$

$$+\frac{S^{RZ}}{j}(\frac{\partial R}{\partial k}\frac{\partial U}{\partial \ell} - \frac{\partial R}{\partial \ell}\frac{\partial U}{\partial k} + \frac{\partial Z}{\partial \ell}\frac{\partial V}{\partial k} - \frac{\partial Z}{\partial k}\frac{\partial V}{\partial \ell}). \quad (6.48)$$

Since all of the dependent variables appearing in Eqs. (6.38)-(6.48) are functions of time and space, we discretize them using different space locations as well as time level.

In Fig. 6.4, the logical meshes (i.e. K and L) are plotted on the physical plane (i.e. R and Z coordinates). The cell centered quantities, i.e. point 5, are M, j, ρ, ϵ and S^{ij} with the vertex quantities, i.e. point 6, U, V, R and Z as shown in Fig. 6.5. Each quadrilateral zone is further divided into four triangular subzones with pressure P and artificial viscosity Q defined at the centroid of each triangle, i.e. point 1, 2, 3 and 4. Therefore, for each quadrilateral cell, we have four P's and Q's.

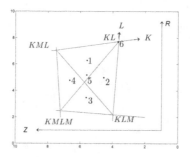

Fig. 6.5 M, j, ρ, ϵ and S^{ij} are defined at cell center point 5, while U, V, R, Z are defined at vertex point 6, pressure and artificial viscosity are at points 1, 2, 3 and 4.

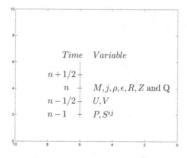

Fig. 6.6 Variables are defined at different time level.

For time level n, we define $M, j, \rho, \epsilon, R, Z$ and Q at t^n and U and V at $t^{n+1/2}$. But pressure P and deviator stress S^{ij} are defined at t^{n-1} as shown in Fig. 6.6. In summary, we have:

cell-centered quantities

$$M_{k-\frac{1}{2},\ell-\frac{1}{2}}, j^n_{k-\frac{1}{2},\ell-\frac{1}{2}}, \rho^n_{k-\frac{1}{2},\ell-\frac{1}{2}}, (S^{ij})^{n-1}_{k-\frac{1}{2},\ell-\frac{1}{2}}, (G)^{n-1}_{k-\frac{1}{2},\ell-\frac{1}{2}}, (Y)^{n-1}_{k-\frac{1}{2},\ell-\frac{1}{2}}$$

and $(\epsilon)^{n-1}_{k-\frac{1}{2},\ell-\frac{1}{2}}$

cell vertex quantities

$(U)^{n-1/2}_{k,\ell}, (V)^{n-1/2}_{k,\ell}, (R)^n_{k,\ell},$ and $(Z)^n_{k,\ell}$

and triangular subzone quantities

$_1P^{n-1}_{k,\ell}, _2P^{n-1}_{k,\ell}, _3P^{n-1}_{k,\ell}, _4P^{n-1}_{k,\ell}, _1Q^n_{k,\ell}, _2Q^n_{k,\ell}, _3Q^n_{k,\ell},$ and $_4Q^n_{k,\ell}$.

Based on the known dependent variables at time $t = t^n$, the following steps are taken logically:

Step 1. Calculate U_R, U_Z, V_R and V_Z.
Step 2. If the sliding interface treatment is required, then update $U^{n+1/2}$, $V^{n+1/2}$, $R^{n+1/2}$ and $Z^{n+1/2}$ due to the effects of sliding.
Step 3. Calculate ρ^{n+1} and ρ_t, from R^n, Z^n and M.
Step 4. Calculate G and Y from Chapter 4.
Step 5. Calculate $(S^{ij})^n$ from Eqs. (6.38) through (6.41) for pure elastic deformation.
Step 6. Check von Mises yield condition.
Step 7. Calculate $(S^{ij})^n$ with Prandtl–Reuss plastic flow term, i.e. Eq. (6.42).
Step 8. Calculate equivalent plastic strain.
Step 9. Add the rigid body rotation correction terms to the deviatoric stresses.
Step 10. Calculate the principle stresses to check the fracture conditions.
Step 11. Calculate velocities $U^{n+1/2}$ and $V^{n+1/2}$ from Eqs. (6.46) and (6.47).
Step 12. Calculate the artificial viscosity $_1Q^n_{k,\ell}, _2Q^n_{k,\ell}, _3Q^n_{k,\ell}$ and $_4Q^n_{k,\ell}$.
Step 13. Calculate energy ϵ^{n+1} from Eq. (6.48).
Step 14. Calculate pressure P^n from equation of state, i.e. $P = f(\rho, \epsilon)$. In our case, it is a table lookup.

Steps 1–14 complete one time step calculations, and we go back to Step 1 for the next time increment computations.

For finite difference, we can solve Eq. (6.38) (i.e. Step 5) by

$$(\dot{S}^{RR})_{k-\frac{1}{2},\ell-\frac{1}{2}} = 2G_{k-\frac{1}{2},\ell-\frac{1}{2}} \left[\frac{1}{j}(Z_\ell \frac{\partial U}{\partial k} - Z_k \frac{\partial U}{\partial \ell}) + \frac{\rho_t}{3\rho} \right]^n_{k-\frac{1}{2},\ell-\frac{1}{2}}, \qquad (6.49)$$

to obtain the deviator stress for pure elastic deformation.

Similar formula are used for computing $(\dot{S}^{ZZ})_{k-\frac{1}{2},\ell-\frac{1}{2}}$ and $(\dot{S}^{RZ})_{k-\frac{1}{2},\ell-\frac{1}{2}}$ from Eqs. (6.39) and (6.40). Once the elastic regime is finished, we can add the Prandtl–Reuss plastic flow and the correction due to rigid body rotations.

The finite difference formula for the momentum equations is more complicated. For the sake of simplicity, we like to rewrite the momentum equation in R component, i.e. Eq. (6.46) as

$$U_t =_1 U_t +_2 U_t, \qquad (6.50)$$

where

$$_1U_t = -\frac{1}{M}\frac{\Delta}{\Delta k}\left(RQ_A \frac{Z_\ell U_k - R_\ell V_k}{U_k^2 + V_k^2} U_k\right)$$

$$+ \frac{1}{M}\frac{\Delta}{\Delta \ell}\left(RQ_B \frac{Z_k U_\ell - R_k V_\ell}{U_\ell^2 + V_\ell^2} U_\ell\right), \qquad (6.51)$$

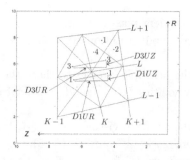

Fig. 6.7 Each cell has four triangular subzones where pressure and artificial viscosity are defined. The momentum components $D1UZ, D1UR, D3UZ$ and $D3UR$ are also shown.

and

$$_2U_t = -\frac{1}{\rho j}\left(Z_\ell \frac{\Delta P}{\Delta k} - Z_k \frac{\Delta P}{\Delta \ell} + Z_k \frac{\Delta S^{RR}}{\Delta \ell} - Z_\ell \frac{\Delta S^{RR}}{\Delta k}\right)$$
$$+\frac{1}{\rho j}\left(R_k \frac{\Delta S^{RZ}}{\Delta \ell} - R_\ell \frac{\Delta S^{RZ}}{\Delta k}\right) + \frac{1}{\rho R}(S^{RR} - S^{\theta\theta}). \qquad (6.52)$$

Since there are four Q's in one cell; namely, $_1Q_{k,\ell}^n, _2Q_{k,\ell}^n, _3Q_{k,\ell}^n$ and $_4Q_{k,\ell}^n$ as calculated in Eqs. (4.27)-(4.30) of Ref. [6.4] (see Fig. 6.5), we divide the right-hand side of Eq. (6.51) into four parts. The first two parts are associated with Q_A, i.e. $Q_A = f(_1Q, _3Q)$ and the last two parts are with Q_B, i.e. $Q_B = f(_2Q, _4Q)$. Figure 6.7 shows four cells; each has four triangular subzones. Within each subzone, we assign one pressure and one artificial viscosity at the centroid of the triangle.

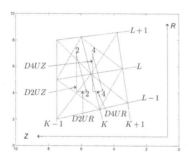

Fig. 6.8 The momentum components $D2UZ, D2UR, D4UZ$ and $D4UR$ are shown.

When we solve the momentum equations for the vertex node (K, L), we connect point 1-1 and point 3-3 as shown in Fig. 6.7. The vertical component of line 1-1 is called $D1UZ$, while the horizontal component is $D1UR$. The vertical and horizontal components for line 3-3 are D3UZ and $D3UR$. The same idea is applied to point 2 and point 4 as shown in Fig. 6.8, which gives $D2UZ, D2UR, D4UZ$ and $D4UR$. The finite difference expressions using the notations shown in Fig. 6.4 are

$$D1UZ = \tfrac{1}{M_S}\{[(Z_{KL} - Z_{KLM})(U_{KPL} - U_{KL}) - (R_{KL} - R_{KLM})$$
$$\times (V_{KPL} - V_{KL})] \cdot \tfrac{(R_1Q)_{KPL}(U_{KPL} - U_{KL})}{(U_{KPL} - U_{KL})^2 + (V_{KPL} - V_{KL})^2}$$
$$+[(R_{KL} - R_{KLM})(V_{KL} - V_{KML}) - (Z_{KL} - Z_{KLM})(U_{KL} - U_{KML})]$$

$$\times \frac{(R_1Q)_{KL}(U_{KL} - U_{KML})}{(U_{KL} - U_{KML})^2 + (V_{KL} - V_{KML})^2}\}, \qquad (6.53)$$

$D3UZ = \frac{1}{M_N}\{[(Z_{KLP} - Z_{KL})(U_{KPL} - U_{KL}) - (R_{KLP} - R_{KL})$

$\times (V_{KPL} - V_{KL})] \cdot \frac{(R_3Q)_{KPLP}(U_{KPL} - U_{KL})}{(U_{KPL} - U_{KL})^2 + (V_{KPL} - V_{KL})^2}$

$+[(R_{KLP} - R_{KL})(V_{KL} - V_{KML}) - (Z_{KLP} - Z_{KL})(U_{KL} - U_{KML})]$

$$\times \frac{(R_3Q)_{KLP}(U_{KL} - U_{KML})}{(U_{KL} - U_{KML})^2 + (V_{KL} - V_{KML})^2}\}, \qquad (6.54)$$

$D2UZ = \frac{1}{M_W}\{[(Z_{KL} - Z_{KML})(U_{KLP} - U_{KL}) - (R_{KL} - R_{KML})$

$\times (V_{KPL} - V_{KL})] \cdot \frac{(R_2Q)_{KLP}(U_{KLP} - U_{KL})}{(U_{KLP} - U_{KL})^2 + (V_{KPL} - V_{KL})^2}$

$+[(R_{KL} - R_{KML})(V_{KL} - V_{KLM}) - (Z_{KL} - Z_{KML})(U_{KL} - U_{KLM})]$

$$\times \frac{(R_2Q)_{KL}(U_{KL} - U_{KLM})}{(U_{KL} - U_{KLM})^2 + (V_{KL} - V_{KLM})^2}\}, \qquad (6.55)$$

$D4UZ = \frac{1}{M_E}\{[(Z_{KPL} - Z_{KL})(U_{KLP} - U_{KL}) - (R_{KPL} - R_{KL})$

$\times (V_{KLP} - V_{KL})] \cdot \frac{(R_4Q)_{KPLP}(U_{KLP} - U_{KL})}{(U_{KLP} - U_{KL})^2 + (V_{KLP} - V_{KL})^2}$

$+[(R_{KPL} - R_{KPLM})(V_{KL} - V_{KLM}) - (Z_{KPL} - Z_{KL})(U_{KL} - U_{KLM})]$

$$\times \frac{(R_4Q)_{KPL}(U_{KL} - U_{KLM})}{(U_{KL} - U_{KLM})^2 + (V_{KL} - V_{KLM})^2}\}, \qquad (6.56)$$

$D1UR = \frac{-1}{M_S}\{[(Z_{KL} - Z_{KLM})(U_{KPL} - U_{KL}) - (R_{KL} - R_{KLM})$

$\times (V_{KPL} - V_{KL})] \cdot \frac{(R_1Q)_{KPL}(V_{KPL} - V_{KL})}{(U_{KPL} - U_{KL})^2 + (V_{KPL} - V_{KL})^2}$

$$+[(R_{KL} - R_{KLM})(V_{KL} - V_{KML}) - (Z_{KL} - Z_{KLM})(U_{KL} - U_{KML})]$$

$$\times \frac{(R_1Q)_{KL}(V_{KL} - V_{KML})}{(U_{KL} - U_{KML})^2 + (V_{KL} - V_{KML})^2}\}, \qquad (6.57)$$

$$D3UR = \frac{-1}{M_N}\{[(Z_{KLP} - Z_{KL})(U_{KPL} - U_{KL}) - (R_{KLP} - R_{KL})$$

$$\times(V_{KPL} - V_{KL})] \cdot \frac{(R_3Q)_{KPLP}(V_{KPL} - V_{KL})}{(U_{KPL} - U_{KL})^2 + (V_{KPL} - V_{KL})^2}$$

$$+[(R_{KLP} - R_{KL})(V_{KL} - V_{KML}) - (Z_{KLP} - Z_{KL})(U_{KL} - U_{KML})]$$

$$\times \frac{(R_3Q)_{KLP}(V_{KL} - V_{KML})}{(U_{KL} - U_{KML})^2 + (V_{KL} - V_{KML})^2}\}, \qquad (6.58)$$

$$D2UR = \frac{-1}{M_W}\{[(Z_{KL} - Z_{KML})(U_{KLP} - U_{KL}) - (R_{KL} - R_{KML})$$

$$\times(V_{KPL} - V_{KL})] \cdot \frac{(R_2Q)_{KLP}(V_{KLP} - V_{KL})}{(U_{KLP} - U_{KL})^2 + (V_{KLP} - V_{KL})^2}$$

$$+[(R_{KL} - R_{KML})(V_{KL} - V_{KLM}) - (Z_{KL} - Z_{KML})(U_{KL} - U_{KLM})]$$

$$\times \frac{(R_2Q)_{KL}(V_{KL} - V_{KLM})}{(U_{KL} - U_{KLM})^2 + (V_{KL} - V_{KLM})^2}\}, \qquad (6.59)$$

and

$$D4UR = \frac{-1}{M_E}\{[(Z_{KPL} - Z_{KL})(U_{KLP} - U_{KL}) - (R_{KPL} - R_{KL})$$

$$\times(V_{KLP} - V_{KL})] \cdot \frac{(R_4Q)_{KPLP}(V_{KLP} - V_{KL})}{(U_{KLP} - U_{KL})^2 + (V_{KLP} - V_{KL})^2}$$

$$+[(R_{KPL} - R_{KPLM})(V_{KL} - V_{KLM}) - (Z_{KPL} - Z_{KL})(U_{KL} - U_{KLM})]$$

$$\times \frac{(R_4Q)_{KPL}(V_{KL} - V_{KLM})}{(U_{KL} - U_{KLM})^2 + (V_{KL} - V_{KLM})^2}\}. \qquad (6.60)$$

In Eqs. (6.53)-(6.60), we use $M_S = (\rho j)_{k,\ell-\frac{1}{2}}$, $M_N = (\rho j)_{k,\ell+\frac{1}{2}}$, $M_W = (\rho j)_{k-\frac{1}{2},\ell}$, and $M_E = (\rho j)_{k+\frac{1}{2},\ell}$. The R in the parentheses $(R_1Q), (R_2Q)$...etc. means the average of R coordinates, e.g. $(R_1Q)_{KPL} = 0.25(R_{KPL} + R_{KL} + R_{KLM} + R_{KPLM}) \cdot (_1Q)_{KPL}$.

Since there are four pressures in one cell, the finite difference of $\partial P/\partial k$ and $\partial P/\partial \ell$ will have similar formulations as in $(\partial Q/\partial k)$ and $(\partial Q/\partial \ell)$. For the reason of simplicity, we just write Eq. (6.52) as

$$(_2U_t)_{K,L} = \{(-\frac{1}{\rho j})[Z_\ell \frac{\partial P}{\partial k} - Z_k \frac{\partial P}{\partial \ell} + Z_k S_\ell^{RR} - Z_\ell S_k^{RR}]$$

$$+(\frac{1}{\rho j})(R_k S_\ell^{RZ} - R_\ell S_k^{RZ}) + \frac{1}{\rho R}(S^{RR} - S^{\theta\theta})\}_{K,L}^n. \qquad (6.61)$$

Using the Eqs. (6.53) through (6.61), we get

$$U_{K,L}^{n+1} = U_{K,L}^n + \Delta t\{D1UZ - D2UZ + D3UZ - D4UZ + (_2U_t)_{K,L}\}, \qquad (6.62)$$

for the new time velocity in the R direction, and

$$V_{K,L}^{n+1} = V_{K,L}^n + \Delta t\{D1UR - D2UR + D3UR - D4UR + (_2V_t)_{K,L}\}, \qquad (6.63)$$

for that in Z direction where $(_2V_t)_{K,L}$ is similar to Eq. (6.52).

The internal energy equation can be rewritten as

$$\frac{\partial \epsilon}{\partial t} = -P\frac{\partial}{\partial t}(\frac{1}{\rho}) - \frac{Q_A(Z_\ell U_k - R_\ell V_k)}{\rho j} - \frac{Q_B(-Z_\ell U_\ell + R_k V_\ell)}{\rho j}$$

$$+\frac{S^{RR}}{\rho j}(Z_\ell \frac{\Delta U}{\Delta k} - Z_k \frac{\Delta U}{\Delta \ell}) + \frac{S^{\theta\theta}}{\rho}(\frac{U}{R})$$

$$+\frac{S^{ZZ}}{\rho j}(R_k \frac{\Delta V}{\Delta \ell} - R_\ell \frac{\Delta V}{\Delta k})$$

$$+\frac{S^{RZ}}{\rho j}(R_k \frac{\Delta U}{\Delta \ell} - R_\ell \frac{\Delta U}{\Delta k} + Z_\ell \frac{\Delta V}{\Delta k} - Z_k \frac{\Delta V}{\Delta \ell}). \qquad (6.64)$$

Let the energy source due to stresses be W, then

$$W = [\frac{S^{RR}}{\rho j}(Z_\ell \frac{\Delta U}{\Delta k} - Z_k \frac{\Delta U}{\Delta \ell}) + \frac{S^{\theta\theta}}{\rho}(\frac{U}{R}) + \frac{S^{ZZ}}{\rho j}(R_k \frac{\Delta V}{\Delta \ell} - R_\ell \frac{\Delta V}{\Delta k})$$

$$+\frac{S^{RZ}}{\rho j}(R_k \frac{\Delta U}{\Delta \ell} - R_\ell \frac{\Delta U}{\Delta k} + Z_\ell \frac{\Delta V}{\Delta k} - Z_k \frac{\Delta V}{\Delta \ell})]_{K,L}^n. \qquad (6.65)$$

Therefore, the new time energy is

$$\epsilon^{n+1} = \epsilon^n - P_{K,L}^{n-1}\left(\frac{1}{\rho^{n+1}} - \frac{1}{\rho^n}\right)_{K,L} - (\Delta t)\left(\frac{1}{\rho j}\right)_{K,L}^{n+1}$$

$$\times [Q_A(Z_\ell U_k - R_\ell V_k) + Q_B(-Z_k U_\ell - R_k V_\ell)]_{K,L}^n + (\Delta t)W. \qquad (6.66)$$

With new energy ϵ^{n+1} and density ρ^{n+1}, one can obtain new pressure P^{n+1} by Equation of State, i.e., $P^{n+1} = f(\rho^{n+1}, \epsilon^{n+1})$.

6.6 Two-Dimensional Lagrangian Method for Radiation Diffusion Problems

6.6.1 *Introduction*

This section describes the two-dimensional Lagrangian method for solving the radiation diffusion equations by using the finite difference approximation and the Monte Carlo sampling.

The radiation diffusion equations comprise the energy balances of the radiation energy and the material energy. We use implicit scheme to solve the partial differential equations of the radiation equations. The system of the algebraic equations from the finite difference approximation are then solved by the matrix inversion method.

The second part of Section 6.6 describes the Monte Carlo method for solving the radiation diffusion equation. We create a large number of particles to represent the total energy (including radiation and material energy) inside each zone and, then, we calculate the movement of each particle until the particle either deposits their energy inside the zone or leaves the boundary and disappeared.

6.6.2 *Finite Difference Approximation for the Radiation Diffusion Equation*

The basic radiation diffusion equations are

$$\frac{\partial E}{\partial t} + \nabla \cdot \vec{F} = c\sigma_a a T^4 - c\sigma_a E, \qquad (6.67)$$

and

$$\rho C_v \frac{\partial T}{\partial t} = c\sigma_a E - c\sigma_a a T^4, \qquad (6.68)$$

where E is the radiation energy density in $(jerks/cm^3)$, F the radiation flux, T the material temperature in (K), c the speed of light in (cm/sec), σ_a the absorption cross section in $(1/cm)$, $a = 4\sigma/c = 7.5657 \times 10^{-16} j/(m^3 \cdot K^4)$, $\sigma = 5.67 \times 10^{-8} j/(m^2 \cdot sec \cdot K^4)$ the Stephan-Boltzmann constant, and C_v the specific heat at constant volume in $[jerks/(g \cdot K)]$. In Eqs. (6.67) and (6.68), σ_a can be replaced by $f\sigma_a$, and

$$f = \frac{1}{1 + \beta c \Delta t \sigma_a}, \qquad (6.69)$$

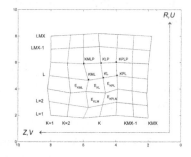

Fig. 6.9 The zone index $KMLP, KLP, KPLP, KML, KL$ and KPL are shown with the zone radiation energy density $E_{KML}, E_{KL}, E_{KPL}, E_{KLM}$ and E_{KPLM}.

where

$$\beta = \frac{4aT^3}{\rho C_v}. \tag{6.70}$$

Using the notations given in Section 6.2, the radiation flux term in Eq. (6.67) is

$$\nabla \cdot \vec{F} = (\hat{i}\frac{\partial}{\partial R} + \hat{j}\frac{\partial}{\partial Z}) \cdot (\hat{i}F_R + \hat{j}F_Z), \tag{6.71}$$

$$= \frac{\partial F_R}{\partial R} + \frac{\partial F_Z}{\partial Z}. \tag{6.72}$$

From Eqs. (6.3), (6.4), (6.7) and (6.8) in Section 6.2, we get

$$\nabla \cdot \vec{F} = \frac{Z_\ell}{j}\frac{\partial F_R}{\partial k} - \frac{Z_k}{j}\frac{\partial F_R}{\partial \ell} - \frac{R_\ell}{j}\frac{\partial F_Z}{\partial k} + \frac{R_k}{j}\frac{\partial F_Z}{\partial \ell}, \tag{6.73}$$

where

$$F_R = -cD \nabla E_R = -cD(E_{k,\ell+1} - E_{k,\ell}), \tag{6.74}$$

and

$$F_Z = -cD \nabla E_Z = cD(E_{k+1,\ell} - E_{k,\ell}). \tag{6.75}$$

In Eqs. (6.73), (6.74) and (6.75), the zone index is shown in Fig. 6.9. For a zone(k, ℓ), the term on the right hand side of Eq. (6.73) are

$$\frac{Z_\ell}{j}\frac{\partial F_R}{\partial k} = \frac{Z_{KL} - Z_{KLM}}{j}\frac{\partial}{\partial k}[-cD(E_{k,\ell+1} - E_{k,\ell})], \tag{6.76}$$

$$= \frac{Z_{KL} - Z_{KLM}}{j}(-cD)(\frac{\partial E_{k,\ell+1}}{\partial k} - \frac{\partial E_{k,\ell}}{\partial k}), \qquad (6.77)$$

$$= (-cD)\frac{Z_{KL} - Z_{KLM}}{j}[E_{k+1,\ell+1} - E_{k,\ell+1} - (E_{k+1,\ell} - E_{k,\ell})], \qquad (6.78)$$

$$\frac{Z_k}{j}\frac{\partial F_R}{\partial \ell} = \frac{Z_{KL} - Z_{KML}}{j}\frac{\partial}{\partial \ell}[-cD(E_{k,\ell+1} - E_{k,\ell})], \qquad (6.79)$$

$$= \frac{Z_{KL} - Z_{KML}}{j}(-cD)(\frac{\partial E_{k,\ell+1}}{\partial \ell} - \frac{\partial E_{k,\ell}}{\partial \ell}), \qquad (6.80)$$

$$= (-cD)\frac{Z_{KL} - Z_{KML}}{j}[E_{k,\ell+1} - E_{k,\ell} - (E_{k,\ell} - E_{k,\ell-1})], \qquad (6.81)$$

$$\frac{R_\ell}{j}\frac{\partial F_Z}{\partial k} = \frac{R_{KL} - R_{KLM}}{j}\frac{\partial}{\partial k}[cD(E_{k+1,\ell} - E_{k,\ell})], \qquad (6.82)$$

$$= \frac{R_{KL} - R_{KLM}}{j}(cD)(\frac{\partial E_{k+1,\ell}}{\partial k} - \frac{\partial E_{k,\ell}}{\partial k}), \qquad (6.83)$$

$$= (cD)\frac{R_{KL} - R_{KLM}}{j}[E_{k+1,\ell} - E_{k,\ell} - (E_{k,\ell} - E_{k-1,\ell})], \qquad (6.84)$$

and

$$\frac{R_k}{j}\frac{\partial F_Z}{\partial \ell} = \frac{R_{KL} - R_{KML}}{j}\frac{\partial}{\partial \ell}[cD(E_{k,\ell+1} - E_{k,\ell})], \qquad (6.85)$$

$$= \frac{R_{KL} - R_{KML}}{j}(cD)(\frac{\partial E_{k,\ell+1}}{\partial \ell} - \frac{\partial E_{k,\ell}}{\partial \ell}), \qquad (6.86)$$

$$= (cD)\frac{R_{KL} - R_{KML}}{j}[E_{k+1,\ell} - E_{k+1,\ell-1} - (E_{k,\ell} - E_{k,\ell-1})]. \qquad (6.87)$$

Substituting Eqs. (6.78), (6.81), (6.84) and (6.87) into Eq. (6.73), one obtains

$$\nabla \cdot \vec{F} = \frac{cD}{j}[(Z_{KLM} - Z_{KL})(E_{k+1,\ell+1} - E_{k,\ell+1} - E_{k+1,\ell} + E_{k,\ell})$$

$$+(Z_{KL} - Z_{KML})(E_{k,\ell+1} - 2E_{k,\ell} + E_{k,\ell-1})$$

$$+(R_{KLM} - R_{KL})(E_{k+1,\ell} - 2E_{k,\ell} + E_{k-1,\ell})$$

$$+(R_{KL} - R_{KML})(E_{k+1,\ell} - E_{k+1,\ell-1} - E_{k,\ell} + E_{k,\ell-1})]. \quad (6.88)$$

Let

$$S_1 = \frac{cD}{j}(Z_{KLM} - Z_{KL}), \quad (6.89)$$

$$S_2 = \frac{cD}{j}(Z_{KL} - Z_{KML}), \quad (6.90)$$

$$S_3 = \frac{cD}{j}(R_{KLM} - R_{KL}), \quad (6.91)$$

and

$$S_4 = \frac{cD}{j}(R_{KL} - R_{KML}). \quad (6.92)$$

Substituting Eqs. (6.89)–(6.92) into Eq. (6.88), one gets

$$\nabla \cdot \overrightarrow{F} = S_1(E_{k+1,\ell+1} - E_{k,\ell+1} - E_{k+1,\ell} + E_{k,\ell})$$

$$+S_2(E_{k,\ell+1} - 2E_{k,\ell} + E_{k,\ell-1}) + S_3(E_{k+1,\ell} - 2E_{k,\ell} + E_{k-1,\ell})$$

$$+S_4(E_{k+1,\ell} - E_{k+1,\ell-1} - E_{k,\ell} + E_{k,\ell-1})]. \quad (6.93)$$

If one replaces σ_a by $f\sigma_a$ in Eqs. (6.67) and (6.68), then, one will get

$$\frac{\partial E}{\partial t} + \nabla \cdot \overrightarrow{F} = cf\sigma_a a T^4 - cf\sigma_a E, \quad (6.94)$$

and

$$\rho C_v \frac{\partial T}{\partial t} = cf\sigma_a E - cf\sigma_a a T^4, \quad (6.95)$$

where f is defined in Eq. (6.69).
Let

$$S_5 = cf\sigma_a a, \quad (6.96)$$

$$S_6 = cf\sigma_a, \quad (6.97)$$

and
$$S_7 = \rho C_v, \tag{6.98}$$

then, Eqs. (6.94) and (6.95) become

$$\frac{\partial E}{\partial t} + \nabla \cdot \vec{F} = S_5 T^4 - S_6 E, \tag{6.99}$$

and

$$S_7 \frac{\partial T}{\partial t} = S_6 E - S_5 T^4. \tag{6.100}$$

The finite difference form of Eq. (6.99) is

$$\frac{E^{n+1} - E^n}{\Delta t} + \nabla \cdot \vec{F} = S_5(T^n)^4 - S_6 E^{n+1}, \tag{6.101}$$

or

$$E^{n+1} - E^n + (\Delta t)\nabla \cdot \vec{F} = S_5(\Delta t)(T^n)^4 - S_6(\Delta t)E^{n+1}, \tag{6.102}$$

or

$$[1 + S_6(\Delta t)]E^{n+1} + (\Delta t)\nabla \cdot \vec{F} = E^n + S_5(\Delta t)(T^n)^4. \tag{6.103}$$

Using the index in k and ℓ, Eq. (6.103) becomes

$$[1 + S_6(\Delta t)]E^{n+1}_{k,\ell} + (\Delta t)\nabla \cdot \vec{F} = E^n_{k,\ell} + S_5(\Delta t)(T^n_{k,\ell})^4. \tag{6.104}$$

In the above equation, $\nabla \cdot \vec{F}$ is evaluated at time level $n+1$ that is assumed all of the E on the right hand side of Eq. (6.93) are set at $t = t^{n+1}$. Therefore, the coefficient for $E^{n+1}_{k,\ell}$ in the left hand side of Eq. (6.103) is S_8, which is

$$S_8 = 1 + (\Delta t)(S_6 + S_1 - 2S_2 - 2S_3 - S_4). \tag{6.105}$$

The coefficient for $E^{n+1}_{k-1,\ell}$ is

$$S_9 = S_3(\Delta t). \tag{6.106}$$

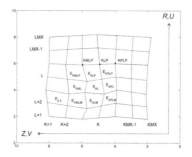

Fig. 6.10 The zone index $KMLP, KLP$ and $KPLP$ are shown with the zone radiation energy density $E_{2,3}, E_{KML}, E_{KL}, E_{KPL}, E_{KMLP}, E_{KLP}, E_{KPLP}, E_{KMLM}, E_{KLM}$ and E_{KPLM}.

The coefficient for $E^{n+1}_{k+1,\ell}$ is

$$S_{10} = (\Delta t)(S_3 - S_1 + S_4). \tag{6.107}$$

The coefficient for $E^{n+1}_{k,\ell+1}$ is

$$S_{11} = (\Delta t)(S_2 - S_1). \tag{6.108}$$

The coefficient for $E^{n+1}_{k,\ell-1}$ is

$$S_{12} = (\Delta t)(S_2 + S_4). \tag{6.109}$$

The coefficient for $E^{n+1}_{k+1,\ell+1}$ is

$$S_{13} = (\Delta t)(S_1). \tag{6.110}$$

The coefficient for $E^{n+1}_{k+1,\ell-1}$ is

$$S_{14} = (\Delta t)(-S_4). \tag{6.111}$$

In Eqs. (6.110) and (6.111), the $E^{n+1}_{k+1,\ell+1}$ and $E^{n+1}_{k+1,\ell-1}$ are obtained from finite difference approximation of $\frac{\partial F_R}{\partial k}$ and $\frac{\partial F_Z}{\partial \ell}$ as shown in Eqs. (6.78) and (6.85).

The interior zones are confined in ($K = 2$ to $K = KMX - 1$) and ($L = 2$ to $L = LMX - 1$) as shown in Fig. 6.9. For the interior zones, Eq. (6.103) becomes

$$S_8 E^{n+1}_{k,\ell} + S_9 E^{n+1}_{k-1,\ell} + S_{10} E^{n+1}_{k+1,\ell} + S_{11} E^{n+1}_{k,\ell+1} + S_{12} E^{n+1}_{k,\ell-1}$$
$$+ S_{13} E^{n+1}_{k+1,\ell+1} + S_{14} E^{n+1}_{k+1,\ell-1} = E^n_{k,\ell} + S_5(\Delta t)(T^n_{k,\ell})^4. \tag{6.112}$$

On the left hand side of Eq. (6.112) there are seven E^{n+1} at different zone centers that will be used to calculate $E_{k,\ell}^{n+1}$. On the right hand side of Eq. (6.112) both terms are already known since they are at the old time level n. Figure 6.9 shows the grids at KML, KL and KPL and the corresponding zone values of radiation energy densities E_{KML}, E_{KL} and E_{KPL}. These seven E^{n+1} on the left hand side of Eq. (6.112) are shown in Fig. 6.10 as $E_{KML}, E_{KL}, E_{KPL}, E_{KLP}, E_{KLM}, E_{KPLP}$ and E_{KPLM}.

The boundary zones that are bounded by $K = 1$ and $K = 2$ for $L = 2$ to $L = LMX - 1$ will be calculated by Eq. (6.112) with $S_9 = 0$, that is

$$S_8 E_{k,\ell}^{n+1} + S_{10} E_{k+1,\ell}^{n+1} + S_{11} E_{k,\ell+1}^{n+1} + S_{12} E_{k,\ell-1}^{n+1}$$

$$+ S_{13} E_{k+1,\ell+1}^{n+1} + S_{14} E_{k+1,\ell-1}^{n+1} = E_{k,\ell}^n + S_5 (\Delta t)(T_{k,\ell}^n)^4 . \quad (6.113)$$

The boundary zones that are bounded by $L = 1$ and $L = 2$ for $K = 2$ to $K = KMX - 1$ will be calculated by Eq. (6.112) with $S_{12} = 0$, that is

$$S_8 E_{k,\ell}^{n+1} + S_9 E_{k-1,\ell}^{n+1} + S_{10} E_{k+1,\ell}^{n+1} + S_{11} E_{k,\ell+1}^{n+1}$$

$$+ S_{13} E_{k+1,\ell+1}^{n+1} + S_{14} E_{k+1,\ell-1}^{n+1} = E_{k,\ell}^n + S_5 (\Delta t)(T_{k,\ell}^n)^4 . \quad (6.114)$$

The boundary zones that are bounded by $K = KMX - 1$ and $K = KMX$ for $L = 2$ to $L = LMX - 1$ will be calculated by Eq. (6.112) with $S_{10} = 0$, that is

$$S_8 E_{k,\ell}^{n+1} + S_9 E_{k-1,\ell}^{n+1} + S_{11} E_{k,\ell+1}^{n+1} + S_{12} E_{k,\ell-1}^{n+1}$$

$$+ S_{13} E_{k+1,\ell+1}^{n+1} + S_{14} E_{k+1,\ell-1}^{n+1} = E_{k,\ell}^n + S_5 (\Delta t)(T_{k,\ell}^n)^4 . \quad (6.115)$$

The top boundary zones that are bounded by $L = LMX - 1$ and $L = LMX$ for $K = 2$ to $K = KMX - 1$ will be calculated by Eq. (6.112) with $S_{11} = 0$, that is

$$S_8 E_{k,\ell}^{n+1} + S_9 E_{k-1,\ell}^{n+1} + S_{10} E_{k+1,\ell}^{n+1} + S_{12} E_{k,\ell-1}^{n+1}$$

$$+ S_{13} E_{k+1,\ell+1}^{n+1} + S_{14} E_{k+1,\ell-1}^{n+1} = E_{k,\ell}^n + S_5 (\Delta t)(T_{k,\ell}^n)^4 . \quad (6.116)$$

The left-bottom corner cell that are bounded by $K = 2$, $L = 2$, $K = 1$, and $L = 1$ has to be calculated by a special treatment. A possible scheme is by setting $S_9 = 0$ and $S_{12} = 0$ from Eq. (6.112), that is

$$S_8 E_{k,\ell}^{n+1} + S_{10} E_{k+1,\ell}^{n+1} + S_{11} E_{k,\ell+1}^{n+1}$$
$$+ S_{13} E_{k+1,\ell+1}^{n+1} + S_{14} E_{k+1,\ell-1}^{n+1} = E_{k,\ell}^n + S_5 (\Delta t)(T_{k,\ell}^n)^4 . \qquad (6.117)$$

The left-bottom corner cell is calculated by

$$S_8 E_{k,\ell}^{n+1} + S_9 E_{k-1,\ell}^{n+1} + S_{11} E_{k,\ell+1}^{n+1}$$
$$+ S_{13} E_{k+1,\ell+1}^{n+1} + S_{14} E_{k+1,\ell-1}^{n+1} = E_{k,\ell}^n + S_5 (\Delta t)(T_{k,\ell}^n)^4 . \qquad (6.118)$$

The left-top corner cell is calculated by

$$S_8 E_{k,\ell}^{n+1} + S_{10} E_{k+1,\ell}^{n+1} + S_{12} E_{k,\ell-1}^{n+1}$$
$$+ S_{13} E_{k+1,\ell+1}^{n+1} + S_{14} E_{k+1,\ell-1}^{n+1} = E_{k,\ell}^n + S_5 (\Delta t)(T_{k,\ell}^n)^4 . \qquad (6.119)$$

The right-top corner cell is calculated by

$$S_8 E_{k,\ell}^{n+1} + S_9 E_{k-1,\ell}^{n+1} + S_{12} E_{k,\ell-1}^{n+1}$$
$$+ S_{13} E_{k+1,\ell+1}^{n+1} + S_{14} E_{k+1,\ell-1}^{n+1} = E_{k,\ell}^n + S_5 (\Delta t)(T_{k,\ell}^n)^4 . \qquad (6.120)$$

There are 36 cells shown in Fig. 6.10, that is, $KMX = 7$ and $LMX = 7$. if one will apply the Eqs. (6.112)–(6.115) to those 36 cells, one may obtain 36 equations which comprise the 36 radiation energy densities at new time t^{n+1}. These 36 simultaneous algebraic equations can be solved for E^{n+1} by a matrix solver. There are many methods for solving simultaneous algebraic equations; we recommend the Incomplete Cholesky-Conjugate Gradient (ICCG) method as given in Ref. [6.6]. Once we have the radiation energy density E^{n+1} for every cells we can use Eq. (6.68) to obtain the material temperature T^{n+1}. The finite difference for Eq. (6.68) is

$$\rho C_v \frac{T_{k,\ell}^{n+1} - T_{k,\ell}^n}{\Delta t} = c\sigma_a E_{k,\ell}^{n+1} - c\sigma_a a (T_{k,\ell}^n)^4 , \qquad (6.121)$$

or

$$T_{k,\ell}^{n+1} = T_{k,\ell}^n + \frac{\Delta t}{\rho C_v} c\sigma_a [E_{k,\ell}^{n+1} - a(T_{k,\ell}^n)^4] . \qquad (6.122)$$

Equation (6.122) is valid only when C_v=constant. Equation (6.68) can be written as

$$\frac{\partial E_m}{\partial t} = c\sigma_a E - c\sigma_a a T^4, \qquad (6.123)$$

Let

$$E_{absorbed} = c(\Delta t)\sigma_a E, \qquad (6.124)$$

which is the energy absorbed by the matter from the radiation field. Therefore, Eq. (6.123) becomes

$$(E_m^{n+1})_{k,\ell} = (E_m^n)_{k,\ell} + (E_{absorbed})_{k,\ell} - (\Delta t)c\sigma_a a(T_{k,\ell}^n)^4. \qquad (6.125)$$

Once we have $(E_m^{n+1})_{k,\ell}$, the temperature can be solved through EOS, that is

$$T_{k,\ell}^{n+1} = (E_m^{n+1})_{k,\ell}/C_v. \qquad (6.126)$$

6.6.3 Monte Carlo Method

In this section, the Monte Carlo method will be used to solve Eq. (6.67). In the previous section, it has been shown that Eq. (6.112) is the finite difference approximation of Eq. (6.67) for the interior zones. From Eq. (6.112), let us define

$$\hat{d}_{k,\ell} = S_8 = 1 + (\Delta t)(S_6 + S_1 - 2S_2 - 2S_3 - S_4), \qquad (6.127)$$

$$\hat{f}_{k,\ell}^+ = S_{10}/S_8 = S_{10}/(\hat{d}_{k,\ell}) \quad \text{for zone (k+1,ℓ)}, \qquad (6.128)$$

$$\hat{f}_{k,\ell}^- = S_9/S_8 = S_9/(\hat{d}_{k,\ell}) \quad \text{for zone (k-1,ℓ)}, \qquad (6.129)$$

$$\hat{f}_{k,\ell}^{up} = S_{11}/S_8 = S_{11}/(\hat{d}_{k,\ell}) \quad \text{for zone (k,$\ell+1$)}, \qquad (6.130)$$

$$\hat{f}_{k,\ell}^{dn} = S_{12}/S_8 = S_{12}/(\hat{d}_{k,\ell}) \quad \text{for zone (k,$\ell-1$)}, \qquad (6.131)$$

$$\hat{f}_{k,\ell}^{upR} = S_{13}/S_8 = S_{13}/(\hat{d}_{k,\ell}) \quad \text{for zone (k+1,$\ell+1$)}, \qquad (6.132)$$

$$\hat{f}_{k,\ell}^{dnR} = S_{14}/S_8 = S_{14}/(\hat{d}_{k,\ell}) \quad \text{for zone (k+1,$\ell-1$)}. \qquad (6.133)$$

Therefore, the probability of the particle in zone (k, ℓ) will jump to zone $(k+1, \ell)$ is

$$p_{k,\ell}^+ = \hat{f}_{k,\ell}^+ / (\hat{d}_{k,\ell}). \tag{6.134}$$

The probability of the particle in zone (k, ℓ) will jump to zone $(k-1, \ell)$ is

$$p_{k,\ell}^- = \hat{f}_{k,\ell}^- / (\hat{d}_{k,\ell}). \tag{6.135}$$

The probability of the particle in zone (k, ℓ) will jump to zone $(k, \ell+1)$ is

$$p_{k,\ell}^{up} = \hat{f}_{k,\ell}^{up} / (\hat{d}_{k,\ell}). \tag{6.136}$$

The probability of the particle in zone (k, ℓ) will jump to zone $(k, \ell-1)$ is

$$p_{k,\ell}^{dn} = \hat{f}_{k,\ell}^{dn} / (\hat{d}_{k,\ell}). \tag{6.137}$$

The probability of the particle in zone (k, ℓ) will jump to zone $(k+1, \ell+1)$ is

$$p_{k,\ell}^{upR} = \hat{f}_{k,\ell}^{upR} / (\hat{d}_{k,\ell}). \tag{6.138}$$

The probability of the particle in zone (k, ℓ) will jump to zone $(k+1, \ell-1)$ is

$$p_{k,\ell}^{dnR} = \hat{f}_{k,\ell}^{dnR} / (\hat{d}_{k,\ell}). \tag{6.139}$$

The probability that the energy of a particle will be tallied (resulted) into the array representing the solution of $E_{k,\ell}^{n+1}$ is

$$p_{k,\ell}^c = 1/(\hat{d}_{k,\ell}). \tag{6.140}$$

The probability that the energy of a particle will be tallied (resulted) into the array representing the material energy $(E_m^{n+1})_{k,\ell}$ is

$$p_{k,\ell}^a = \frac{c(\Delta t)\sigma_a}{\hat{d}_{k,\ell}}, \tag{6.141}$$

or

$$p^a_{k,\ell} = 1 - p^+_{k,\ell} - p^-_{k,\ell} - p^{up}_{k,\ell} - p^{dn}_{k,\ell} - p^{upR}_{k,\ell} - p^{dnR}_{k,\ell} - p^c_{k,\ell}. \qquad (6.142)$$

The probabilities presented in Eqs. (6.134)–(6.142) will be used for calculating the particle movement for the particle resided in the interior zones as shown on Fig. 6.10. The interior zones are confined in $K = 2, K = KMX - 1, L = 2$ and $L = LMX - 1$.

For the boundary zones, for example, the particle movement of these zones bounded by $K = 1$ and $K = 2$ for $L = 2$ to $L = LMX - 1$ will be calculated using the probabilities obtained from Eq. (6.113) similar to the probabilities as shown in Eqs. (6.134)–(6.142) for the interior zones. From Eq. (6.113), we define

$$\hat{d}_{k,\ell} = S_8 = 1 + (\Delta t)(S_6 + S_1 - 2S_2 - 2S_3 - S_4), \qquad (6.143)$$

$$\hat{f}^+_{k,\ell} = S_{10}/S_8 = S_{10}/(\hat{d}_{k,\ell}) \quad \text{for zone}(k+1,\ell), \qquad (6.144)$$

$$\hat{f}^{up}_{k,\ell} = S_{11}/S_8 = S_{11}/(\hat{d}_{k,\ell}) \quad \text{for zone}(k,\ell+1), \qquad (6.145)$$

$$\hat{f}^{dn}_{k,\ell} = S_{12}/S_8 = S_{12}/(\hat{d}_{k,\ell}) \quad \text{for zone}(k,\ell-1), \qquad (6.146)$$

$$\hat{f}^{upR}_{k,\ell} = S_{13}/S_8 = S_{13}/(\hat{d}_{k,\ell}) \quad \text{for zone}(k+1,\ell+1), \qquad (6.147)$$

and

$$\hat{f}^{dnR}_{k,\ell} = S_{14}/S_8 = S_{14}/(\hat{d}_{k,\ell}) \quad \text{for zone}(k+1,\ell-1). \qquad (6.148)$$

Therefore, the probability of the particle in zone (k,ℓ) will jump to zone $(k+1,\ell)$ is

$$p^+_{k,\ell} = \hat{f}^+_{k,\ell}/(\hat{d}_{k,\ell}). \qquad (6.149)$$

The probability of the particle in zone (k, ℓ) will jump to zone $(k, \ell+1)$ is

$$p_{k,\ell}^{up} = \hat{f}_{k,\ell}^{up}/(\hat{d}_{k,\ell}). \tag{6.150}$$

The probability of the particle in zone (k, ℓ) will jump to zone $(k, \ell-1)$ is

$$p_{k,\ell}^{dn} = \hat{f}_{k,\ell}^{dn}/(\hat{d}_{k,\ell}). \tag{6.151}$$

The probability of the particle in zone (k, ℓ) will jump to zone $(k+1, \ell+1)$ is

$$p_{k,\ell}^{upR} = \hat{f}_{k,\ell}^{upR}/(\hat{d}_{k,\ell}). \tag{6.152}$$

The probability of the particle in zone (k, ℓ) will jump to zone $(k+1, \ell-1)$ is

$$p_{k,\ell}^{dnR} = \hat{f}_{k,\ell}^{dnR}/(\hat{d}_{k,\ell}). \tag{6.153}$$

The probability that the energy of a particle will be tallied (resulted) into the array representing the solution of $E_{k,\ell}^{n+1}$ is

$$p_{k,\ell}^{c} = 1/(\hat{d}_{k,\ell}). \tag{6.154}$$

The probability that the energy of a particle will be tallied (resulted) into the array representing the material energy $(E_m^{n+1})_{k,\ell}$ is

$$p_{k,\ell}^{a} = \frac{c(\Delta t)\sigma_a}{\hat{d}_{k,\ell}}, \tag{6.155}$$

or

$$p_{k,\ell}^{a} = 1 - p_{k,\ell}^{+} - p_{k,\ell}^{up} - p_{k,\ell}^{dn} - p_{k,\ell}^{upR} - p_{k,\ell}^{dnR} - p_{k,\ell}^{c}. \tag{6.156}$$

If the boundary zones are inflow-boundary condition, then, we have to derive a new equation similar to Eq. (6.112) for these boundary zones. In general, the left boundary has an inflow of radiation energy, that means the zones on the left side of $K = 1$ have known radiation energy densities. The rest three boundaries, $L = LMX$, $L = 1$, and $K = KMX$ have non-flow boundaries.

6.6.4 Monte Carlo Procedure

To describe the Monte Carlo method for two-dimensional problems, let us assume that we want to calculate the radiation transport with the problem geometry as shown in Fig. 6.10. The left boundary has the strongest radiation energy density E_r, while the interior zones are with very low radiation energy initially. At time $t = 0$, we calculate the radiation energy source for each zone as

$$(E_{source}^{n=0})_{k,\ell} = E_{k,\ell}^{n=0} + (\Delta t)c(\sigma_a)_{k,\ell}(a)(T_{k,\ell}^{n=0})^4 . \qquad (6.157)$$

The total radiation energy source is

$$E_{total} = \sum_{k=2}^{kmx} \sum_{\ell=2}^{\ell mx} (E_{source}^{n=0})_{k,\ell} . \qquad (6.158)$$

Let N be the total number of particle in the whole problem, for example, say $N = 1000$, then, the radiation energy density of each particle is

$$E_{particle} = \frac{E_{total}}{N} . \qquad (6.159)$$

The number of particle in each zone is

$$N_{k,\ell} = \frac{(E_{source}^{n=0})_{k,\ell}}{E_{particle}} . \qquad (6.160)$$

If $N_{k,\ell} < 1$, then, we set $N_{k,\ell} = 0$. To avoid the $N_{k,\ell} < 1$, it is recommended to choose a large number N. Once the $N_{k,\ell}$ for each zone has been decided we can recompute the total number N, that is

$$N^* = \sum_{k=2}^{kmx} \sum_{\ell=2}^{\ell mx} N_{k,\ell} . \qquad (6.161)$$

We hope that N^* is very close to N appeared in Eq. (6.159). If N^* is not close to N, then, we should reset $N > 1000$ and repeat the calculations of Eqs. (6.160) and (6.161) until $N^* \approx N$. Now, we will start the Monte Carlo calculations.

Assume that we begin the calculation with the zone ($k = 2, \ell = 3$) which has the radiation energy density $E_{2,3}$ as shown in Fig. 6.10. Also, let $N_{2,3} = 5$ which means there are five particles in zone ($k = 2, \ell = 3$).

For calculating the motion of the particle, we will use Eqs. (6.149)–(6.156) to compute the probability for tracing the particle movement. Since Eqs. (6.149)–(6.156) are used for the boundary zones including zone ($k = 2, \ell = 3$). Let us pick one particle and choose a random number from the random number generator function routine. Then, we will compare the random number (between 1 and 0) with $p_{k,\ell}^{+}, p_{k,\ell}^{up}, p_{k,\ell}^{dn}, p_{k,\ell}^{upR}, p_{k,\ell}^{dnR}, p_{k,\ell}^{c}$ and $p_{k,\ell}^{a}$.

Let us assume that the random number is close to $p_{k,\ell}^{c}$, then, we will deposit the total energy of this particle into the array $E_{2,3}$, and this particle is disappeared.

Now, let us pick the second particle from zone ($k = 2, \ell = 3$) and draw another number from the random number generator program. This time, assume the drawn random number is close to $p_{k,\ell}^{upR}$, then, we will put this particle into the zone ($k = 3, \ell = 4$). Since zone ($k = 3, \ell = 4$) belongs to the interior zones, therefore, we will use the probabilities described in Eqs. (6.134)–(6.142) to decide the movement of this second particle. Now, we will draw a number from the random number generator program. Assume the drawn number is close to the probability presented in Eq. (6.142) which is $p_{3,4}^{a}$, then, we will put the total energy of this second particle into the array $(E_m)_{3,4}$ and the second particle is also disappeared.

Now, we will pick the third particle from zone ($k = 2, \ell = 3$) and trace the movement of this particle. Let us assume that the random number for this third particle is close to p^{+}, then, we will move this particle to zone ($k = 3, \ell = 3$). Since zone ($k = 3, \ell = 3$) belongs to the interior zones, we will use the probabilities from Eqs. (6.134)–(6.142) to decide the motion of this third particle in zone ($k = 3, \ell = 3$). At this moment, we choose a random number again. If this random number is close to p^{dnR}, then, we will move this particle to zone ($k = 4, \ell = 2$). Since zone ($k = 4, \ell = 2$) belongs to the lower boundary zone, we will use different set of probabilities to compute the motion of this particle inside zone ($k = 4, \ell = 2$).

We will use the same procedure as described previously to trace the motion of this third particle from zone ($k = 2, \ell = 3$), until this particle disappeared. That means this particle will be absorbed by either E or E_m, or runs outside the boundary and leaves the problem domain.

At this moment, we still have two particle remained inside the zone ($k = 2, \ell = 3$). We will use the same procedure to follow the motions of the last two particles until they are disappeared.

Now, we will calculate the motion of the particles inside the zone ($k = 3, \ell = 2$). Assume that there are seven particles inside the zone ($k = 3, \ell = 2$). We will use exactly the same method to handle these seven particles until they are all absorbed by either E or E_m or run out of the boundary zone and disappeared.

Now, we will trace the movement of the particles inside the zone ($k = 2, \ell = 4$) until all of the particles are gone. Using the same method, we will trace all 1000 particles (if $N = 1000$) until all of them disappeared. At this moment, we have finished the calculations for one time step which means we are now at $t = \Delta t$.

For the next time step, i.e., $t = \Delta t + \Delta t = 2\Delta t$ or $t = t^{n=2}$, the physical parameters E_r, E, E_m, T_m etc. at time level $t = t^{n=1}$ are already known. From the known variables $E_r^{n=1}, E^{n=1}, E_m^{n=1}, T_m^{n=1}$ etc., one will start to compute the next time-step variables $E_r^{n=2}, E^{n=2}, E_m^{n=2}, T_m^{n=2}$ etc. The energy source for each zone is available from

$$(E_{source}^{n=1})_{k,\ell} = E_{k,\ell}^{n=1} + (\Delta t)c(\sigma_a)_{k,\ell}(a)(T_{k,\ell}^{n=1})^4 . \qquad (6.162)$$

The total radiation energy for the whole problem is

$$E_{total} = \sum_{k=2}^{kmx} \sum_{\ell=2}^{\ell mx} (E_{source}^{n=1})_{k,\ell} . \qquad (6.163)$$

If one will choose the total number of particles for this second time step as $N = 500$, then, each particle has the energy

$$E_{particle} = \frac{E_{total}}{N} = \frac{E_{total}}{500} . \qquad (6.164)$$

From $(E_{source}^{n=1})_{k,\ell}$ and $E_{particle}$, one can decide the number of particle in each zone, that is

$$N_{k,\ell} = \frac{(E_{source}^{n=1})_{k,\ell}}{E_{particle}} , \qquad (6.165)$$

which is the same equation as given by Eq. (6.160). Next, one will obtain N^* from

$$N^* = \sum_{k=2}^{kmx} \sum_{\ell=2}^{\ell mx} N_{k,\ell} , \qquad (6.166)$$

until $N^* \approx N$ as described previously. Let the new N be $N = 600$, one will use the same procedure as described previously to calculate each particle which may deposit its energy into either E_r or E_m or go outside the problem domain and disappeared. Once all of these 600 particles have been computed the second time step is finished. One will repeats the same procedure until $t = t_{maximum}$ which is the termination time for the problem.

Bibliography

6.1 Schulz, WD. (1964). Two-dimensional Lagrangian hydrodynamic difference-equation, *Method Comput. Phys.*, 3, pp. 1-45.
6.2 Wallick, KB. (1987). *REZONE: A method for automatic rezoning in two-dimensional Lagrangian hydrodynamic problems*. Los Alamos National Laboratory report LA- 10829-MS.
6.3 Wilson, HL and Pomraning, GC. (1974). *Theoretical and numerical radiation hydrodynamics*, SAI-73-525-LJ, AD-530919.
6.4 Lee, WH. (2006). *Computer simulation of shaped charge problems*, World Scientific Publishing Co., Singapore.
6.5 Schulz, WD. (1963). *Two-dimensional Lagrangian hydrodynamic difference equation*, Lawrence Radiation Laboratory report UCRL-6776.
6.6 Kershaw, DS. (1978). The incomplete Cholesky-conjugate gradient method for the iteration solution of systems of linear equations, *J. Comput. Phys.*, 26, pp. 43-65.

Appendix A

Rezone for Two-Dimensional Lagrangian Hydrodynamic Code

Notations

c speed of light $(3 \times 10^{10} cm/s)$
E^m material internal energy per unit volume $(jerks/cm^3)$
$\dot{e}^{ij}$ strain rate deviator tensor $(1/shake)$
$\vec{F}$ radiation flux vector $(jerks/cm^2/shake)$
G shear modulus of elasticity $(jerks/cm^3)$
$I(\vec{r}, \nu, \vec{\Omega}, t)$ specific intensity of the radiation field defined as the rate of energy flow per unit frequency and solid angle across a unit area oriented normal to the direction of propagation at point $\vec{r}$, frequency ν, in the direction Ω, at time t $(jerks/cm^2)$
J volume Jacobian (cm^3)
j Jacobian of the transformation between (R, Z) and (k, ℓ) (cm^2)
K Lagrangian coordinate used in figures and difference equation
k Lagrangian coordinate
L Lagrangian coordinate used in figures and difference equation
ℓ Lagrangian coordinate
M mass constant (g)
P^m material pressure $(jerks/cm^3)$
$(P^r)^{ij}$ radiation energy tensor $(jerks/cm^3)$
Q artificial viscosity $(jerks/cm^3)$
$\widetilde{R}$ normal vector to $\vec{r}$, i.e., $\widetilde{R} = (Z, -R)$
$\hat{R} = R$ for cylindrical coordinate and $\hat{R} = 1$ for Cartesian coordinate
$\vec{r}$ position vector (R, Z)
S^{ij} stress deviator tensor $(jerks/cm^3)$
$\dot{S}^{ij}$ stress rate deviator tensor $(jerks/cm^3/shake)$
$S^{RR}, S^{ZZ}, S^{RZ}, S^{\theta\theta}$ stress deviator components $(jerks/cm^3)$
t time $(shake)$

u absolute value of the velocity vector, i.e., $u = |\vec{u}|$ ($cm/shake$)
$\vec{u}$ velocity vector (U, V) ($cm/shake$)
W energy source ($jerks/g/shake$)
$\dot{W}$ rate of energy source ($jerks/g/shake^2$)
Y^0 yield stress in simple tension ($jerks/cm^3$)

Greek letters

γ ratio of the specific heat, i.e., $\gamma = C_P/C_V$ (no unit)
δ^{RR}, δ^{ZZ} correction terms for the rigid body rotation ($jerks/cm^3/shake$)
ϵ specific internal energy per mass ($jerks/g$)
η weighting function defined in Ref. [6.1]
ξ weighting function defined in Ref. [6.1]
ρ density ($\frac{g}{cm^3}$)
τ specific volume ($\frac{cm^3}{g}$)
ϕ azimuthal angle (radian)
$\vec{\Omega}$ unit vector in the direction of the photon transport

Subscripts

0 initial value
R derivative with respect to R coordinate
t derivative with respect to time
Z derivative with respect to Z coordinate

Superscript

n time at n time-step, i.e., $t^n = t_0 + n \cdot \Delta t$

A.1 Introduction

In the Lagrangian method, a mesh is imbedded in and moves with the fluid, so a given zone always contain the same material. Lagrangian methods maintain good identification of materials, but often become difficult, if not impossible, to calculate when the mesh or fluid distorts. What is need in the Lagrangian calculation is some kind of rezoning or mixed Lagrangian/Eulerian calculation that can smooth out the distortions of the mesh without stopping the problem and that will disturb the final results as little as possible.

The hydrodynamics problem is run for a time with the pure Lagrangian code. While the problem is running, the mesh is monitored for distortion.

When the tests indicate that the mesh is distorted sufficiently to require a rezone, the hydrodynamics calculation is stopped. The REZONE code then takes over to smooth out the mesh with no time change during the rezoning operations. Only those points in the mesh that show distortion are rezoned. The new mesh is passed back to the hydrodynamics code for more time-dependent Lagrangian calculations until more distortion appears. All this is done automatically.

This rezone method does not attempt to reconstruct an entire mesh or to keep the zoning to some theoretical optimum shape. The principal object is to keep problems running while disturbing the normal calculation as little as possible. The overall shape of the mesh will still indicate the natural flow of the hydrodynamics problem.

This method does not reconstruct a mesh that is already in bad shape. The rezone operates by making small modifications in selected areas of the mesh during the entire run of the problem. It is not always necessary for the rezone to be used in the entire mesh of a problem. Often, using the rezone in selected areas of a mesh is very effective. Partial use of the rezone also will reduce any possible disturbance of the normal calculation of the hydrodynamics problem.

A.2   The REZONE Model

A.2.1   Zone and Point Model

The general model is that of a mesh consisting of quadrilateral zones imbedded in the material to be studied. About each vertex we have four zones and four adjacent vertices as shown in Fig. A.1. In this model the kinetic energy equation is applied to the points and the internal energy equation to the zones.

Thus, associated with the points we have the coordinates: $r_x, z_x; r_A, z_A; r_B, z_B; \ldots$ and the velocities: $\dot{r}_x, \dot{z}_x; \dot{r}_A, \dot{z}_A; \dot{r}_B, \dot{z}_B; \ldots$. For the zones we have internal energies: $\epsilon_1, \epsilon_2, \ldots$ and densities: $\rho_1, \rho_2, \ldots$.

A.2.2   The Mass Model

Mass is a doubly used quantity that must be associated with the points in someway and with the zones in some different but consistent way. To do this we adopt a model in which each zone is thought of as divided into four "sub-zones" obtained by joining a "point 8" within the zone to the midpoints of the sides. Point 8 is the average of the four corner points of the zone.

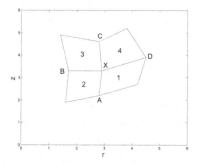

Fig. A.1 The zone model with vertex x surrounded by zones 1, 2, 3, and 4 and vertices A, B, C, and D.

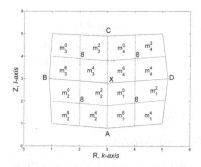

Fig. A.2 The mass model with vertex X surrounded by zones 1, 2, 3, and 4 and each zone has four sub-zones.

Quantities relating to the four sub-zones are denoted by superscripts 0, 2, 4, 6 numbered clockwise about the zone as shown in Fig. A.2. The masses of the sub-zones are calculated from the original mesh and are carried along in the problem. They will change when the REZONE code is used. For velocity and kinetic energy calculations, the sub-zones are considered to be associated with the adjacent vertex. For internal energy and pressure calculations, the sub-zones are considered to belong to the zone in which they lie.

For example, in Fig. A.2, the masses $m_1^0, m_2^2, m_3^4, m_4^6$, are assumed to have the velocities $\dot{r}_x, \dot{z}_x$ of the vertex X, while the masses $m_1^0, m_1^2, m_1^4, m_1^6$ are assumed to have the internal energy ϵ and density ρ of zone 1. The

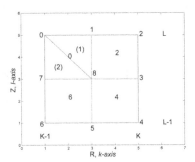

Fig. A.3 Sub-zone definition with points 1, 3, 5, and 7 the midpoint of the sides of the zone and point 8 the average of the four corners.

total energy of a given sub-zone is then given by

$$TE = m[0.5(\dot{r}^2 + \dot{z}^2) + \epsilon]. \qquad (A.1)$$

A.2.3 Sub-zone Definition

A.2.3.1 Description

For the REZONE code, each zone in the mesh is divided into four sub-zones, as shown in Fig. A.3.

Points 1, 3, 5, and 7 are the midpoints of the sides of the zone. Point 8 is the average of the four corners. At setup time the sub-zone masses are calculated and stored. These masses are used for the entire run of the problem, subject to changes made by the REZONE code. During the run, sub-zone volume are calculated for each zone in the entire mesh every time a rezone is called.

A.2.3.2 Equations

Sub-zone 0 (see Fig. A.3) is used as an example. Each sub-zone is divided into two triangles for calculations.

$$\text{Vol. tri. } (1) = \pi[(z_0 - z_8)(r_1 - r_8) - (z_1 - z_8)(r_0 - r_8)]\frac{(r_0 + r_1 + r_8)}{3},$$
$$(A.2)$$

$$\text{Vol. tri. } (2) = \pi[(z_0 - z_7)(r_8 - r_7) - (z_8 - z_7)(r_0 - r_7)]\frac{(r_0 + r_7 + r_8)}{3},$$
$$(A.3)$$

$$\text{Vol. subzone } 0 = \text{Vol. tri.}(1) + \text{Vol. tri.}(2), \tag{A.4}$$

$$\text{Mass subzone } 0 = \rho_{K-\frac{1}{2}, L-\frac{1}{2}} \times (\text{Vol. subzone } 0). \tag{A.5}$$

A.3 A Brief Description of the Rezone Method

A.3.1 The REZONE Code

The REZONE code is organized into several passes as follows:
1st pass: the displacement pass.
2nd pass: the expansion pass.
3rd pass: the vertex pass.
4th pass: the midpoint pass.
5th pass: the point 8 pass.
6th pass: the velocity adjustment pass.
7th pass: the averaging pass.

The displacement pass goes through the entire mesh, calculating the new position for all the points to be rezoned. For each vertex K, L and the related zone $K - 1/2, L - 1/2$, i.e., Zone 2 in Fig. A.1, and its sub-zones (see Fig. A.3), we will need the following quantities:
The "old" coordinates: $r_{K,L}, z_{K,L}$.
The sub-zone masses: $m_2^0, m_2^2, m_2^4, m_2^6$.
The midpoint coordinates: $(r_1)_2, (z_1)_2, (r_3)_2, (z_3)_2, (r_5)_2, (z_5)_2, (r_7)_2, (z_7)_2$.
The point 8 coordinates: $(r_8)_2, (z_8)_2$.
The sub-zone volumes: $v_2^0, v_2^2, v_2^4, v_2^6$.
The sub-zone internal energies: $E_2^0, E_2^2, E_2^4, E_2^6$.
The sub-zone directed kinetic energies($=\frac{2KE}{m}$):
$\dot{r}_2^0|\dot{r}_2^0|, \dot{z}_2^0|\dot{z}_2^0|, \dot{r}_2^2|\dot{r}_2^2|, \dot{z}_2^2|\dot{z}_2^2|, \dot{r}_2^4|\dot{r}_2^4|, \dot{z}_2^4|\dot{z}_2^4|, \dot{r}_2^6|\dot{r}_2^6|, \dot{z}_2^6|\dot{z}_2^6|$.
The "new" coordinates: $r^*_{K,L}, z^*_{K,L}$. The definition of directed kinetic energy is described in Section A.15.

A.3.2 The Displacement Pass

The purpose of this pass is to obtain the new coordinates r^*, z^* for each point that is to be moved. There are several ways to find this new position, but the basic objective is to displace some or all of the vertices in the mesh in such a way as to reduce the distortion.

For example, in the simple case shown in Fig. A.4, it appears that if the vertex X were to be displaced to some point X^* the mesh would be less distorted.

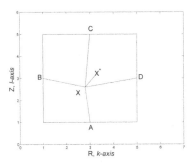

Fig. A.4 Simple displacement example with X^* the new vertex point.

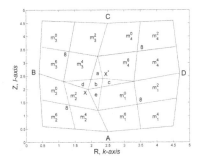

Fig. A.5 Vertex pass with X^* the new vertex point.

A.3.3 The Expansion Pass

The purpose of this pass is to obtain the quantities list in Section A.3.1. Most are obtained directly, and some require calculation, but all can be obtained logically if the model is kept in mind.

A.3.4 The Vertex Pass

The purpose of this and the next two passes is to shift the mesh through the fluid while keeping exact account of the volume, mass, and energy that will be transferred between sub-zones in the mesh. For this discussion, it will be assumed that mass, internal energy, and directed kinetic energy will be conserved.

In changing the mesh we are changing the sub-zones in which the various masses and energies are located, but we are not moving the fluid. Consider the simple example shown in Fig. A.4 with the various sub-zones shown in Fig. A.5. If the new vertex, X^*, is joined to the midpoints of the sides which meet at X, a number of triangles and quadrilaterals are defined. Denote the volumes of these elements (which can be calculated exactly) by $\Delta v_a, \Delta v_b, \Delta v_c, \Delta v_d$, and Δv_e. If the vertex X is shifted to X^* with the midpoints, T and W, of the sides held constant, the volumes Δv will be shifted from one sub-zone to another. For example, Δv_a will be shifted from sub-zone (4, 6) in zone 4 to sub-zone (3, 4) in zone 3.

On this pass all transfers will be among the four sub-zones (1, 0), (2, 2), (3, 4), and (4, 6). It is now possible to write equations conserving volume, mass, internal energy, and directed kinetic energy. The equations for volume transfer will be as follows. (Denote quantities before transfer by -, those after by +.)

$$(v_1^0)^+ = (v_1^0)^- + \Delta v_c - \Delta v_e, \tag{A.6}$$

$$(v_2^2)^+ = (v_2^2)^- + \Delta v_b + \Delta v_d + \Delta v_e, \tag{A.7}$$

$$(v_3^4)^+ = (v_3^4)^- + \Delta v_a - \Delta v_d, \tag{A.8}$$

$$(v_4^6)^+ = (v_4^6)^- - \Delta v_a - \Delta v_b - \Delta v_c. \tag{A.9}$$

The corresponding masses transferred will be given by

$$\Delta m_a = (m_4^6)^- \frac{\Delta v_a}{(v_4^6)^-}, \tag{A.10}$$

etc., and the mass transfer equations (which are just like v equations) by

$$(m_1^0)^+ = (m_1^0)^- + \Delta m_c - \Delta m_e, \tag{A.11}$$

etc. The internal energies transferred will be given by

$$\Delta E_a = (E_4^6)^- \frac{\Delta v_a}{(v_4^6)^-}, \tag{A.12}$$

etc., and the corresponding energy transfer equations, which again are like the v equation will be given by

$$(E_1^0)^+ = (E_1^0)^- + \Delta E_c - \Delta E_e, \tag{A.13}$$

etc. Assuming that when a given mass, Δm, is transferred out of a sub-zone with the velocities $\dot{r}, \dot{z}$, the corresponding directed kinetic energy losses are $\Delta m(\dot{r}|\dot{r}|), \Delta m(\dot{z}|\dot{z}|)$, we can write the equations of conservation of the r component of the kinetic energy by

$$(m_1^0)^+(\dot{r}_1^0|\dot{r}_1^0|)^+ = (m_1^0)^-(\dot{r}_1^0|\dot{r}_1^0|)^- + \Delta m_c(\dot{r}_4^6|\dot{r}_4^6|)^- - \Delta m_e(\dot{r}_1^0|\dot{r}_1^0|)^-, \quad (A.14)$$

$$(m_2^2)^+(\dot{r}_2^2|\dot{r}_2^2|)^+ = (m_2^2)^-(\dot{r}_2^2|\dot{r}_2^2|)^- + \Delta m_b(\dot{r}_4^6|\dot{r}_4^6|)^-$$

$$+ \Delta m_d(\dot{r}_3^4|\dot{r}_3^4|)^- + \Delta m_e(\dot{r}_1^0|\dot{r}_1^0|)^-, \quad (A.15)$$

$$(m_3^4)^+(\dot{r}_3^4|\dot{r}_3^4|)^+ = (m_3^4)^-(\dot{r}_3^4|\dot{r}_3^4|)^- + \Delta m_a(\dot{r}_4^6|\dot{r}_4^6|)^- - \Delta m_d(\dot{r}_3^4|\dot{r}_3^4|)^-, \quad (A.16)$$

$$(m_4^6)^+(\dot{r}_4^6|\dot{r}_4^6|)^+ = (m_4^6)^-(\dot{r}_4^6|\dot{r}_4^6|)^- - \Delta m_a(\dot{r}_4^6|\dot{r}_4^6|)^-$$

$$- \Delta m_b(\dot{r}_4^6|\dot{r}_4^6|)^- - \Delta m_c(\dot{r}_4^6|\dot{r}_4^6|)^-. \quad (A.17)$$

These equations can be solved for the $\dot{r}|\dot{r}|$'s. There is a similar set of equations for the $\dot{z}$ component with $\dot{r}$ replaced by $\dot{z}$. For the vertex pass, it is not necessary to calculate directed kinetic energy conservation since the four sub-zones concerned all have the same sub-zone velocity. This

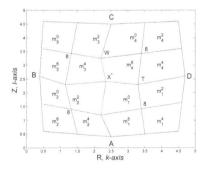

Fig. A.6 Midpoint pass with W and T the midpoints.

completes the calculation of the vertex pass. We have considered only the simple case shown in Fig. A.4. There are a number of other cases, depending on the relative orientation of XX^* with the sides.

A.3.5 The Midpoint Pass

After the vertex pass, the starting mesh of Fig. A.5 will now take on a shape like that represented in Fig. A.6. The lines joining the vertices will no longer be straight but will consist of two line segments joining the vertices to the old midpoints. The purpose of the midpoint pass is to straighten these lines (that is, move "old" midpoints to new midpoints.)

Since, in this pass, we deal with four sub-zones around any given midpoint, we can use exactly the same equations and code that were used on the vertex pass to transfer elements of volume, mass, and energy.

A.3.6 The Point 8 Pass

After the midpoint pass, the mesh will again consist of straight lines joining the vertices with less distortion than before the vertex and midpoint passes. However, the point 8 is no longer the average of the four corner points, and the distribution of the total mass of the zone among the sub-zones may be unrealistic. Therefore, we now compute a new point 8 and move mass, etc., between the sub-zones of each zone. Since each point 8 is surrounded by four sub-zones, it is possible to use the same transfer equations and code that were used for the vertex pass and the midpoint pass.

A.3.7 The Velocity Adjustment Pass

When directed kinetic energy conservation is chosen, velocities of the sub-zones will be changed. Having points with their movement constrained in some manner will result in incorrect velocities. The velocity adjustment pass check for these constraints and where necessary replaces the $\dot{r}|\dot{r}|$ or $\dot{z}|\dot{z}|$ of the sub-zones with the appropriate values from the constrained points. This pass is not used if velocity interpolation is chosen.

A.3.8 The Averaging Pass

After the previous passes, the mesh surrounding the moved mesh point has been completely redefined with new positions for the vertices, midpoint, and point 8's. However, the sub-zones are still represented as entities.

Returning to the hydrodynamics code after the rezone is complete requires most of the sub-zone quantities to be averaged to fit them back into the model for the main hydrodynamics code. For the zones, a new ρ, ϵ, and P are required. As an example, for zone 1, the new density is given by

$$\rho_1 = \frac{m_1^0 + m_1^2 + m_1^4 + m_1^6}{v_1^0 + v_1^2 + v_1^4 + v_1^6} . \qquad (A.18)$$

The new energy is giving by

$$\epsilon_1 = \frac{E_1^0 + E_1^2 + E_1^4 + E_1^6}{m_1^0 + m_1^2 + m_1^4 + m_1^6} . \qquad (A.19)$$

Pressure is then obtained from an equation of state. The new zone mass can be obtained from the new zone density and the new zone volume. New velocities are needed for the vertices, which are obtained from the $\dot{r}|\dot{r}|$'s and $\dot{z}|\dot{z}|$'s as follows:

$$\dot{r}_{X^*}|\dot{r}X^*| = \frac{m_1^0 \dot{r}_1^0 |\dot{r}_1^0| + m_2^2 \dot{r}_2^2 |\dot{r}_2^2| + m_3^4 \dot{r}_3^4 |\dot{r}_3^4| + m_4^6 \dot{r}_4^6 |\dot{r}_4^6|}{m_1^0 + m_1^2 + m_1^4 + m_1^6} . \qquad (A.20)$$

There is a similar equation for $\dot{z}_{X^*}|\dot{z}X^*|$, and

$$\dot{r}_{X^*} = |\dot{r}_{X^*}|\dot{r}X^*||^{1/2} , \qquad (A.21)$$

with the sign of $\dot{r}_{X^*}|\dot{r}X^*|$. Similarly for $\dot{z}_{X^*}$.

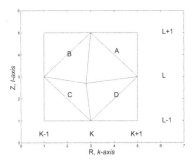

Fig. A.7   ADOT tests-case 1.

A.4 Testing for a Rezone

A.4.1 *Philosophy*

All rezoning disturbs a mesh, so rezoning should be done only when absolutely necessary and only on points that really need it. That is, sections of a mesh that are in reasonably good shape should not be rezoned even though rezoning would improve the appearance of the mesh. The test for rezoning should be designed with this in mind. The tests used in this rezone are designed to look for areas of a mesh that are distorting badly and for areas where there is a probability of points crossing.

A major purpose of this testing method is to keep the number of necessary rezones to a minimum. When a rezone is done, all points where there is any indication of trouble are included.

A.4.2 *Test Details — General Case*

During each calculation cycle of the hydrodynamics code, all active points are monitored for distortion. When the tests indicate that a rezone is needed, the hydrodynamics code is stopped. The REZONE code then takes over to smooth out the mesh in the indicated region. Several tests are used to located the distortions. Probably a number of additional tests might be used. Other tests may be added in future versions of the code.

A.4.2.1 *The ADOT Tests*

a. The areas of the four triangles A, B, C, and D (as shown in Fig. A.7) surrounding point K, L are tested to see how fast the areas are going to zero.

Triangle A will be used as an example. The other triangles are calculated similarly. Let the area of triangle A $= A$ and $\frac{dA}{dt} = \dot{A}$.

$$2A = (z_{K,L+1} - z_{K,L})(r_{K+1,L} - r_{K,L}) - (z_{K+1,L} - z_{K,L})(r_{K,L+1} - r_{K,L}), \quad \text{(A.22)}$$

$$2\dot{A} = (z_{K,L+1} - z_{K,L})(\dot{r}_{K+1,L} - \dot{r}_{K,L}) + (\dot{z}_{K,L+1} - \dot{z}_{K,L})(r_{K+1,L} - r_{K,L})$$
$$- (z_{K+1,L} - z_{K,L})(\dot{r}_{K,L+1} - \dot{r}_{K,L})$$
$$- (\dot{z}_{K+1,L} - \dot{z}_{K,L})(r_{K,L+1} - r_{K,L}). \quad \text{(A.23)}$$

If $A > 0$ and $\dot{A} < 0$, the time it would be until A goes to zero with the current velocities would be

$$\Delta t_A = \frac{-A}{\dot{A}}. \quad \text{(A.24)}$$

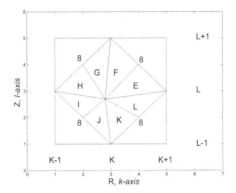

Fig. A.8 ADOT tests-case 2.

The number of cycles required would be

$$k = \frac{\Delta t_A}{\Delta t_H}, \qquad (A.25)$$

where Δt_H is the problem time step.

If $k \leq 10$: the point is marked for rezoning if a rezone is called on that cycle. If $k \leq 4$: the point is marked for rezoning and a flag is set to force a rezone on that cycle. Note: If $\dot{A} > 0$ the triangle is expanding and the test does not apply. If $A \leq 0$, the test can not be used and the point is automatically marked for rezoning.

b. The areas of the eight triangles E, F, G, H, I, J, K, and L (as shown in Fig. A.8) surrounding the point K, L are tested to see how fast the areas are going to zero. Point 8 in each zone is the centroid of the zone with the coordinates

$$r^8_{K+1/2,L+1/2} = 0.25(r_{K,L} + r_{K+1,L} + r_{K+1,L+1} + r_{K,L+1}), \qquad (A.26)$$

$$z^8_{K+1/2,L+1/2} = 0.25(z_{K,L} + z_{K+1,L} + z_{K+1,L+1} + z_{K,L+1}), \qquad (A.27)$$

etc., similarly for $\dot{r}_8$ and $\dot{z}_8$.

The number of cycles to go to zero, k, is calculated similarly to that described in Eq. A.25. If $k \leq 9$: the point is marked for rezoning if a rezone is called on that cycle. If $k \leq 4$: the point is marked for rezoning and a flag is set to force a rezone on that cycle.

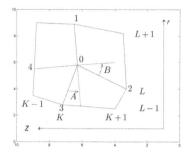

Fig. A.9 The 180° test.

A.4.2.2 The 180° Tests

This tests how far opposite sides have moved from 180°. Given a general point in the mesh surrounded by four zones (see Fig. A.9), we wish to test the size of angles A and B. For angle A

$$sin(1,3) = \frac{2|A_{013}|}{R_1 R_3}, \tag{A.28}$$

where

$$R_1 = [(r_1 - r_0)^2 + (z_1 - z_0)^2]^{\frac{1}{2}}, \tag{A.29}$$

$$R_3 = [(r_3 - r_0)^2 + (z_3 - z_0)^2]^{\frac{1}{2}}, \tag{A.30}$$

and

$$2A_{013} = (z_3 - z_0)(r_1 - r_0) - (z_1 - z_0)(r_3 - r_0). \tag{A.31}$$

For angle B,

$$sin(2,4) = \frac{2|A_{042}|}{R_2 R_4}, \tag{A.32}$$

where

$$R_2 = [(r_2 - r_0)^2 + (z_2 - z_0)^2]^{\frac{1}{2}}, \tag{A.33}$$

$$R_4 = [(r_4 - r_0)^2 + (z_4 - z_0)^2]^{\frac{1}{2}}, \tag{A.34}$$

and

$$2A_{042} = (z_2 - z_0)(r_4 - r_0) - (z_4 - z_0)(r_2 - r_0). \tag{A.35}$$

If $sin(1,3) > sin(\text{Anglemark})$ or $sin(2,4) > sin(\text{Anglemark})$, the point is marked for rezoning and will be repositioned if a rezone is done on that cycle. Anglemark may be a variable. The value that seems to give the best result is 30°. The 180° test is not used to force a rezone.

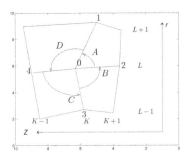

Fig. A.10   The 90° test.

A.4.2.3 The 90° Test

This tests how far the four angles adjacent to a point are from 90°. This test is very useful but can also cause considerable trouble. In problems where the initial mesh has nonorthogonal areas, points will be rezoned before there is any need. This can give a mesh that will not run well. It is found that the test gives the best results if it is used only on points where at least one of the four adjacent zones has an aspect ratio of greater than 30 to 1. Given a general point in the mesh surrounded by four zones (as shown in Fig. A.10), we wish to test the size of angles A, B, C, and D.

For angle A

$$sin(A) = \frac{2|A_{012}|}{R_1 R_2}, \tag{A.36}$$

where

$$R_1 = [(r_1 - r_0)^2 + (z_1 - z_0)^2]^{\frac{1}{2}}, \tag{A.37}$$

$$R_2 = [(r_2 - r_0)^2 + (z_2 - z_0)^2]^{\frac{1}{2}}, \tag{A.38}$$

and

$$2A_{012} = (z_1 - z_0)(r_2 - r_0) - (z_2 - z_0)(r_1 - r_0), \tag{A.39}$$

similarly for angles B, C, and D. If $sin(A), sin(B), sin(C)$, or $sin(D)$ is less than $sin(90° - \text{Animark})$, the point is marked for rezoning and will be repositioned if a rezone is done on that cycle. Animark may be a variable. The value that seems to give the best results is 40°. The 90° test is not used to force a rezone.

A.4.3 Tests on Boundaries

A.4.3.1 ADOT Tests

The ADOT tests are used on all boundary points. The calculation is the same as for the general case except that the tests are limited to the appropriate existing triangles inside the mesh.

A.4.3.2 The 180° Tests

a. Constrained Points. A point outside the mesh is assumed for points with a constant r or constant z constraint. The point reflects the point inside the mesh. The calculations for the test are the same as in the general case.
b. Free Surface Points. For points on a free surface, only the angle formed by the point and the two adjacent surface points is tested.

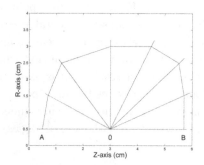

Fig. A.11 The center of mass test.

A.4.3.3 The 90° Test

The 90° test is not used on boundary points.

A.4.3.4 Test for Center of Mass Points

None of the standard tests will work for points at a center of mass. A special test based on the relative positions of the points adjacent to the center of mass points is used. The lengths of the sides AO and BO (as shown in Fig. A.11) are calculated and compared.

If $\dfrac{\text{length of } AO}{\text{length of } BO}$ or $\dfrac{\text{length of } BO}{\text{length of } AO}$ is greater than 2.0, the center of mass is marked to be rezoned if a rezone is done on that cycle.

If $\frac{\text{length of } AO}{\text{length of } BO}$ or $\frac{\text{length of } BO}{\text{length of } AO}$ is greater than 6.0, the center of mass is marked to be rezoned and a rezone is forced for that cycle.

A.4.4 Additions to the Tests

After all points have been tested and marked, the four zones adjacent to each marked point are examined. If any of these zones have an aspect ratio greater than 20 to 1, the two points adjacent to the marked point in the direction of the smallest zone dimension also marked for rezoning. This is also used on boundaries where only the existing zones and points are involved.

A.4.5 Limitations on the Testing

In many problems the ADOT tests will indicate the need for a rezone when there is no necessary for it. For example, in a spherical problem with a strong shock, the ADOT test will frequently give a result within the

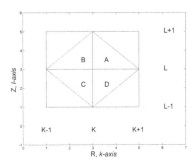

Fig. A.12   Relative ADOT test.

limits for forcing a rezone when there is no real distortion. This may cause perturbations in what should be a perfectly smooth run. Some method of preventing this must be included in the testing. The following checks are being used.

A.4.5.1 Relative ADOT Test

On each point, after the tests have been completed, the $\dot{A}$'s for the triangles A, B, C, and D (as shown in Fig. A.12) are compared in pairs to see if they

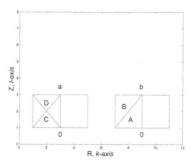

Fig. A.13   Relative area test.

are near the same value. We calculate

$$\text{PCTAB} = \left| \frac{\dot{A}_A - \dot{A}_B}{\dot{A}_A} \right|, \tag{A.40}$$

$$\text{PCTCD} = \left| \frac{\dot{A}_C - \dot{A}_D}{\dot{A}_C} \right|, \tag{A.41}$$

$$\text{PCTBC} = \left| \frac{\dot{A}_B - \dot{A}_C}{\dot{A}_B} \right|, \tag{A.42}$$

$$\text{PCTAD} = \left| \frac{\dot{A}_A - \dot{A}_D}{\dot{A}_A} \right|. \tag{A.43}$$

If PCTAB and PCTCD are both < PCADOT or PCTBC and PCTAD are both < PCADOT, then the results of the ADOT and 180° tests are ignored. The 90° test is not affected. The value used for PCADOT is 0.225. This check is good only for points interior to the mesh.

A.4.5.2   *Relative Area Test*

Since the relative ADOT test does not work for boundary points, a different approach must be used. For these points we compare relative areas. There are two tests. In Fig. A.13 we are testing point 0 for a rezone. When area A is tested by the ADOT test we compare the size of areas A and B (see Fig. A.13a). If these areas are within 10% of each other, we ignore the results of the ADOT test for that triangle. The same is true for the other triangles tested. In testing area C by the ADOT test, we compare the size of area C with area D (see Fig. A.13b). If these areas are within 10% of each other, we ignore the results of the ADOT tests for that triangle. The same is true for the other triangles tested.

A.4.6 Changes in Test Values

The values used in all of the tests described have been chosen as a compromise to work well with the implosion problems. For a different class of problems, a completely different set of standard values may be needed.

A.4.7 General Remarks

The tests for a rezone are very important. If too few or too many points are included or if rezones are done too often or too seldom, the entire method may fail. It is necessary that only the points that really need adjustment be included. The tests described in this section are certainly not the only methods that could be used. Other tests may be added in the future and

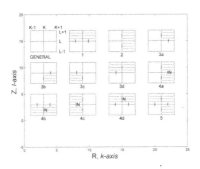

Fig. A.14 Displacement cases.

some of the current tests may be modified or abandoned. Finding the proper tests has been the most difficult and time-consuming part of developing this rezone method.

A.5   The Displacement Pass

The displacement pass, the first stage of the actual rezone process, calculates the displacement (new position) of the points that were marked for rezoning. A sweep is made through the entire mesh. The displacements are calculated for all points that have been marked for rezoning. All displacements are calculated before any other part of the rezone code is called. All displacement calculations are made on the basis of the current mesh positions of the neighboring points, taking into account any neighbors that have already been moved.

A.5.1 *Displacement Cases for the Interior Points*

At the start of each run the mesh is examined and a flag is set to identify the various cases for the displacement calculation (as shown in Fig. A.14). The flag is based on the changes of materials in the zones surrounding the point.

When an interface point is moved, the distortion of the interface should be minimized. Therefore, the general rule for interface displacements is that the new point must lie on the old interface.

A.5.1.1 *General Case*

For the general case the displacement is not constrained.

A.5.1.2 *Case 1 and Case 2*

These cases are standard interfaces. The displacement calculation is constrained. The new position of the point must lie on one of the surfaces marked I.

A.5.1.3 *Cases 3a, 3b, 3c, and 3d*

For these cases it is impossible to avoid distortion of the interface. The best results were obtained in most situations by allowing the displacement to be calculated as in the general case. However, there are some types of meshes where this calculation will cause trouble. These cases still need to be studied to find the best way to calculate the displacement.

A.5.1.4 *Cases 4a, 4b, 4c, and 4d*

For these cases, three surfaces are interface, those marked I and the one marked IN. The least distortion of the interfaces will occur if the new position of the point is constrained to lie on one of the surfaces marked I. This works well for all situations, but there still may be a better way to handle these cases.

A.5.1.5 Case 5

This case has interfaces on all four surfaces surrounding the point. There is no way of moving the point without distorting the interface. The case is treated the same as the general case.

A.5.2 Displacement on Boundary Points

Except for points on a free surface and on an exterior corner, all boundary points use the same displacement calculation. They are flagged as boundary points during the internal examination of the mesh.

A.5.3 The Displacement Method

The displacement used in this rezone method was first proposed and developed by Browne [A.1].

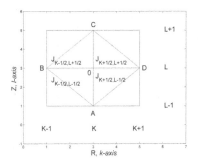

Fig. A.15 Displacement notation.

A.5.3.1 General Case

We compute a displacement by trying to minimize a sum of quantities

$$\frac{L^2}{J} = \frac{(z_j^2 + r_j^2) + (z_i^2 + r_i^2)}{J}. \tag{A.44}$$

a. Notation. Point $A, B, C,$ and D are the end points of the sides meeting at the point to be moved. $J_{K-\frac{1}{2}, L-\frac{1}{2}}$, etc., are the areas of the triangles adjacent to point 0 (as shown in Fig. A.15).

b. Calculation. For a general point as shown in Fig. A.15, we will calculate L^2/J for the adjacent triangles. The terms associated with K, L

in the summation to be minimized are

$$\left[\frac{L^2}{J}\right]_{K,L}^{K+\frac{1}{2},L+\frac{1}{2}}, \tag{A.45}$$

$$\left[\frac{L^2}{J}\right]_{K,L}^{K-\frac{1}{2},L+\frac{1}{2}}, \tag{A.46}$$

$$\left[\frac{L^2}{J}\right]_{K,L}^{K-\frac{1}{2},L-\frac{1}{2}}, \tag{A.47}$$

and

$$\left[\frac{L^2}{J}\right]_{K,L}^{K+\frac{1}{2},L-\frac{1}{2}}, \tag{A.48}$$

where

$$\left[\frac{L^2}{J}\right]_{K,L}^{K+\frac{1}{2},L+\frac{1}{2}} = \frac{\text{sum of squares of lengths of sides meeting at } K, L}{\text{Area of triangle in zone}(K+\frac{1}{2}, L+\frac{1}{2})\text{next to pt.}K, L}, \tag{A.49}$$

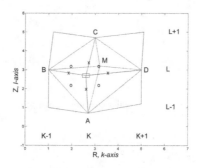

Fig. A.16 Test points for displacement.

or

$$\left[\frac{L^2}{J}\right]_{K,L}^{K+\frac{1}{2},L+\frac{1}{2}} = \frac{(r_C - r_0)^2 + (z_C - z_0)^2 + (r_D - r_0)^2 + (z_D - z_0)^2}{0.5[(z_C - z_0)(r_D - r_0) + (z_D - z_0)(r_C - r_0)]}. \tag{A.50}$$

The same is true for the other terms. To calculate the displacement, we do the following.

Step 1. Starting with the point K, L ($\square$) and the four adjacent zones (as shown in Fig. A.16), select eight neighboring points as follow. Four points (X) are chosen along the four sides meeting at K, L.

They are set a fractional distance f from K, L out to the next point in the mesh. f has an initial value of $f = 0.1$. Four more points (0) are chosen so that they lie midway between the X points.

Step 2. For each point of this cluster of 9 points (1'$\square$', 4'X', 4'0') the value of $\sum(L^2/J)$ is calculated using that point as the vertex K, L. (See above.) This gives nine values of $\sum(L^2/J)$.

Step 3. The case of $J \leq 0$. For any of the nine points, J may be ≤ 0 for any or all of the four surrounding triangles. Any point where this occurs must be ruled out, since use of $J > 0$ will probably make the corresponding $\sum$ a minimum when compared to $\sum(L^2/J)$ of the other points. So, for any $\sum(L^2/J)$ where $J \leq 0$ the value of 10^{200} is substituted for the $\sum(L^2/J)$. This eliminates this $\sum$ from consideration as a minimum. If a $J \leq 0$ occurs for all nine points, we set $f = f + 0.1$ and return to step 1. f is limited to ≤ 0.9. If $f > 0.9$ the point under consideration can not be moved.

Step 4. We find the minimum $\sum(L^2/J)$. If the minimum $\sum(L^2/J)$ occurs on any of the points x or o, that point is taken as a new $\square$ point and we return to step 1. If the minimum value of $\sum(L^2/J)$ occurs on

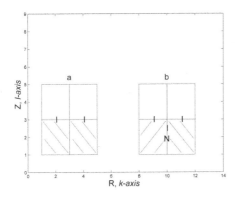

Fig. A.17 Configurations for cases 1, 2, 4a, 4b, 4c, and 4d.

the $\square$ point, the factor f is divided by 2 and we return to step 1. When $f < 5.0 \times 10^{-3}$ the minimum has been found and the current point $\square$ is taken as the new position of point K, L.

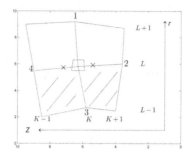

Fig. A.18   Test points for interfaces.

A.5.3.2 *Interface Cases*

In Section A.5.1, 12 displacement cases are described. The general case and cases 3a, 3b, 3c, 3d, and 5 use the general case displacement described above.

The remaining six cases are calculated as follows. Cases 1, 2, 4a, 4b, 4c, and 4d are basically the same. The basic configuration is as in either Fig. A.17a or Fig. A.17b with the new position of the point required to be on

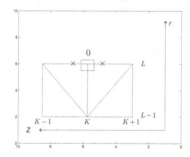

Fig. A.19   Two-zone case.

one of the surfaces marked I. The method of calculating the displacement is the same as described in the general case, except that instead of the nine points used in the general case, three points are used (as shown in Fig. A.18).

The rest of the procedure for minimizing $\sum(L^2/J)$ is as in the general case, except that as we progress to the new position all of the test points must be on the old interface.

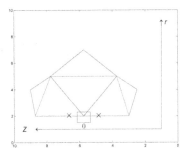

Fig. A.20 Three-zone case.

A.5.3.3 *Boundary Cases*

a. Constant r, constant z, and slide angle. Exterior corners are a separate case and are not included in this section. These boundary cases use the same method as the general case. The only difference is in how many triangles are involved in the $\sum(L^2/J)$ and in the number of test points used.

* Two-zone case. For the two-zone case we have two triangles adjacent to the point.

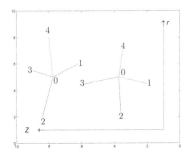

Fig. A.21 Free surface case.

We pick the initial three points as shown in Fig. A.19. On subsequent passes all of the points must lie on the boundary.

* Three-zone case. For the three-zone case we have three triangles

adjacent to the point. The initial points are as shown in Fig. A.20. Again, on the subsequent passes the test points must be on the boundary.

b. Free surface points. On free surface points (as shown in Fig. A.21) we construct a side 4 bisecting the angle on the free surface (side 1 and 3) the length of side 2. If either of the adjacent zones next to the free surface has an aspect ratio greater than 10 to 1, we set the length of side 4 equal to three times the length of side 2. After the fourth side has been set, we calculate the displacement by using the method of the general case.

c. Exterior corners. Rezoning on exterior corners is limited to corners that have one side constrained (constant r or constant z). For example, in Fig. A.22, side 1 is fixed to constant r and side 2 is free. We construct side 3 as a continuation of side 1 with length equal to that of side 1. If zone A has an aspect ratio greater than 10 to 1, we set the length of side 3 equal to three times that of side 1. The displacement is calculated using the two-zone boundary case method. All other corners are calculated similarly.

d. Center of mass points. The center of mass consists of a number of superimposed points, which must all be moved together. For the center of mass case we use the basic method described above. The initial three test

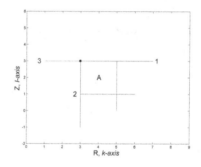

Fig. A.22   Corner case.

points are as shown in Fig. A.23. On subsequent passes all of the test points must lie on the boundary. We minimize

$$\sum_{\ell} \frac{L_\ell^2 + L_{\ell+1}^2}{J_{\ell+\frac{1}{2}}}, \qquad (A.51)$$

where the J's are the whole zone areas and the L^2's are the length squared of the adjacent sides.

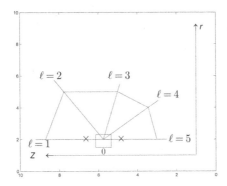

Fig. A.23 Center of mass case.

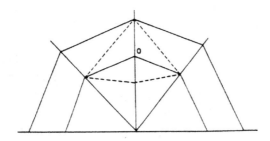

Fig. A.24 Special centroid calculation.

A.5.3.4 *Special Centroid Calculation*

When points next to a center of mass are rezoned, the points tend to collapse inward. This frequently will cause great difficulty in running as the zones in the center become smaller. For these points, we replace the end point of the line that joins the center of mass with a point $1/3$ of the distance from point 0 to the real end point. This point is then used to calculate the triangular areas (as shown in Fig. A.24) for the $\sum(L^2/J)$ calculation. The rest of the calculation is as in the general case.

A.5.4 *Limitations on the Displacement Calculations*

In addition to minimizing $\sum(L^2/J)$, the new position must satisfy a number of conditions.

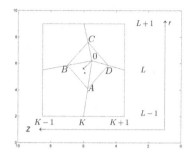

Fig. A.25 Displacement size test — general case.

A.5.4.1 *Displacement Size Test*

The new position of the point must be inside the quadrilateral formed by joining the midpoints of the sides that meet at the point being moved (as shown in Fig. A.25). The areas of the triangles *AB,* BC,* CD, and *DA are calculated. If all are greater than zero, the new point is acceptable. If any of the areas are less than zero, the distance 0* is divided by 2 and the test is started again from the beginning. This process is repeated up to 10 times if necessary. If, after cutting the size of the displacement 10 times, a negative area still exists, the point is set back to its original position and the rezone mark is removed from the points. The point will then be ignored for the rest of the rezone.

A.5.4.2 *Displacement Size Test on Boundary*

This is for the two-zone and three-zone cases. The principle of the size test on boundary points is the same as for the general case. The only difference is the number of triangles. For the two-zone case there are two triangles (Fig. A.26) to test. For the three-zone case there are three triangles (Fig. A.27). The free surface case is similar to the normal two-zone case, except that the displacement point * is not confined to the boundary. The principle is the same for the exterior corner case.

A.5.4.3 *Interface Points on Boundary*

If a point on a boundary is an interface point, it is first tested as described in Section A.5.2. There is then an additional test as follows:

a. The area formed by the points *OB is calculated to determine where point * is located (Fig. A.26). If the area is less than 0, the point * is on

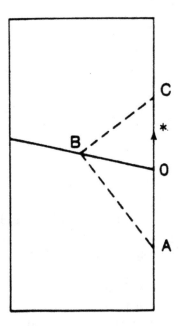

Fig. A.26 Displacement size test — two-zone case.

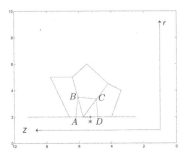

Fig. A.27 Displacement size test — three-zone case.

OC. If the area is greater than 0, the point * is on OA.
b. Suppose point * is on OC. Calculate the area formed by points OCB.
c. Calculate the area formed by point *CB.

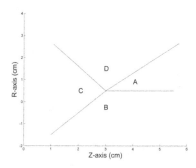

Fig. A.28 Angle change test.

d. If the area of $*CB < 0.75\times$ area of OCB, the displacement is divided by 2 and the process is repeated from step 2. This is repeated up to 10 times. If after 10 iterations, the condition is still not met, the point is set back to its original position and rezone mark is removed from the point. The same is true if the point $*$ is on OA. This test limits the displacement of interface point on a boundary such that the adjacent triangular area is reduced by less than 25%. This test is used only on boundary points constrained by constant r or constant z.

A.5.4.4 Bowtie Test

When all of the above conditions are met, we test to see whether the new position of the point will create a bowtie or boomerang in relation to the surrounding points. If either is formed, the displacement is divided by 2 and the test is repeated. This will be done up to 10 times. If after 10 tries a bowtie or a boomerang is still formed, the point is returned to its original position and the rezone mark is removed from the point.

A.5.4.5 Angle Change Test

For each point where a displacement is to be calculated the four angles formed by the sides meeting at the point are examined. The angle farthest from 90° is located. For example, in Fig. A.28 it would be angle A. After the new position of the point has been determined, the new angle at that position is examined. If the new angle is farther from 90° than it was before moving the point, the displacement of the point is divided by 2 and the new angle tested. This is done up to 10 times. If after 10 iterations, the angle is still farther from 90° than the initial angle, the point is put back to its original position and the rezone mark is removed. When all tests are satisfied, we have the new position of the point and are ready to proceed with the rezone process.

A.5.5 *Remarks*

This displacement method works very well for most situations. However, there are some types of mesh configurations where improvements are needed. The tendency for spherical sections to move inward needs to be modified and there often are difficulties on long thin zones, particularly when they are near interface. These situations are being studied.

A.6 Expansion Pass

Since the rezone method is based on the use of sub-zones, it is necessary to expand the normal mesh to a sub-zone format before starting the rezone process. The expansion pass create this sub-zone format.

A.6.1 *Introduction*

The marked points are processed one at a time by the rezone code. In the expansion pass, the zones surrounding the point to be moved are expanded into a 4 K line by 4 L line mesh. The point K, L being moved is located at position (2,2) in the sub-mesh. The sub-zone masses are carried throughout the run. The sub-zone volumes for the entire mesh are calculated each time the REZONE code is called and are carried until the rezone is finished. This sub-mesh is for the basic hydrodynamics quantities. Other quantities that may need to be mapped are treated in a similar manner.

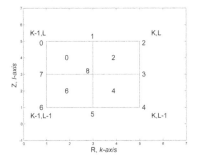

Fig. A.29   Sub-zone velocity definition.

A.6.2 Sub-mesh Storage

A.6.2.1 Zone Quantities

For each K, L point in the sub-mesh the following zone quantities are stored.
a. The K, L point r and z coordinates.
b. The material ID.
c. The zone internal energy ($\epsilon_{K,L}$).

A.6.2.2 Sub-Zone Quantities

For each K, L point in the sub-mesh the following sub-zone quantities are stored.

a. The 4 sub-zone masses associated with the point (from the full mesh array carried permanently).
b. The 4 sub-zone volumes associated with the point (from the full mesh array calculated at the start of the rezone).
c. The sub-zone-directed kinetic energies.

Each sub-zone (Fig. A.29) is assigned the velocities of the adjacent mesh point. These velocities are associated with the sub-zone mass for the duration of the rezone calculation. The directed kinetic energies are calculated for each sub-zone. For convenience in calculation, we carry only the velocity terms for each sub-zone in the sub-mesh.

$$\dot{r}^0_{K,L}|\dot{r}^0_{K,L}| = \dot{r}_{K-1,L}|\dot{r}_{K-1,L}|, \tag{A.52}$$

$$\dot{z}^0_{K,L}|\dot{z}^0_{K,L}| = \dot{z}_{K-1,L}|\dot{z}_{K-1,L}|, \tag{A.53}$$

$$\dot{r}^2_{K,L}|\dot{r}^2_{K,L}| = \dot{r}_{K,L}|\dot{r}_{K,L}|, \tag{A.54}$$

$$\dot{z}^2_{K,L}|\dot{z}^2_{K,L}| = \dot{z}_{K,L}|\dot{z}_{K,L}|, \tag{A.55}$$

$$\dot{r}^4_{K,L}|\dot{r}^4_{K,L}| = \dot{r}_{K,L-1}|\dot{r}_{K,L-1}|, \tag{A.56}$$

$$\dot{z}^4_{K,L}|\dot{z}^4_{K,L}| = \dot{z}_{K,L-1}|\dot{z}_{K,L-1}|, \tag{A.57}$$

$$\dot{r}^6_{K,L}|\dot{r}^6_{K,L}| = \dot{r}_{K-1,L-1}|\dot{r}_{K-1,L-1}|, \tag{A.58}$$

$$\dot{z}^6_{K,L}|\dot{z}^6_{K,L}| = \dot{z}_{K-1,L-1}|\dot{z}_{K-1,L-1}|. \tag{A.59}$$

A.6.2.3 The Midpoints of the Sides

The midpoints 1 and 3 (Fig. A.29) are calculated for each K, L point. Points 5 and 7 are points 1 and 3 from adjacent zones.

$$r^1_{K,L} = 0.5(r_{K-1,L} + r_{K,L}),\tag{A.60}$$

$$z^1_{K,L} = 0.5(z_{K-1,L} + z_{K,L}),\tag{A.61}$$

$$r^3_{K,L} = 0.5(r_{K,L} + r_{K,L-1}),\tag{A.62}$$

$$z^3_{K,L} = 0.5(z_{K,L} + z_{K,L-1}).\tag{A.63}$$

A.6.2.4 The Point 8 of the Zone (Fig. A.29)

$$r^8_{K,L} = 0.5(r^3_{K,L} + r^3_{K-1,L}),\tag{A.64}$$

$$z^8_{K,L} = 0.5(z^3_{K,L} + z^3_{K-1,L}).\tag{A.65}$$

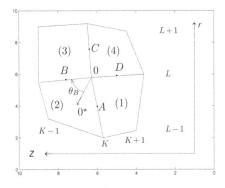

Fig. A.30 System definition.

A.6.2.5 The Sub-zone Internal Energies

Internal energy is conserved for both directed kinetic energy conservation and velocity interpolation.

$$E^0_{K,L} = \epsilon_{K,L} m^0_{K,L}, \qquad (A.66)$$

$$E^2_{K,L} = \epsilon_{K,L} m^2_{K,L}, \qquad (A.67)$$

$$E^4_{K,L} = \epsilon_{K,L} m^4_{K,L}, \qquad (A.68)$$

$$E^6_{K,L} = \epsilon_{K,L} m^6_{K,L}. \qquad (A.69)$$

A.6.2.6 Other Sub-zone Quantities

If there are other quantities that must be mapped, they also must be expanded to a sub-zone mesh such that fractions of the quantities may be moved from sub-zone to sub-zone.

A.7 Rezone Method — General Case

Before continuing with the description of the rezone passes, an explanation of the rezone calculation is appropriate. This discussion will assume that the vertex pass is being calculated. The other passes use the same method.

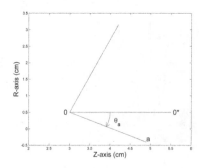

Fig. A.31   Correct angle selection (A and B).

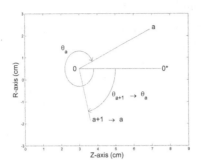

Fig. A.32 Correct angle selection (C and D).

A.7.1 Preparation — Definition of the System

Suppose that at a vertex 0 we are given a displacement 00* and the midpoints of the four sides meeting at 0, denote by A, B, C, and D. (Fig. A.30.) The zones surrounding point 0 are denoted by (1), (2), (3), and (4). We will denote by θ_A, θ_B, θ_C, and θ_D, the positive angles that $0A$, $0B$, $0C$, and $0D$ make with respect to 00* when measured clockwise. We now get the relative location for all points to r_0, z_0 by subtracting r_0, z_0 from all coordinates.

A.7.2 Find Orientation of the System

The number of possible cases for calculating the intersections and volumes will be reduced if all cases are given the same logical orientation with respect to 00*. To accomplish this we do as follows.

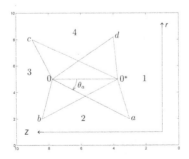

Fig. A.33 System with notation change.

A.7.2.1 Find the Smallest θ

a. We wish to find the smallest θ_i for $i = A, B, C,$ and D. We calculate the sines and cosines of the angles

$$\cos\theta_i = \frac{r^* r_i + z^* z_i}{[(r^*)^2 + (z^*)^2]^{\frac{1}{2}}[(r_i)^2 + (z_i)^2]^{\frac{1}{2}}}, \tag{A.70}$$

$$\sin\theta_i = \frac{z^* r_i - r^* z_i}{[(r^*)^2 + (z^*)^2]^{\frac{1}{2}}[(r_i)^2 + (z_i)^2]^{\frac{1}{2}}}. \tag{A.71}$$

b. From the θ_i select the largest $\cos\theta_i$ (first in case of two). Call this $\cos\theta_a$. Now examine $\sin\theta_a$. If $\sin\theta_a \geq 0$, the θ_a is the smallest positive θ_i (Fig. A.31). If $\sin\theta_a \leq 0$, then choose $\theta_{a+1} \to \theta_a$ (Fig. A.32).

A.7.2.2 Change the Notation of the System

We now change the notation of the system and reorient it logically, so that as we start from 00^* and move clockwise around the vertex we will see the sides in order $a, b, c,$ and d (Fig. A.33) that is, θ_a is the smallest $\theta_i \geq 0$. The coordinates of $a, b, c,$ and d are stored in the new order. For the surrounding sub-zones 1, 2, 3, and 4 we store in order of the new orientation $v^1, v^2, v^3,$ and v^4, the sub-zone volumes. We now proceed to join the new vertex 0^* to the midpoints of the sides $a, b, c,$ and d (Fig. A.33). These lines are the new sides of the sub-zones. These lines and their intersections with the old sides will define a number of triangular and quadrilateral elements used to calculate delta volume (Δv) to be moved from sub-zone to sub-zone. There are several possible cases, depending on the orientation of sides $a, b, c,$ and d with 00^*.

A.7.3 The Rezone Process

A.7.3.1 Selecting the Proper Case

There are five possible cases. The case number $= \sum -$ signs of the sines $+ 1$ (calculated above). It is possible that one of the sides $a, b, c,$ and d may be of zero length. Certain of these cases is legal for the rezone, but others mean that rezoning for that point must be terminated. The following are legal: Side $c = 0$ for case 3; Side $d = 0$ for case 4.

Table A.1 The 13-word buffer for holding the Δv's to be moved from sub-zone to sub-zone.

Buffer#	Definition of Sub-zone Move
1	Sub-zone 1 to sub-zone 2
2	Sub-zone 1 to sub-zone 3
3	Sub-zone 1 to sub-zone 4
4	Sub-zone 2 to sub-zone 1
5	Sub-zone 2 to sub-zone 3
6	Sub-zone 2 to sub-zone 4
7	Sub-zone 3 to sub-zone 1
8	Sub-zone 3 to sub-zone 2
9	Sub-zone 3 to sub-zone 4
10	Sub-zone 4 to sub-zone 1
11	Sub-zone 4 to sub-zone 2
12	Sub-zone 4 to sub-zone 3
13	Temporary

A.7.3.2 General Procedure

a. Intersection calculations and Δv calculation. After the proper case has been chosen, we proceed to calculate the intersections and the volume changes (Δv's). There are 12 possible intersections for the five cases. For each case we form a set of indices telling which point to use, which intersections to calculate, and where to store the resulting coordinates of the intersections. There is a 13-word buffer for holding the Δv's to be moved from sub-zone to sub-zone. (Table A.1.) For each case there are 13 sets of indices, each giving the location of the corner points of a triangular torus for volume calculations. The Δv's are now calculated and stored in the appropriate buffer location. Most of the Δv's are calculated directly, but some must have additional calculation to obtain the actual volume to be moved. The reason for this is that triangular elements are calculated while some of the elements to be transferred are quadrilaterals. These are obtained easily by sums or differences of the triangular elements. For each of the Δv's obtained, a fraction ($fr = \Delta v/v$) of the volume to be moved from one sub-zone to another is calculated. These fractions are reoriented to the original system and used to map the data from sub-zone to sub-zone. After finishing the proper rezone case we calculate the following for each buffer where the fr for that buffer is not equal to 0. For this discussion we will assume that directed kinetic energy is conserved.

$$\Delta \text{volume} = \Delta v_i = v_i \times fr_i \qquad \text{cell}_i \to \text{cell}j, \qquad (A.72)$$

$$\Delta \text{mass} = \Delta m_i = m_i \times fr_i \qquad \text{cell}_i \to \text{cell}j, \qquad (A.73)$$

$$\Delta \text{energy} = \Delta E_i = E_i \times fr_i \qquad \text{cell}_i \to \text{cell}j, \qquad (A.74)$$

$$\Delta \text{KE}_R = \Delta m_i \dot{r}_i |\dot{r}_i| \qquad \text{cell}_i \to \text{cell}j, \qquad (A.75)$$

$$\Delta \text{KE}_z = \Delta m_i \dot{z}_i |\dot{z}_i| \qquad \text{cell}_i \to \text{cell}j. \qquad (A.76)$$

For the vertex pass, Eqs. (A.75) and (A.76) are omitted since all four sub-zones involved have the same velocities. The values calculated are stored in the same order as the fractions.

b. Completing the mapping. All of the required data changes have been calculated. Now we calculate the kinetic energy for each sub-zone if not on the vertex pass. The factor of 0.5 is left out since it cancels out during calculation.

$$\Delta \text{KE}_{r_i} = \Delta m_i \dot{r}_i |\dot{r}_i| \qquad \text{cell}_i, \qquad (A.77)$$

$$\Delta \text{KE}_{z_i} = \Delta m_i \dot{z}_i |\dot{z}_i| \qquad \text{cell}_i. \qquad (A.78)$$

At this point we have stored all of the data to be moved, so each item is identified with the moves between sub-zones. It is now a straightforward process to do the actual mapping. All that is necessary is to subtract the Δ quantities from zone i and add to zone j for each fraction. We now move volume, mass, internal energy, and directed kinetic energy from cell to cell with the following exceptions.

(1) Only volume is moved across interfaces.
(2) On the vertex pass, no kinetic energy is moved.

c. Clean up.
(1) Calculate the new $\dot{r}_i |\dot{r}_i|$ and $\dot{z}_i |\dot{z}_i|$ for the sub-zones involved except on the vertex pass.

$$\dot{r}_i |\dot{r}_i| = \frac{\text{KE}_{r_i}}{m_i}, \qquad (A.79)$$

$$\dot{z}_i |\dot{z}_i| = \frac{\text{KE}_{z_i}}{m_i}. \qquad (A.80)$$

(2) Store all new sub-zone quantities in the sub-zone mesh.

Table A.2 The rezone case 1 with the notations shown in Fig. A.34.

The New Sides and the Intersections	Form the Triangles	We Want the Volume of
ab	a_1	a_1
ac	a_2	a_2
ad	a_3	a_3
bc	b_{123}	b_1
bd	b_{23}	b_2
cd	b_3	b_3
	c_{12}	c_1
	c_2	c_2
	d_1	d_1

A.7.4 The Rezone Cases

A.7.4.1 Case 1

We have the configuration in Fig. A.34. The rezone case 1 is described in Table A.2 with the notations shown in Fig. A.34. We calculate the triangular volumes. The $\bar{r}$'s are in terms of coordinates with respect to point 0, so it is necessary to add the r_0 coordinate to obtain correct volumes. In the following equations, $\bar{r}$ is the centroid of triangle and A the area of triangle.

$$\Delta v_{a1} = 2\pi(r_0 + \bar{r}_{a1})A_{a1}, \tag{A.81}$$

$$\Delta v_{a2} = 2\pi(r_0 + \bar{r}_{a2})A_{a2}, \tag{A.82}$$

$$\Delta v_{a3} = 2\pi(r_0 + \bar{r}_{a3})A_{a3}, \tag{A.83}$$

$$\Delta v_{b123} = 2\pi(r_0 + \bar{r}_{b123})A_{b123}, \tag{A.84}$$

$$\Delta v_{b123} = 2\pi(r_0 + \bar{r}_{b123})A_{b123}, \tag{A.85}$$

$$\Delta v_{b23} = 2\pi(r_0 + \bar{r}_{b23})A_{b23}, \tag{A.86}$$

$$\Delta v_{b3} = 2\pi(r_0 + \bar{r}_{b3})A_{b3}, \tag{A.87}$$

Table A.3 The moving of Δv from one sub-zone to another sub-zone.

Δv goes	From Sub-zone	To Sub-zone
a_1	1	2
a_2	1	3
a_3	1	4
b_1	2	3
b_2	2	4
b_3	2	1
c_1	3	4
c_2	3	1
d_1	4	1

$$\Delta v_{c12} = 2\pi(r_0 + \overline{r}_{c12})A_{c12}, \tag{A.88}$$

$$\Delta v_{c2} = 2\pi(r_0 + \overline{r}_{c2})A_{c2}, \tag{A.89}$$

and

$$\Delta v_{d1} = 2\pi(r_0 + \overline{r}_{d1})A_{d1}. \tag{A.90}$$

We need $\Delta v_{b1}, \Delta v_{b2}$, and Δv_{c1}. They are

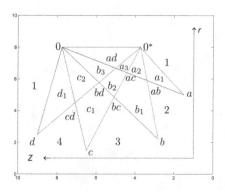

Fig. A.34 Rezone case 1.

$$\Delta v_{b1} = \Delta v_{b123} - \Delta v_{b23}, \tag{A.91}$$

$$\Delta v_{b2} = \Delta v_{b23} - \Delta v_{b3}, \tag{A.92}$$

and
$$\Delta v_{c1} = \Delta v_{c12} - \Delta v_{c2}. \tag{A.93}$$

Now consider what shifting of volume takes place. The moving of Δv from one sub-zone to another sub-zone is described in Table A.3. The equations for volume would be: (+ indicates after move, − before)

$$v_+^1 = v_-^1 - \Delta v_{a1} - \Delta v_{a2} - \Delta v_{a3} + \Delta v_{b3} + \Delta v_{c2} + \Delta v_{d1}, \tag{A.94}$$

$$v_+^2 = v_-^2 - \Delta v_{b1} - \Delta v_{b2} - \Delta v_{b3} + \Delta v_{a1}, \tag{A.95}$$

$$v_+^3 = v_-^3 - \Delta v_{c1} - \Delta v_{c2} + \Delta v_{a2} + \Delta v_{b1}, \tag{A.96}$$

$$v_+^4 = v_-^4 - \Delta v_{d1} + \Delta v_{a3} + \Delta v_{b2} + \Delta v_{c1}. \tag{A.97}$$

A.7.4.2 Case 2

We have the configuration in Fig. A.35. The triangles formed by the new sides and the intersection are given in Table A.4. We calculate the triangular volumes $\Delta v_{a1}, \Delta v_{a2}, \Delta v_{a3}, \Delta v_{b12}, \Delta v_{b2}, \Delta v_{c1}$, and Δv_{d1} as in case 1. We need

$$\Delta v_{b1} = \Delta v_{b12} - \Delta v_{b2}. \tag{A.98}$$

Now consider what shifting of volume takes place. The shifting of volumes from one sub-zone to another sub-zone is described in Table A.5.

Since d_1 and a_3 both move from 1 to 4, we need

$$\Delta v_{a3d1} = \Delta v_{a3} + \Delta v_{d1}. \tag{A.99}$$

The equations for volume would be

$$v_+^1 = v_-^1 - \Delta v_{a1} - \Delta v_{a2} - \Delta v_{a3d1}, \tag{A.100}$$

$$v_+^2 = v_-^2 - \Delta v_{b1} - \Delta v_{b2} + \Delta v_{a1}, \tag{A.101}$$

$$v_+^3 = v_-^3 - \Delta v_{c1} + \Delta v_{a2} + \Delta v_{b1}, \tag{A.102}$$

$$v_+^4 = v_-^4 + \Delta v_{a3d1} + \Delta v_{b2} + \Delta v_{c1}. \tag{A.103}$$

Table A.4 The triangles formed by the new sides and the intersections.

The New Sides and the Intersections	Form the Triangles	We Want the Volume of
ab	a_1	a_1
ac	a_2	a_2
bc	a_3	a_3
	b_{12}	b_1
	b_2	b_2
	c_1	c_1
	d_1	d_1

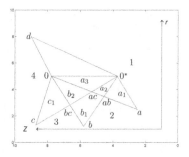

Fig. A.35 Rezone case 2.

A.7.4.3 Case 3

We have the configuration shown in Fig. A.36. The triangles formed by the new sides and the intersection are described in Table A.6. We calculate the triangular volumes $\Delta v_{a1}, \Delta v_{a2}, \Delta v_{b1}, \Delta v_{c1}, \Delta v_{d1}$, and Δv_{d2} as in case 1. Now consider what shifting of volume takes place. The shifting of volumes from one sub-zone to another sub-zone is described in Table A.7. Since Δv_{a2} and Δv_{d2} both move from 1 to 3, we need

$$\Delta v_{a2d2} = \Delta v_{a2} + \Delta v_{d2}. \tag{A.104}$$

The equations for volume would be

$$v_+^1 = v_-^1 - \Delta v_{a1} - \Delta v_{a2d2} - \Delta v_{d1}, \tag{A.105}$$

$$v_+^2 = v_-^2 - \Delta v_{b1} + \Delta v_{a1}, \tag{A.106}$$

Table A.5 The shifting of volumes from one sub-zone to another sub-zone.

The New Sides and the Intersections	Form the Triangles	We Want the Volume of
a_1	1	2
a_2	1	3
a_3	1	4
b_1	2	3
b_2	2	4
c_1	3	4
d_1	1	4

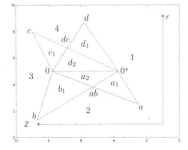

Fig. A.36 Rezone case 3.

Table A.6 The rezone case 3 with the notations shown in Fig. A.35.

The New Sides and the Intersections	Form the Triangles	We Want the Volume of
ab	a_1	a_1
dc	a_2	a_2
	b_1	b_1
	c_1	c_1
	d_1	d_1
	d_2	d_2

$$v_+^3 = v_-^3 + \Delta v_{a2d2} + \Delta v_{b1} + \Delta v_{c1}, \qquad (A.107)$$

$$v_+^4 = v_-^4 - \Delta v_{c1} + \Delta v_{d1}. \qquad (A.108)$$

Table A.7 The moving of Δv from one sub-zone to another sub-zone.

Δv goes	From Sub-zone	To Sub-zone
a_1	1	2
a_2	1	3
b_1	2	3
c_1	4	3
d_1	1	4
d_2	1	3

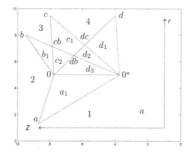

Fig. A.37   Rezone case 4.

A.7.4.4   *Case 4*

We have the configuration in Fig. A.37. The triangles formed by the new sides and the intersection are described in Table A.8. We calculate the triangular volumes $\Delta v_{a1}, \Delta v_{b1}, \Delta v_{c12}, \Delta v_{c2}, \Delta v_{d1}, \Delta v_{d2}$, and Δv_{d3} as in case 1. we need

$$\Delta v_{c1} = \Delta v_{c12} - \Delta v_{c2}. \qquad (A.109)$$

Now consider what shifting of volume takes place. The shifting of volumes from one sub-zone to another sub-zone is described in Table A.9. Since Δv_{a1} and Δv_{d3} both move from 1 to 2, we need

$$\Delta v_{a1d3} = \Delta v_{a1} + \Delta v_{d3}. \qquad (A.110)$$

The equations for volume would be

$$v_+^1 = v_-^1 - \Delta v_{a1d3} - \Delta v_{d1} - \Delta v_{d2}, \qquad (A.111)$$

Table A.8 The rezone case 4 with the notations shown in Fig. A.37.

The New Sides and the Intersections	Form the Triangles	We Want the Volume of
dc	a_1	a_1
db	b_1	b_1
cb	c_{12}	c_1
	c_2	c_2
	d_1	d_1
	d_2	d_2
	d_3	d_3

Table A.9 The moving of Δv from one sub-zone to another sub-zone.

Δv goes	From Sub-zone	To Sub-zone
a_1	1	2
b_1	3	2
c_1	4	3
c_2	4	2
d_1	1	4
d_2	1	3
d_3	1	2

Table A.10 The rezone case 5 with the notations shown in Fig. A.38.

The New Sides and the Intersections	Form the Triangles	We Want the Volume of
ba	a_1	a_1
ca	b_{12}	b_1
cb	b_2	b_2
da	c_{123}	c_1
db	c_{23}	c_2
dc	c_3	c_3
	d_1	d_1
	d_2	d_2
	d_3	d_3

$$v_+^2 = v_-^2 + \Delta v_{a1d3} + \Delta v_{b1} + \Delta v_{c2}, \quad (A.112)$$

$$v_+^3 = v_-^3 - \Delta v_{b1} + \Delta v_{c1} + \Delta v_{d2}, \quad (A.113)$$

$$v_+^4 = v_-^4 - \Delta v_{c1} - \Delta v_{c2} + \Delta v_{d1}. \quad (A.114)$$

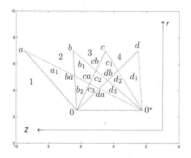

Fig. A.38 Rezone case 5.

A.7.4.5 Case 5

We have the configuration in Fig. A.38. The triangles formed by the new sides and the intersection are described in Table A.10. We calculate the triangular volumes $\Delta v_{a1}, \Delta v_{b12}, \Delta v_{b2}, \Delta v_{c123}, \Delta v_{c23}, \Delta v_{c3}, \Delta v_{d1}, \Delta v_{d2}$, and Δv_{d3} as in case 1. we need $\Delta v_{b1}, \Delta v_{c1}$, and Δv_{c2}.

$$\Delta v_{b1} = \Delta v_{b12} - \Delta v_{b2}, \qquad (A.115)$$

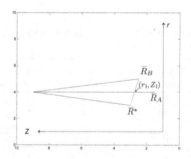

Fig. A.39 Intersection calculation.

$$\Delta v_{c1} = \Delta v_{c123} - \Delta v_{c23}, \qquad (A.116)$$

$$\Delta v_{c2} = \Delta v_{c23} - \Delta v_{c3}. \qquad (A.117)$$

Now consider what shifting of volume takes place. The shifting of volumes from one sub-zone to another sub-zone is described in Table A.11.

Table A.11 The moving of Δv from one sub-zone to another sub-zone.

Δv goes	From Sub-zone	To Sub-zone
a_1	2	1
b_1	3	2
b_2	3	1
c_1	4	3
c_2	4	2
c_3	4	1
d_1	1	4
d_2	1	3
d_3	1	2

The equations for volume would be

$$v_+^1 = v_-^1 - \Delta v_{d1} - \Delta v_{d2} - \Delta v_{d3} + \Delta v_{a1} + \Delta v_{b2} + \Delta v_{c3}, \tag{A.118}$$

$$v_+^2 = v_-^2 - \Delta v_{a1} + \Delta v_{b1} + \Delta v_{c2} + \Delta v_{d3}, \tag{A.119}$$

$$v_+^3 = v_-^3 - \Delta v_{b1} - \Delta v_{b2} + \Delta v_{c1} + \Delta v_{d2}, \tag{A.120}$$

$$v_+^4 = v_-^4 - \Delta v_{c1} - \Delta v_{c2} - \Delta v_{c3} + \Delta v_{d1}. \tag{A.121}$$

A.7.5 *Mapping Other Quantities*

So far the discussion of mapping has been limited to the hydrodynamic quantities. Any other quantities necessary in a given code also may be mapped. All that is required is that the quantities to be moved from sub-zone to sub-zone must be put into a form such that fractions of the quantities may be moved directly.

A.7.6 *Intersection Calculation*

Given a set of vectors as shown in Fig. A.39 calculate

$$k = \frac{r_B z^* - z_B r^*}{r_A(z^* - z_B) - z_A(r^* - r_B)}. \tag{A.122}$$

If the denominator $= 0$ there is no intersection. Rezone is terminated for the point under consideration. If $k > 1$ there is no intersection on the vector $\overline{R}_A$. The displacement $\overline{R}^*$ is reduced and we try again. If $k \leq 1$ then

$$r_I = k r_A, \tag{A.123}$$

$$z_I = k z_A. \tag{A.124}$$

A.8 Rezone Method — Boundary Cases

Boundary rezoning is allowed on boundary points that are constrained to constant r, constant z, and slide angle. It is also allowed on free surface points and center of mass points. The code cannot do rezoning on point that are involved in triangular rezones or on corner points that have unconstrained motion. These feature need to be added to the code.

A.8.1 Constant R, Constant Z, and Slide Angle

A.8.1.1 Two-zone Case

The two-zone case is quite simple, with only two possible cases. The logical configuration of the sub-zones will be either of the two shown in Fig. A.40. We first calculate the volume of element $0A0^* = \Delta v_{a1}$.

$$\Delta v_{a1} = 0.5[(z^* - z_0)(r_A - r_0) - (z_A - z_0)(r^* - r_0)]\frac{r_0 + r_A + r^*}{3}. \quad (A.125)$$

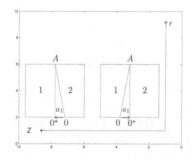

Fig. A.40 Two-zone case. Case 1 (left) and Case 2 (right).

If $\Delta v_{a1} > 0$, we have case 1. If $\Delta v_{a1} < 0$, we have case 2.

a. Case 1 $\Delta v_{a1} > 0$. For this case we move from cell 1 to cell 2. The volume equations would be

$$v_+^1 = v_-^1 - \Delta v_{a1}, \quad (A.126)$$

$$v_+^2 = v_-^2 + \Delta v_{a1}. \quad (A.127)$$

We calculate the fraction of the volume moved from cell 1 to cell 2 ($fr = \Delta v_{a1}/v^1$). The mapping then proceeds as in the general case.

Table A.12 The moving of Δv from one sub-zone to another sub-zone.

Δv goes	From Sub-zone	To Sub-zone
a	1	3
b	2	3
c	1	2

b. Case 2 $\Delta v_{a1} < 0$. For this case we move from cell 2 to cell 1. We use the absolute value of Δv_{a1}. The volume equations would be

$$v_+^1 = v_-^1 + |\Delta v_{a1}|, \qquad (A.128)$$

$$v_+^2 = v_-^2 - |\Delta v_{a1}|. \qquad (A.129)$$

We calculate the fraction of the volume moved from cell 2 to cell 1 ($fr = \Delta v_{a1}/v^2$). The mapping then proceeds as in the general case. Again, for both case 1 and case 2, no directed kinetic energy is moved on the vertex pass and volume is moved only if $0A$ is an interface.

A.8.1.2 Three-zone Case

The three-zone case is a little more complicated than the two-zone case, but there still are only two cases.

a. Case 1. We have the configuration in Fig. A.41. Calculate the intersection of $B0^*$ with $A0$. The new sides and the intersection BA form

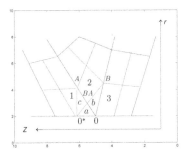

Fig. A.41 Three-zone case. Case 1.

the triangle a, b, and c. We calculate the triangular volumes

$$\Delta v_a = \bar{r}_a A_a, \qquad (A.130)$$

$$\Delta v_b = \bar{r}_b A_b, \qquad (A.131)$$

$$\Delta v_c = \bar{r}_c A_c. \qquad (A.132)$$

Now consider what shifting of volume takes place. The shifting of volumes from one sub-zone to another sub-zone is described in Table A.12. The equations for volume would be

$$v_+^1 = v_-^1 - \Delta v_a - \Delta v_c, \qquad (A.133)$$

$$v_+^2 = v_-^2 - \Delta v_b + \Delta v_c, \qquad (A.134)$$

$$v_+^3 = v_-^3 + \Delta v_a + \Delta v_b. \qquad (A.135)$$

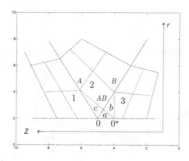

Fig. A.42 Three-zone case. Case 2.

b. Case 2. We have the configuration in Fig. A.42. Calculate the intersection of $A0^*$ with $B0$. The new sides and the intersection AB form the triangles a, b, and c. Calculate the triangular volumes.

$$\Delta v_a = \bar{r}_a A_a, \qquad (A.136)$$

Table A.13 The moving of Δv from one sub-zone to another sub-zone.

Δv goes	From Sub-zone	To Sub-zone
a	3	1
b	3	2
c	2	1

$$\Delta v_b = \overline{r}_b A_b, \qquad (A.137)$$

$$\Delta v_c = \overline{r}_c A_c. \qquad (A.138)$$

Now consider what shifting of volume takes place. The shifting of volumes from one sub-zone to another sub-zone is described in Table A.13. The equations for volume would be

$$v_+^1 = v_-^1 + \Delta v_a + \Delta v_c, \qquad (A.139)$$

$$v_+^2 = v_-^2 - \Delta v_c + \Delta v_b, \qquad (A.140)$$

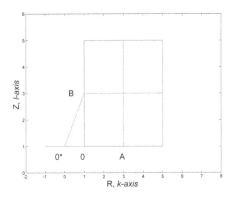

Fig. A.43 Corner case.

$$v_+^3 = v_-^3 - \Delta v_a - \Delta v_b. \qquad (A.141)$$

c. Picking the case for the three-zone case. Referring to Figs. A.41-A.42, we calculate the area of the element with corners $0A0^*$. If the area is positive, we have case 1. If the area is negative, we have case 2. The area is calculated by

$$\text{Area}_{0A0*} = 0.5[(z^* - z_0)(r_A - r_0) - (z_A - z_0)(r^* - r_0)]. \quad (A.142)$$

d. The mapping. For both of the above cases we calculate the fractions of the volume moved from sub-zone to sub-zone. We then do the actual mapping in a manner similar to that of the general case. Again, no directed kinetic energy is moved on the vertex pass, and volume is moved only across an interface.

A.8.1.3 Corner Case

The corner case is very simple. All cases are similar to that shown in Fig. A.43. In the case in Fig. A.43, side $0A$ is constrained to constant r. Side $0B$ is free. The new corner point is 0^*. All other corner cases are handled similarly. There is no change of mass, etc. All that is necessary is to calculate the new volume of the sub-zone.

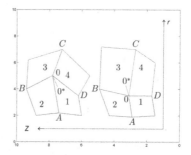

Fig. A.44 Free surface case.

A.8.2 Free Surface Case

For the free surface case the point 0^* may be either inside the mesh or outside the mesh (Fig. A.44). We have the old point 0; the new point 0^*; the three sides A, B, and C; and the two sub-zones 1 and 2. To do the rezoning we first construct a fourth side, C, outside the mesh and create two empty sub-zone 3 and 4, also outside the mesh. We now have sufficient data to use the general case for the rezoning. No mass, energy, or directed kinetic energy is transferred across the free surface. The total volume of the two sub-zones will either decrease or increase, depending on whether the surface is concave or convex. The exact position of point C is unimportant. A convenient way to get it is to construct a side bisecting the angle BOD of a length equal to side A as was done in the displacement calculation. Mapping is done in a manner similar to that of the other cases by calculating the fractions of volume to be moved from sub-zone to sub-zone.

A.8.3 Center of Mass Case

In the center of mass case the mesh will be changing as shown in Fig. A.45. The new point 0^* may be on either side of the old point 0. If the center points are moved directly to the new position as calculated by the displacement calculation there may be difficulty completing the rezone. The midpoint pass may move the side past the old point 8, which would violate the rules for the rezone. To keep this from happening we test that all motion on the vertex, midpoint and point 8 passes for all points in the center of mass will be legal before starting the move. If there will be any trouble anywhere, the size of the displacement is reduced until all calculations can be done correctly. If there is no position of the new points where this is possible, the center of mass is removed from the rezone.

So that we may position the new point at the calculated position, we

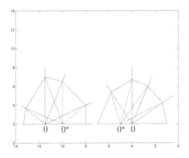

Fig. A.45 Center of mass case.

use an iterative process. We move the center to the new position in small increments as determined by the above tests. Each time we move the points as far as possible and after the full mapping has been completed, we again check to see if the point can now be moved to the calculated position. If it cannot, the points are again moved as far as possible.

This process continues until the calculated position is achieved or there is no more possible motion. For the actual mapping we do not try to move all of the center points simultaneously. Instead we move them one at a time. Figure A.46 shows a sample of an intermediate step in the process. The points to the right have been moved and the points to the left are still at the old position. By moving points one at a time it is possible to use standard boundary cases for the mapping with only minor modifications. As each point is moved, we do the vertex, midpoint, point 8, and averaging passes just as would be done for a normal boundary point. The mapping is done as in the normal boundary cases.

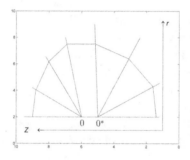

Fig. A.46 Intermediate step-center of mass case.

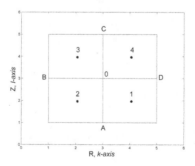

Fig. A.47 General case.

A.9 The Vertex, Midpoint, Point 8, and Velocity Passes

The vertex, midpoint, and point 8 passes all use the same mapping routines. All that is necessary is to supply the mapping routines with the locations of appropriate sub-zones.

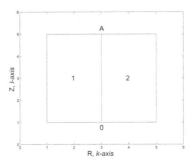

Fig. A.48   Two-zone case.

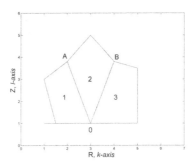

Fig. A.49   Three-zone case.

A.9.1 *Possible Cases*

A.9.1.1 *General Case*

Figure A.47 shows the general configuration of the sub-zones. We have a point 0 surrounded by the four sub-zones 1, 2, 3, 4 and the four points A, B, C, and D.

A.9.1.2 Sliding Cases

a. Two zones. In Fig. A.48 we have the point 0, the two sub-zones 1, 2, and the point A.

b. Three zones. In Fig. A.49 we have the point 0, the three sub-zones 1, 2, 3, and the points A and B.

A.9.1.3 Free Surface Case

In Fig. A.50 we have the point 0, the projected point C, and the real points A, B, D with the real sub-zone 1, 2 and the artificial sub-zones 3 and 4.

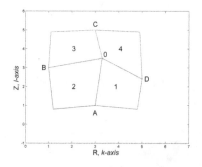

Fig. A.50   Free surface case.

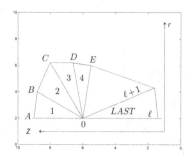

Fig. A.51   Center of mass.

A.9.1.4 Center of Mass

In Fig. A.51 we have the point 0, the sub-zones 1, 2, 3, 4,...last, and the points $A, B, C, D, E, ... \ell - 1, \ell$.

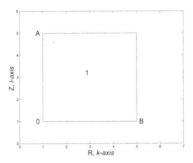

Fig. A.52 Corner point.

A.9.1.5 Corner Point

In Fig. A.52 we have the point 0, the sub-zone 1, and the points A and B.

A.9.2 The Vertex Pass

On the vertex pass we set the new vertex point as defined by the displacement calculation and do the mapping between the adjacent sub-zones. The vertex pass may use any of the cases described in Section A.9.1. No directed kinetic energy is moved on the vertex pass.

A.9.3 The Midpoint Pass

For a point inside the mesh there are four midpoints surrounding the vertex. Each of these must be moved to a new position. The new midpoint is the average of the end points of the line connecting the new point 0 and one of the four surrounding points. We set the midpoint and do the mapping between the adjacent sub-zones. If the vertex point is on a boundary, one or more of the midpoints may not exist and we will skip the calculation for the nonexistent point. The midpoint pass may use the general case, the two-zone slide case, or the free surface case from Section A.9.1.

A.9.4 The Point 8 Pass

For a point inside the mesh there are four point 8's surrounding the vertex. Each of these must be moved to a new position. The new point 8 is the average of the four corner points of the zone in which it is located. We set the new point 8 and do the mapping between the adjacent sub-zones.

If the vertex point is on a boundary, one or more of the point 8's will be nonexistent. Calculation for these points will be skipped. The point 8 pass can use only the general case from Section A.9.1. At the end of each of the above three passes, the sub-zone storage is replaced by the new quantities calculated on that pass.

A.9.5 The Velocity Adjustment Pass

When directed kinetic energy conservation is chosen, kinetic energy is transferred from sub-zone to sub-zone. This will change the velocities of the sub-zones involved. On constrained boundaries this may result in giving velocities to points where the r or z velocities should be zero or in changing velocities on points that are constrained in some other manner. The velocity adjustment pass checks all points and sets $\dot{r}|\dot{r}|$ and/or $\dot{z}|\dot{z}|$ in the sub-zones with appropriate values from the velocities of constrained points. This pass is not called when directed kinetic energy is not being conserved.

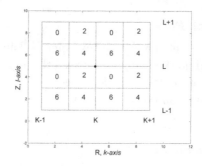

Fig. A.53 Sub-zones surrounding moved point.

A.10 The Averaging Pass

When all mapping is completed for the sub-zones surrounding the point being moved, it is necessary to return the data to the normal mesh from the expanded sub-zone mesh.

A.10.1 Point Quantities

A.10.1.1 Coordinates

The new position of the point is stored in the normal mesh storage (r and z coordinates).

A.10.1.2 Velocities

New velocities are calculated at this point only if directed kinetic energy is being conserved. After the mapping the directed kinetic energy of the sub-zones surrounding a point, K, L in the sub-mesh (Fig. A.53) may all be different from the original values. This means that the velocities associated with the sub-zones have been changed and the velocities of point K, L must be recalculated. We first get the $\dot{r}|\dot{r}|$ and $\dot{z}|\dot{z}|$ for point K, L by calculating the total r and z components of the kinetic energy of sub-zones surrounding the point and dividing by the sum of the masses of the sub-zones.

$$\dot{r}|\dot{r}|_{K,L} =$$

$$\frac{(\dot{r}^0|\dot{r}^0|m^0)_{K+\frac{1}{2},L-\frac{1}{2}} + (\dot{r}^2|\dot{r}^2|m^2)_{K-\frac{1}{2},L-\frac{1}{2}} + (\dot{r}^4|\dot{r}^4|m^4)_{K-\frac{1}{2},L+\frac{1}{2}} + R6}{m^0_{K+\frac{1}{2},L-\frac{1}{2}} + m^2_{K-\frac{1}{2},L-\frac{1}{2}} + m^4_{K-\frac{1}{2},L+\frac{1}{2}} + m^6_{K+\frac{1}{2},L+\frac{1}{2}}},$$
(A.143)

where $R6 = (\dot{r}^6|\dot{r}^6|m^6)_{K+\frac{1}{2},L+\frac{1}{2}}$,

$$\dot{z}|\dot{z}|_{K,L} =$$

$$\frac{(\dot{z}^0|\dot{z}^0|m^0)_{K+\frac{1}{2},L-\frac{1}{2}} + (\dot{z}^2|\dot{z}^2|m^2)_{K-\frac{1}{2},L-\frac{1}{2}} + (\dot{z}^4|\dot{z}^4|m^4)_{K-\frac{1}{2},L+\frac{1}{2}} + Z6}{m^0_{K+\frac{1}{2},L-\frac{1}{2}} + m^2_{K-\frac{1}{2},L-\frac{1}{2}} + m^4_{K-\frac{1}{2},L+\frac{1}{2}} + m^6_{K+\frac{1}{2},L+\frac{1}{2}}},$$
(A.144)

where $Z6 = (\dot{z}^6|\dot{z}^6|m^6)_{K+\frac{1}{2},L+\frac{1}{2}}$.

Then from $\dot{r}|\dot{r}|_{K,L}$ and $\dot{z}|\dot{z}|_{K,L}$, we get the velocities as follows:

$$\dot{r}_{K,L} = (|\dot{r}|\dot{r}||_{K,L})^{1/2} \quad \text{with the sign of } \dot{r}|\dot{r}|_{K,L}, \quad \text{(A.145)}$$

$$\dot{z}_{K,L} = (|\dot{z}|\dot{z}||_{K,L})^{1/2} \quad \text{with the sign of } \dot{z}|\dot{z}|_{K,L}. \quad \text{(A.146)}$$

Since all of the sub-zones in the zones surrounding the moved point have had velocities changed, it is necessary to recalculate point velocities for the eight points adjacent to the moved point in addition to the moved point.

A.10.2 Zone Quantities

A.10.2.1 Zone Density

The zone density is obtained by dividing the sum of masses by the sum of the sub-zone volume.

$$\rho_{K,L} = \frac{m_{K,L}^0 + m_{K,L}^2 + m_{K,L}^4 + m_{K,L}^6}{v_{K,L}^0 + v_{K,L}^2 + v_{K,L}^4 + v_{K,L}^6}. \tag{A.147}$$

A.10.2.2 Zone Energy

The zone energy (ϵ) is obtained by dividing the sum of energies by the sum of the sub-zone mass.

$$\epsilon_{K,L} = \frac{E_{K,L}^0 + E_{K,L}^2 + E_{K,L}^4 + E_{K,L}^6}{m_{K,L}^0 + m_{K,L}^2 + m_{K,L}^4 + m_{K,L}^6}. \tag{A.148}$$

A.10.2.3 Sub-zone Masses and volumes

The new sub-zone masses are taken from the expanded sub-mesh and are replaced in the permanent sub-zone mass storage. The new sub-zone volumes are taken from the expanded sub-mesh and replaced in the temporary sub-zone volume storage.

A.10.2.4 Other Quantities

Any other quantities that are being rezoned must be averaged back to zone quantities and replaced in the full mesh storage. This completes the average pass. We are now ready to go on to the next point to be moved.

A.11 Completing the Rezone

After all marked points have been moved, the mapping completed and the new data replaced in the normal mesh, there is still some cleanup to be done before we can return to the main calculation. We need to complete the calculation of the new zone and point quantities. During the rezone, flags have been set indicating changed zones and points. These flags control the calculation of the final values.

A.11.1 New Velocities

If kinetic energy is being conserved, the new velocities have already been calculated during the rezone process. If kinetic energy is not being conserved, we obtain the new velocities by interpolation on the values in the mesh before the rezone. For the velocity interpolation, only those points which have been moved need new velocities.

A.11.2 New Zone Mass

For zones where the mass has been changed we need to obtain a new zone mass. The four zones adjacent to each moved point will need new masses.

A.11.3 New Pressure

We need new pressures for the zones where energy has changed. Again this means the four zones adjacent to each moved point.

A.11.4 New q Terms (Artificial Viscosity)

Since velocities and positions of points have been changed, the q terms in the zones surrounding these points are no longer correct. Where kinetic energy is not conserved, we need new q terms for the four zones adjacent to the moved point. Where kinetic energy is conserved we need to calculate q terms for the sixteen zones adjacent to the moved point.

A.11.5 Clear Flags

All of the flags that have been set during the rezone are clear to zero. This completes the rezone and we return to the main calculation.

A.12 Summary

This automatic rezone method monitor the mesh and interrupts the hydrodynamic calculation when the mesh needs attention. The mesh is moved through the fluid without changing the physical position or properties of the fluid. Depending on the options chosen, mass, internal energy, and directed kinetic energy will be conserved exactly. There are some area that still need attention. Spherical sections tend to move inward when point are rezoned. Section of the mesh where there are zones with large aspect ratios often do not behave well, particularly when the zones are adjacent to an interface.

A.13 Final Remarks

This automatic rezone method is currently included in a large production code and is in regular use. There are some problems where it does not work well and the situations mentioned in the summary (Section A.12) sometimes cause difficulties. However, the method is being used with considerable success. It is now possible to calculate many problems that previously have been either very difficult because of mesh twisting or have been impossible to do.

This automatic rezone has not yet completely eliminated the need to stop problems and manually reconstruct the mesh. However, the number of times this is necessary has been greatly reduced and the reconstructions are usually much simpler than without rezone. Some problems will now run from start to finish without any intervention, and many problems use less computer time to run when the automatic rezone is used.

The exact methods and parameters described in this appendix have been developed specifically for the type of problem that is calculated most frequently with this code. Other types of problems may required modifications of the methods and the parameters.

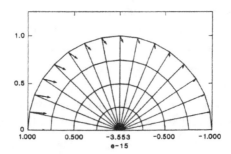

Fig. A.54 The spherical challenge problem at T = 0.0. Initial mesh with velocity vectors.

A.14 Examples

The first example is the spherical challenge problem. This problem is driven by a set of constant velocities imposed on the outer surface. Figure A.54 shows the initial mesh with the velocity vectors. The problem was run with the normal hydrodynamics equations with no use of any mesh-straightening techniques.

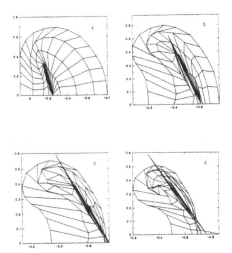

Fig. A.55 The spherical challenge problem run without rezoning. At T = 0.6, cycle 276 (a); T = 0.8, cycle 770 (b); T = 0.9, cycle 1837 (c); T = 1.0, cycle 4535 (d). At T = 0.6 (a) twisting of the mesh has started. At T = 0.8 (b) and T = 0.9 (c) twisting has increased and bow-ties have formed. At T = 1.0 mesh is badly distorted.

The problem first was run without rezoning. The scales in the figures are not all the same. The mesh is kept expanded so that details may be seen more clearly. Figure A.55a shows the mesh at T = 0.6. Twisting of the mesh has started. At T = 0.8 and T = 0.9 (Figs. A.55b and A.55c) the twisting has increased and bow-ties have been formed. Finally at T = 1.0 (Fig. A.55d) the mesh is badly distorted and the run has become meaningless. Figures A.56a, A.56b, A.56c, and A.56d show the mesh at the same times but with the automatic rezone in use. At T = 1.0 (Fig. A.56d) the mesh is still in good shape and meaningful results may be obtained. In Fig. A.56d the tendency mentioned above for spherical sections to be moved inward is apparent. The automatic rezone was called on 83 cycles during the run. It is possible to run this problem past T = 1.0 when automatic rezone is used. The problem was run until the outer points on the axis to the left and right reached the same position. No difficulties were encountered.

The initial mesh of the second example is shown in Fig. A.57. This problem consists of a layer of low-density material, a straight section of fine spaced high density material, a tapered section of high-density material, and another layer of low-density material. The fine spaced layers has zones with high aspect ratios. The problem is driven by feeding energy into the

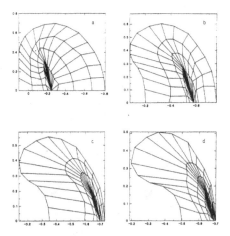

Fig. A.56 The spherical challenge problem run with automatic rezone. At T = 0.6, cycle 290 (a); T = 0.8, cycle 512 (b); T = 0.9, cycle 681 (c); T = 1.0, cycle 1091 (d). At T = 1.0 (d) the mesh is still in good shape. The number of calculation cycles needed for this problem is much smaller than the number needed for the problem without automatic rezone.

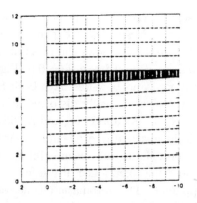

Fig. A.57 Second test problem at T = 0.0. Initial mesh.

top layer of low-density material to produce a nonsymmetric shock. The points on the boundaries are constrained to slide along the boundaries.

The problem was first run without the automatic rezone. At T = 20.0 (Fig. A.58a) the upper section has distorted and bow-ties have formed. As the problem continues to T = 60.0 (Fig. A.58b), the distortion increases

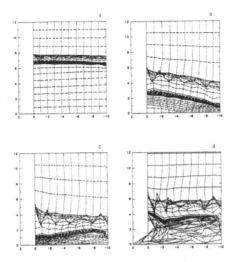

Fig. A.58 Second test problem run without rezoning. At T = 20.0, cycle 603 (a); T = 60.0, cycle 973 (b); T = 80.0, cycle 1259 (c); T = 100.0, cycle 2799 (d). At T = 20.0 (a) the upper section has distorted. At T = 60.0 (b) the distortion has increased. At T = 80.0 (c) the lower section is distorted and at T = 100.0 the mesh is badly distorted.

and by T = 80.0 (Fig. A.58c) the lower section also is distorted. By T = 100.0 (Fig. A.58d) the mesh is badly twisted and has 103 bow-ties. Figures A.59a, A.59b, A.59c, and A.59d show the same time on the problem, but with the automatic rezone in use. There are no bow-ties in the mesh at T = 100.0 (Fig. A.59d). This problem required rezone on 65 cycles.

Both of these examples are simple problems, but they illustrate the general principles and capabilities of this rezone method. Rezoning is done on a small percentage of the total cycles and the mesh shape reflects the hydrodynamic flow.

A.15 Directed Kinetic Energy

The concept of directed kinetic energy suggested by Philip L. Browne, involves a different way of looking at the calculation of kinetic energy in Lagrangian meshes. The usual definition of kinetic energy in a Lagrangian mesh is

$$KE = \frac{1}{2}m(\dot{r}^2 + \dot{z}^2), \qquad (A.149)$$

where $\dot{r}$ and $\dot{z}$ are the r and z velocities associated with the mass m. There are several possible ways of defining the mass to be used and the velocities

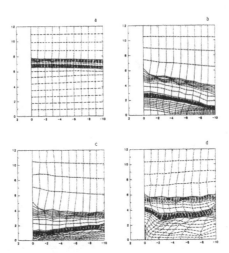

Fig. A.59 Second test problem run with the automatic rezone. At T = 20.0, cycle 638 (a); T = 60.0, cycle 1012 (b); T = 80.0, cycle 1318 (c); T = 100.0, cycle 1495 (d). Note that at T = 100.0 (d) the mesh is still in good shape. The number of calculation cycles needed for this problem is much smaller than the number needed for the problem without automatic rezone.

associated with that mass, but the basic equation is as above. For directed kinetic energy, we divide the kinetic energy into signed r and z components.

$$r \text{ component of } KE/m = 0.5\dot{r}|\dot{r}| \qquad \text{may be } + \text{ or } -, \qquad (A.150)$$

$$z \text{ component of } KE/m = 0.5\dot{z}|\dot{z}| \qquad \text{may be } + \text{ or } -. \qquad (A.151)$$

By this definition each component has a direction as well as a magnitude. In the automatic rezone code, the masses used are the sub-zone masses. Each sub-zone mass is assigned the r and z velocities of the adjacent mesh point. The r and z components of kinetic energy are defined for each sub-zone by using the sub-zone velocities. The r and z components can be associated with the elements of mass (Δm) that are transferred from sub-zone to sub-zone during the rezone process. After all mapping is completed, the new point velocities may be calculated from the directed kinetic energy components of the four sub-zones surrounding each point. With this method of conserving directed kinetic energy, the automatic rezone should conserve mass, internal energy, and total energy.

This Appendix is written based on the Wallick's report [A.2].

Bibliography

A.1 Browne, BG. (1961). *Numerical solution of neutron transport problems,* Reactor theory, Vol.11, Am. Math. Soc. Proc. Symp. Appl. Math, American Mathematical Society, Providence, R.I.

A.2 Wallick, KB. (1987). *REZONE: A method for automatic rezoning in two-dimensional Lagrangian hydrodynamic problems,* Los Alamos National Laboratory report LA-10829-MS.

Appendix B

Eigenvalue Calculations

Notations

$\vec{e}$ unit vector in the direction of the spatial vector $\vec{r}$
E energy, also $E = mv^2/2$
F surface source
$F(E)$ Coulomb scattering term
$\vec{j}$ normalized current
$\vec{J}$ current
m particle mass(g)
$n(E, \vec{\Omega})$ angular particle density
N particle number density
$\vec{r}$ position vector
$S(E, \vec{\Omega})$ external source
$t_E(E)$ Coulomb scattering term
t time

Greek letters

δ' Kronecker delta
μ the cosine of the polar angle
v particle speed
$\Sigma_t(E)$ total nuclear reaction cross-section
$\Sigma(E' \to E, \vec{\Omega}' \to \vec{\Omega})$ nuclear reaction
$\Sigma_c(E)$ Coulomb scattering term
τ Coulomb scattering term
ϕ azimuthal angle
$\Psi_g(\vec{\Omega})$ normalized angular density
$\vec{\Omega}$ unit vector in the direction of the photon transport

Subscripts

0 initial value
g for a particular group g
i Legendre polynomial index
t derivative with respect to time

Superscript

i finite difference at grid location i

B.1 Introduction

The time absorption (α) eigenvalue is one of the implicit eigenvalue search options found in most current transport codes [Refs. B.1-B.5]. Because these codes are designed to treat a variety of implicit eigenvalue problems (critical size search, concentration search), their generalized search methods are less than optimal for a specific type of eigenvalue calculation. Due to the explicit nature of the α eigenvalue problem and the types of nuclear systems of interest at Los Alamos, a number of methods have been developed to considerably improve the efficiency of these codes for this eigenvalue search. This Appendix describes these methods.

We will start with the homogeneous, time- and energy-dependent, transport equation

$$\frac{1}{v}\frac{\partial \Psi}{\partial t} + \nabla \cdot \vec{\Omega}\Psi + \Sigma_t(\vec{r}, E)\Psi(\vec{r}, E, \vec{\Omega}, t)$$
$$= \int dE' \int d\vec{\Omega}' \Sigma_s(\vec{r}, E' \to E, \vec{\Omega} \cdot \vec{\Omega}')\Psi(\vec{r}, E', \vec{\Omega}', t)$$
$$+ \chi(E) \int dE' \int d\vec{\Omega}' \nu\Sigma_f(\vec{r}, E')\Psi(\vec{r}, E', \vec{\Omega}', t), \qquad (B.1)$$

where
$\Psi(\vec{r}, E, \vec{\Omega}, t)$= angular fluxes as a function of the independent variables time (t), energy (E) or velocity (v), angle ($\vec{\Omega}$), and space ($\vec{r}$),
v= diagonal matrix of neutron group speeds,
$\Sigma_t(\vec{r}, E)$=macroscopic total cross section,
$\Sigma_s(\vec{r}, E' \to E, \vec{\Omega} \cdot \vec{\Omega}')$=macroscopic scattering function,
$\chi(E)$=fission spectrum, and
$\nu\Sigma_f(\vec{r}, E')$ =macroscopic prompt fission neutron production cross sections.

Eigenvalue Calculations

By making the time separability ansatz

$$\Psi(\vec{r}, E, \vec{\Omega}, t) = e^{\alpha t}\Psi_\alpha(\vec{r}, E, \vec{\Omega}), \tag{B.2}$$

Eq. (B.1) is transformed to the time-independent eigenvalue problem

$$L\Psi_\alpha + (\Sigma_t + \frac{\alpha}{v})\Psi_\alpha = (S + F)\Psi_\alpha, \tag{B.3}$$

where
L = leakage operator,
S = scattering operator,
F = fission operator, and
Ψ_α and α are the time-absorption eigenfunctions and eigenvalues, respectively.
There is a very rich literature [B.6-B.11] on the eigenvalue spectrum of Eq. (B.3). Duderstadt and Martin [B.11] give a particularly good discussion of this problem, with abundant references. The Larsen and Zweifel [B.9] article is probably the most complete and general of the recent papers.

Of particular interest is the eigen-solution of Eq. (B.3) with the largest real part, α_0, the so-called dominant eigenvalue [B.10]. This solution corresponds to the asymptotic (as $t \to \infty$) solution of the original initial value problem, Eq. (B.1). This dominant eigenvalue is the physical quantity measured in the laboratory by pulsed neutron (die-away) and Rossi-α experiments. Physical consideration would obviously require

$$Im(\alpha_0) = 0, \tag{B.4}$$

and

$$\Psi_\alpha(\vec{r}, E, \vec{\Omega}) \geq 0 \text{ for all } \vec{r}, E, \vec{\Omega}. \tag{B.5}$$

Also of important is the sign of α_0
$\alpha_0 > 0$ supercritical,
$\alpha_0 = 0$ critical, and
$\alpha_0 < 0$ subcritical.

It should be noted that α eigenfunctions of Eq. (B.3) differ from the λ (or k_{eff}) eigenfunction, Ψ_λ [B.6], which satisfy

$$L\Psi_\lambda + \Sigma_t \Psi_\lambda = (S + \frac{1}{\lambda}F)\Psi_\lambda, \tag{B.6}$$

except for the exactly critical case ($\lambda = 1$ and $\alpha = 0$). The solution of Eq. (B.6), the k_{eff} problem, is the calculation most commonly done with steady-state transport codes. Because the α eigenvalue appears in Eq. (B.3) as a $1/v$ absorber (hence the name time absorption), the energy spectra of Ψ_α and Ψ_λ can differ greatly for systems far from exactly critical. Furthermore, the spatial distributions can also differ greatly. For systems far subcritical, the Ψ_α eigenfunction can have a convex, downward spatial distribution, contrary to physical intuition.

For this appendix, it is sufficient to proceed directly to the multi-group [B.6, B.12] form of Eq. (B.3), written as

$$L\Psi + (\Sigma_t + \frac{\alpha}{v})\Psi(\vec{r}, \vec{\Omega}) = (S + \frac{1}{\lambda}F)\Psi, \tag{B.7}$$

where $\Psi(\vec{r}, \vec{\Omega})$ is a vector of multi-group angular fluxes, $\Psi_g(\vec{r}, \vec{\Omega})$, $g = 1$ to IGM, and L, Σ_t, S, and F are the multi-group approximations to the corresponding operators in Eq. (B.3). The intermediate eigenvalue λ has been introduced and the solution, α and Ψ, of Eq. (B.7) is sought such that $\lambda = 1$. The angular variable $\vec{\Omega}$ will be treated with the standard discrete-ordinates approximation [B.12]. For the schemes to be described in this appendix, the actual geometry and the spatial approximation used are immaterial, although the testing was done for two-dimensional geometry with a finite difference (diamond difference [B.12, B.14]) approximation of the spatial variable $\vec{r}$.

The existence of a dominant eigenvalue, α_0, for the multi-group transport Eq. (B.7), for finite media, under very general conditions, has been shown by Larsen.[B.10].

This appendix is organized in the following manner. Section II briefly describes the methods currently used in most current transport codes. Section III and IV outline the application of coarse mesh re-balance (CMR) [B.3] acceleration to the α eigenvalue search. Section V details some modifications to the iteration strategy that can improve computational efficiency. Section VI presents some numerical results for typical test problems to demonstrate the efficiencies effected by these improved methods. Finally,

Section VII summarizes the mathematical literature for the α eigenvalue problem and describes a procedure to make more robust the eigenvalue search for subcritical systems, a particularly difficult calculation. Section VIII provides the coding details of the two α re-balance schemes ($GCCMR$ and $WSGR$). The last section gives the conclusions.

B.2 Standard Methods

The standard iterative methods for solving the multi-group discrete-ordinates transport equation are detailed in the appropriate code manuals [B.1, B.3] and elsewhere. [B.6] The inner iterations for a single energy group are performed on the within-group scatter source. The outer iterations are performed on the fission source, yielding the inter-mediate eigenvalue λ of Eq. (B.3) upon convergence. For $\alpha = 0$, this λ is merely k_{eff} for the system.

The α eigenvalue problem is solved as a sequence of λ eigenvalue problem until an α for $\lambda = 1$ is obtained. This α iteration overlying the outer and inner iterations can be written as

$$L\Psi^k + (\Sigma_t + \alpha^k v^{-1})\Psi^k = (S + \frac{1}{\lambda^k}F)\Psi^k, \qquad (B.8)$$

where k is the α iteration index. The procedure by which the next α^{k+1} is selected, based on α^k, λ^k and Ψ^k, constitute the α eigenvalue search procedure, the subject of this appendix.

The standard eigenvalue search procedure is roughly as follows:
1. Make an initial guess of eigenvalue α^0, usually $= 0$.
2. Solve the transport Eq. (B.8), performing inner and outer iterations until convergence, for λ^0.
3. Obtain the second guess α^1 by adjusting α^0 with the eigenvalue modifier EVM, such that $\alpha^1 = \alpha^0 + EVM$.
4. Solve the transport Eq. (B.8) for λ^1.
5. Using the points (λ^k, α^k) and $(\lambda^{k-1}, \alpha^{k-1})$, perform a linear extrapolation of $\lambda(\alpha)$ to that α^{k+1} for which $\lambda^{k+1}(\alpha^{k+1}) = 1$. Or alternatively, using the points (λ^k, α^k), $(\lambda^{k-1}, \alpha^{k-1})$ and $(\lambda^{k-2}, \alpha^{k-2})$, perform a quadratic extrapolation of $\lambda(\alpha)$ to that α^{k+1} for which $\lambda^{k+1}(\alpha^{k+1}) = 1$.
6. Solve the transport Eq. (B.8) for λ^{k+1}.
7. Repeat (go back to Step 5) until $\lambda^{k+1} = 1$.

The code user is required to provide an initial eigenvalue guess α^0 and an eigenvalue modifier EVM. In many cases, particularly if the neutronics are coupled to a hydrodynamics calculation, the user has little idea of the value, or even the sign, of α^0. The eigenvalue modifier is seldom more than a wild guess. A poor choice of EVM, either in magnitude or sign, can significantly slow the search procedure, if not, in the case of subcritical systems, cause it to fail. In the early α iterations, the root-finding procedure of step 5 can sometimes cause the α search to flounder about.

Further details of the eigenvalue search procedure are described in the code manuals. [B.1-B.3] Additional input parameters are available to make the eigenvalue search proceed more efficiently. However, seldom is enough known about a problem to permit an a priori selection of these parameters, even by the most intelligent code users.

Because of these deficiencies in the α search procedure, it is desirable to devise an alternate procedure, one preferably based on the physics of the problem, that will produce more rapid convergence and require fewer input parameters from the code user.

B.3   Group-Collapse Coarse Mesh Re-balance

The coarse mesh re-balance (CMR) method [B.1, B.3, B.12] has been found to be an efficient means of accelerating the inner and outer iterations of the transport equation solution. This method is based on integrating the transport equation over various coarse mesh spatial regions to obtain a (usually) small system of equations for re-balance factors; factors which, when used to multiply the fluxes, at the end of each iteration force the re-balanced fluxes to satisfy particle conservation over each coarse mesh zone. This procedure has been found to effect a sometimes substantial acceleration of the iterations. For most problems, the coarse mesh used for the re-balance acceleration is simply chosen to be the material mesh. For example, a metal shell may be divided into 5 zones in the radius direction, and, now, the whole metal shell is used as one zone only for the coarse mesh.

Typically, for the outer iteration acceleration, a group-collapse rebalance, obtained by summing over all energy groups, is used to obtain a matrix eigenvalue equation for the intermediate eigenvalue λ. Because the eigenvalue α appears as an explicit scalar variable in Eq. (B.7), similar to λ, the CMR procedure can also be applied to accelerate the α iterations of Eq. (B.8).

To obtain the group-collapse coarse mesh re-balance ($GCCMR$) equations, the solution of Eq. (B.7) is sought for the eigenvalue α with $\lambda = 1$. Multiplying the fluxes in Eq. (B.7) by coarse mesh dependent re-balance factors, f_k, and integrating over all angles, all energy groups (group collapse), and all mesh cells in coarse mesh zone K yields a matrix eigenvalue equation

$$[FL + AB - FS]f = -(\alpha)(FV)f, \qquad (B.9)$$

of size equal to the number of coarse mesh zones KM, for the eigenvalue α and eigenvector of re-balance factors f. The diagonal matrices AB, FS, and FV are

$$AB_K = \int_K dV \int dE \Sigma_a \phi = \text{absorption}, \qquad (B.10)$$

$$FS_K = \int_K dV \int dE \nu \Sigma_f \phi = \text{fission neutron production}, \qquad (B.11)$$

and

$$FV_K = \int_K dV \int dE \frac{1}{V} \phi = \text{total neutrons}. \qquad (B.12)$$

Here, ϕ is the spatial and energy-dependent scalar flux, ν is the average neutron number per fission, Σ_f is the fission cross section, and Σ_a is the macroscopic effective absorption cross section [B.1, B.3]

$$\Sigma_a(E) = \Sigma_t(E) - \int dE' \int d\mu_0 \Sigma_s(E \to E', \mu_0), \qquad (B.13)$$

where $\mu_0 = \vec{\Omega} \cdot \vec{\Omega}' =$ scattering angle, Σ_t is the total cross section and Σ_s is the macroscopic scattering cross section. In practice, the integrals of Eqs. (B.10-B.12) are carried out by sums over spatial mesh cells and energy groups in the discretized space.

The coarse mesh flow matrix, FL, consists of the group-summed outflows from zone K on the diagonal and negative inflows from zone I into zone J on the off-diagonal elements:

$$FL = \begin{bmatrix} OF_1 & -IF_{2\to1} & \ldots & -IF_{KM\to1} \\ -IF_{1\to2} & OF_2 & \ldots & \cdot \\ \cdot & \cdot & \ldots & \cdot \\ \cdot & & \ldots & \cdot \\ \cdot & & \ldots & \cdot \\ -IF_{1\to KM} & \cdot & \ldots & OF_{KM} \end{bmatrix} \quad (B.14)$$

For one-dimensional geometries, this flow matrix will be tri-diagonal. In general, for multidimensional, orthogonal and nonorthogonal meshes, it will be a full, nonsymmetric, diagonally dominant matrix. It can be noted that

$$OF_K \geq \Sigma_{I=1}^{KM} IF_{K\to I} \text{ for } I \neq K, \quad (B.15)$$

with the equality holding when coarse mesh zone K has no surfaces on the outer boundary. This implies the L_1 norm $\parallel FL \parallel > 0$.

In general, Eq. (B.9) permits the existence of complex eigenvalues, necessitating the use of a generalized eigenvalue-eigenvector routine for its solution if all the eigenvalues are to be obtained. Experience on a variety of problems has shown the eigenvalue spectrum of Eq. (B.9) to range from negative eigenvalues, large in magnitude, occasionally in complex conjugate pairs, to a most positive, dominant eigenvalue corresponding to α_0 of Eq. (B.4 and B.5). It can, in fact, be shown [B.15] that there exists a dominant eigenvalue of Eq. (B.9), with zero imaginary component, which corresponds to an eigenvector with entirely positive components.

Because of the large negative eigenvalues, the simple power iteration [B.1] which is used to solve the outer CMR equations for λ cannot be used for Eq. (B.9) inasmuch as it converges to the largest eigenvalue in magnitude, not the most positive one. However, the inverse power iteration [B.16] (Wielandts method of fractional iteration) can be applied to obtain the dominant eigenvalue of Eq. (B.9). This method involves choosing some estimate of an eigenvalue and this modified power method then converges to the eigenvalue nearest that guess. By choosing an initial guess of the dominant eigenvalue sufficiently large, this inverse power iteration will converge to that desired, most positive eigenvalue. In typical problems, this inverse power iteration is found to converge quite rapidly, usually in from 3 to 8 iterations.

The $GCCMR$ is performed as follows: During the inner iterations for each group, when the angular flux is being calculated, the flows between each region $IF_{I \to J}$ and the outflows for each region OF_K are computed. At the completion of the inners for that group, these flows are accumulated (group summed) into the flow array, FL. Upon convergence of the outer iterations (for a particular α guess), the three flux integrals of Eqs. (B.10-B.12) are computed for each coarse mesh zone. The matrix Eq. (B.9) is then solved for a dominant eigenvalue, which will then be used as the next α guess. The re-balance factors, f, of Eq. (B.9) are applied to the scalar flux and moments and the next α iteration is then begun.

This $GCCMR$ α search procedure eliminates the need for the eigenvalue modifier and the root-finding procedure of the standard method. In fact, the intermediate eigenvalue λ plays no role in this search procedure. In practice, it has been found that $GCCMR$ gives remarkably good estimates of α for the early iterations, but that it converges much more slowly than the root-finding procedure in the later stages. Thus, the recommended procedure is to use $GCCMR$ for the first few (three or four) α estimates, then switch over to a root-finding procedure for the final convergence.

A simple whole system variant of the α CMR can be derived. Integrating Eq. (B.8) over all energy groups and all space points yields the balance equation

$$NL^k + AB^k + \alpha^k FV^k = \frac{1}{\lambda^k} FS^k, \qquad (B.16)$$

for the kth α iteration, where NL is the total net leakage of neutrons from the system. We would like the $k+1$st iteration to satisfy the balance equation with $\lambda = 1$

$$NL^{k+1} + AB^{k+1} + \alpha^{k+1} FV^{k+1} = FS^{k+1}. \qquad (B.17)$$

Subtracting Eq. (B.17) from Eq. (B.16) and assuming $\phi^{k+1} \cong \phi^k$ (so that $NL^{k+l} \cong NL^k$, $AB^{k+l} \cong AB^k$, etc.) yield

$$\alpha^{k+1} = \alpha^k + (1 - \frac{1}{\lambda^k}) \frac{FS^k}{FV^k}, \qquad (B.18)$$

as the next estimate for the eigenvalue. Equation (B.9) for the case of a material region can be shown equivalent to Eq. (B.18).

The $GCCMR$ Eq. (B.9) yields not only the next guess for the eigenvalue α, but also the vector of re-balance factors f which are applied to all the fluxes in each coarse mesh zone. Since the magnitude of this eigenvector is arbitrary, normalization is typically chosen to maintain a total fission neutron source of unity

$$FT = \Sigma_{k=1}^{KM} FS_k = 1. \tag{B.19}$$

At the completion of the first α iteration, the re-balance factors for the outer re-balance equation

$$[FL^0 + AB^0]f^0_{outer} = \frac{1}{\lambda^0} FS^0 f^0_{outer}, \tag{B.20}$$

will be identically unity ($\alpha^0 = 0$ is assumed), where $\lambda^0 = k_{eff}$. The re-balance factors from the first α re-balance equation

$$[FL^0 + AB^0 - FS^0]f^0_\alpha = -\alpha^1 FV^0 f^0_\alpha, \tag{B.21}$$

will be those coarse mesh re-balance factors that convert the spatial distribution of the k_{eff} solution fluxes to the approximate spatial distribution of the converged α eigenvalue fluxes. Clearly, these re-balance factors will differ the most from unity (and, hence, provide the most acceleration) for problems in which the α spatial distribution differs greatly from the k_{eff} spatial distribution.

This will occur for problems far removed from critical ($\alpha = 0$) and with much spatial inhomogeneity. For spatially homogeneous problems (where $KM = 1$), there is no acceleration from the re-balance factor ($f^0_\alpha = 1$), only from the improved estimate of α^1. In practice, only a modest amount of the acceleration from $GCCMR$ comes from the re-balance factors; most of the acceleration comes from the improved estimates of α. For a typical test problem (#4) described in Section VI, which contains considerable spatial structure, these re-balance factors for f^0_α ranged from 1.43 in the innermost material zone, to 0.25 in the outermost zone.

B.4 Whole System Group-wise Re-balance

The α eigenvalue appears in Eq. (B.3) as a $1/v$ absorber. For problems with many energy groups and a broad range of neutron speeds, the α eigenvalue can greatly change the spectral distribution from the k_{eff} solution. This suggests a re-balance that does not integrate out the energy dependence (no group collapse), but that maintains a group-wise dependence, might provide an effective α acceleration.

One such procedure would be to eliminate the integration over all energies in deriving the CMR equations in Section III. This would yield a matrix equation of size $KM * IGM$, the number of coarse mesh regions times the number of energy groups. For many problems, this can be of size 50 to 100 or larger, resulting in a rather large re-balance equation to be solved, and greatly increasing the usually negligible overhead to perform the re-balance acceleration.

A second, more feasible, procedure to obtain re-balance equations would be to eliminate the integration over all energy groups, but integrate over all space, rather than each coarse mesh zone. This whole system group-wise re-balance ($WSGR$) matrix equation will be only of size IGM, the number of energy groups.

To obtain the $WSGR$ equations, we start with the transport equation, Eq. (B.7) for one energy group g with the condition of $\lambda = 1$.

$$L\Psi_g + (\Sigma_t + \frac{\alpha}{V_g})\Psi_g(\vec{r},\vec{\Omega}) = \Sigma_{g'=1}^{g}\Sigma_{s,g'\to g}\phi_{g'} + \chi_g \Sigma_{g'=1}^{IGM}\nu\Sigma_{f,g'}\phi_{g'}. \quad (B.22)$$

In Eq. (B.22) we have assumed no up-scattering, although the method can just as easily treat problems with up-scatter. Multiplying the fluxes in Eq. (B.22) by group-wise dependent re-balance factors, f_g, and integrating over all angles and all mesh cells yields the $WSGR$ matrix eigenvalue equation

$$[NL + C - S - FS]f = -(\alpha)(FV)f, \quad (B.23)$$

of size equal to the number of energy groups IGM, for the eigenvalue α and the eigenvector of re-balance factors f. In Eq. (B.23), S is the scattering neutron source, FS the fission production neutron source, NL neutron sink due to leakage and C the neutron sink from collision.

The diagonal matrices NL, C, and FV are

$$NL_g = \text{net leakage from the system for group } g, \quad (B.24)$$

$$C_g = \int dV \Sigma_{t,g}\phi_g = \text{within-group total collision rate in group } g, \quad (B.25)$$

$$FV_g = \int dV \frac{1}{V_g} \phi_g = \text{total neutrons in group } g. \qquad (B.26)$$

Again, in practice, the integrals over all space in Eqs. (B.24-B.26) are carried out as sums over spatial-mesh cells in the discretized space. The lower triangular (for down-scatter only) matrix S of group-to-group scattering rates is of the form

$$[S] = \begin{bmatrix} \int dV \Sigma_{1\to 1}\phi_1 & & & & \cdot & & \cdot \cdot & & & \cdot \\ \int dV \Sigma_{1\to 2}\phi_1 & \int dV \Sigma_{2\to 2}\phi_2 & & \cdot \cdot & & & & & \cdot \\ \cdot & & \cdot & & \cdot \cdot & & & \cdot & \\ \cdot & & & \cdot & & \cdot \cdot & & & \cdot \\ \int dV \Sigma_{1\to IGM}\phi_1 & \int dV \Sigma_{2\to IGM}\phi_2 & \cdot \cdot & \int dV \Sigma_{IGM\to IGM}\phi_{IGM} \end{bmatrix}, \qquad (B.27)$$

or

$$[S_{gg'}] = [\int dV \Sigma_{g'\to g}\phi_{g'}]. \qquad (B.28)$$

The full matrix FS of fission production rates is of the form

$$[FS] = \begin{bmatrix} \chi_1 & \cdot & \cdot \cdot & & \cdot \\ & \cdot & \chi_2 & \cdot \cdot & & \cdot \\ & \cdot & \cdot & \cdot \cdot & \cdot \\ & \cdot & \cdot & \cdot \cdot & \cdot \\ & \cdot & \cdot \cdot & \chi_{IGM} \end{bmatrix}$$

$$\times \begin{bmatrix} \int dV \nu \Sigma_{f1}\phi_1 & \int dV \nu \Sigma_{f2}\phi_2 & \cdots & \int dV \nu \Sigma_{fIGM}\phi_{IGM} \\ \cdot & \cdot & \cdots & \cdot \\ \cdot & \cdot & \cdots & \cdot \\ \cdot & \cdot & \cdots & \cdot \\ \cdot & \cdot & \cdots & \cdot \end{bmatrix}, \qquad (B.29)$$

or

$$[FS_{gg'}] = [\chi_g \int dV \nu \Sigma_{fg'}\phi_{g'}], \qquad (B.30)$$

where χ_g is the fission neutron distribution function. For one-group problems, Eq. (B.23) can be shown identical to Eq. (B.18). As in $GCCMR$, the non-symmetry of Eq. (B.23) admits complex eigenvalues. Test calculations have shown the eigenvalue spectrum range from negative eigenvalues, large in magnitude, occasionally in complex conjugate pairs, to the most positive, dominant eigenvalue, corresponding to α_0. By the same proof as for $GCCMR$ [B.15], the existence of a dominant eigenvalue, with zero imaginary component, corresponding to a eigenvector with entirely positive components, can be shown. Implementation of the $WSGR$ is done exactly as the $GCCMR$, with the inverse power iteration being used to solve Eq. (B.23) for the dominant eigenvalue.

For a typical test problem (#4) described in Section VI, with 12 energy groups, the re-balance factors of Eq. (B.23) range from 2.27 for the first group down to 1.56×10^{-5} for the bottom group, indicating that a large spectral effect is, in fact, present.

Between the two α re-balance schemes, $GCCMR$ or $WSGR$, it is not clear, a priori, which is the better. In practice, the scheme of choice is found to be problem dependent. Since most code users would seldom have the knowledge to correctly choose between the two schemes, this suggests that a hybrid re-balance scheme, using first $GCCMR$ and then $WSGR$, might have some merit. Since convergence of the $\alpha^n \to \alpha_0$ is often monotonic in the early α iterations, the next α guess can be chosen (from between the two different re-balance α's) to produce the largest change from the current α guess. However, both sets of re-balance factors, from $GCCMR$ and $WSGR$, can be applied to the scalar flux and moments. By using both sets of re-balance factors, it is hoped that a serendipitous acceleration of the α iterations is achieved. In practice, this does not occur and the hybrid re-balance merely results in the most effective re-balance scheme being automatically used.

B.5 Variable Convergence Precision and Iteration Strategies

In this section, we discuss some of the modifications to the iteration strategy that can be made to improve the overall efficiency of the calculation.

The standard method for α eigenvalue problems consists of a set of λ (or k_{eff}) eigenvalue problems and use of a root-finding procedure, linear or quadratic, to estimate, usually by an extrapolation, the next α guess. Because the root-finding procedure can often lead to wild extrapolations, especially for subcritical systems, it is important that each λ calculation, be fully converged.

If a re-balance method, as opposed to a root-finding method, is utilized to provide the next α estimate, it may be possible to converge rather loosely the early λ calculations, when the estimate of α is poor, inasmuch as the re-balance is a function only of the fluxes.

One scheme for permitting a loose convergence early in the α calculation is termed variable convergence precision. In typical discrete-ordinates transport codes, there are several different convergence precisions, fixed at the start of the calculation, that are used to determine convergence of the various iterations. Two of these convergence precision are on the inner iterations, ϵ_i, so that

$$Max \left\{ \frac{|\phi_{ijg}^{\ell} - \phi_{ijg}^{\ell-1}|}{\phi_{ijg}^{\ell}} \right\}_{ij} \leq \epsilon_i, \qquad (B.31)$$

where ℓ = inner iteration index, ij = spatial mesh index; and on the outer iterations, ϵ_0, so that

$$|\lambda^n - \lambda^{n-1}| \leq \epsilon_0, \qquad (B.32)$$

where n = outer iteration index. The α calculation is continued until the final convergence test is satisfied

$$|\lambda^k - 1| \leq \epsilon_{\text{final}}, \qquad (B.33)$$

Table B.3 The two re-balance estimates of α.

Re-balance	α^1
GCCMR	16.79 ($\alpha_0 = 18.32$)
WSGR	8.42 ($\alpha_0 = 18.32$)

Table B.4 The complete eigenvalue spectrum and eigenvectors of re-balance factors of $GCCMR$ scheme.

Eigenvalues	$GCCMR$ Re-balance Factors by Material Region
$\alpha_0^1 = 16.79$ (Mat 1)	(1.738, 0.2615) (eigenvector)
$\alpha_1^1 = -3.23$ (Mat 2)	(-0.1838, 2.1830)

Table B.5 The eigenvalue spectrum and eigenvectors of re-balance factors of $GCCMR$ scheme after the third α iteration.

Eigenvalues	$GCCMR$ Re-balance Factors by Material Region
$\alpha_1^3 = 18.32$ (Mat 1)	(1.0004, 0.9996) (eigenvector)
$\alpha_1^3 = -53.69$ (Mat 2)	(-0.0076, 2.0076)

b. the $GCCMR$ is the more effective re-balance, which one would expect for a one-group problem.

c. no acceleration is coming from the re-balance factors.

d. the best overall improvement ($MXOUT5$ scheme) results in 1/5 the computational effort of the standard scheme.

Problem #2

This is another small, hypothetical, two-group, two-material, spherical test problem. The material and spatial meshes are the same as for Problem #1. The two group macroscopic cross sections (cm^{-1}) are shown in Table B.6.

The results for this problem are shown in Table B.7.

At the completion of the first α iteration (for which $k_{eff} = 1.816$), the two re-balance estimates of α are shown in Table B.8.

Table B.6 The macroscopic cross sections (cm^{-1}) for each material.

Mat	Group	Σ_a	$\nu\Sigma_f$	Σ_t	$\Sigma_{g \to g}$
1	1	2.0	4.0	3.0	1.0
	2	1.0	2.0	1.5	0.5
2	1	0.1	0	2.1	2.0
	2	0.2	0	4.2	4.0

$\chi = (g_1, g_2)$ $v = (g_1, g_2) cm/sh$

$\chi = (0.7, 0.3)$ $v = (10.0, 0.1) cm/sh$

$\Sigma_{1 \to 2}$ (no down-scatter)

Table B.7 The total number of inner iterations for the schemes.

Scheme	Total Inners
STD	No convergence
VCP	773
GCCMR	714
WSGR	531 (best re-balance)
HYB	534
HYBNOF	557 (re-balance factors accelerate)
MXOUT5	266

For the $GCCMR$ at the end of the first α iteration, the complete eigenvalue spectrum and eigenvectors of re-balance factors are shown in Table B.9.

For the $WSGR$ at the end of the first α iteration, the complete eigenvalue spectrum and eigenvectors of re-balance factors are shown in Table B.10.

Table B.8 The two re-balance estimates of α.

Re-balance	α^1
GCCMR	0.276 ($\alpha_0 = 6.367$)
WSGR	3.284 ($\alpha_0 = 6.367$)

Table B.9 The complete eigenvalue spectrum and eigenvectors of re-balance factors of $GCCMR$ scheme.

Eigenvalues	GCCMR Re-balance Factors by Material Region
$\alpha_0^1 = 0.276$ (Mat 1)	(1.648, 0.3520) (eigenvector)
$\alpha_1^1 = -0.105$ (Mat 2)	(-0.459, 2.459) (eigenvector)

Table B.10 The complete eigenvalue spectrum and eigenvectors of re-balance factors of $WSGR$ scheme.

Eigenvalues	WSGR Re-balance Factors by Material Region
$\alpha_0^1 = 3.284$ (Mat 1)	(1.955, 0.0450) (eigenvector)
$\alpha_1^1 = -0.150$ (Mat 2)	(5.267, -0.267) (eigenvector)

From the Table B.10 of total inners, we conclude

a. the $WSGR$ is the more effective re-balance, which one might expect because of the strong spectral skewing between the two material regions.
b. less than 5% of the acceleration is due to the re-balance factors.
c. the best overall improvement results from the $MXOUT5$ scheme, in one-third the computational effort of the worst case (VCP scheme).

Problem #3

This is a more realistic problem of a small central sphere ($R = 5\,cm$) of H_2O, surrounded by a sphere of pure U^{235} ($R = 20\,cm$). All materials are at nominal density. A mesh spacing is used of 40 cells in the azimuthal direction and 20 cells in the radial direction (5 in the H_2O, 15 in the U^{235}), as shown in Fig. B.2.

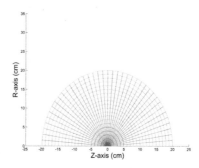

Fig. B.2 Problem #3 material mesh and spatial mesh.

Table B.11 The total number of inner iterations for the schemes.

Scheme	Total Inners
STD	3511
VCP	1175
GCCMR	1147
WSGR	984 (best re-balance)
HYB	983
HYBNOF	1127 (re-balance factors accelerate)
MXOUT5	847

A twelve-group, fast neutron cross-section set is used, with velocities ranging from $v_1 = 51.93\,cm/sh$ down to $v_{12} = 0.00333\,cm/sh$. The results for this problem are shown in Table B.11.

At the completion of the first α iteration (for which $k_{eff} = 1.706$), the two re-balance estimates of α are shown in Table B.12.

For the $GCCMR$ at the end of the first α iteration, the complete eigenvalue spectrum and eigenvectors of re-balance factors are shown in Table B.13.

For the $WSGR$ at the complete convergence of the problem, the eigenvalue spectrum for all 12 groups is shown in Fig. B.3.

Table B.12 The two re-balance estimates of α.

Re-balance	α^1
GCCMR	0.479 ($\alpha_0 = 0.844$)
WSGR	0.792 ($\alpha_0 = 0.844$)

Table B.13 The complete eigenvalue spectrum and eigenvectors of re-balance factors of $GCCMR$ scheme.

Eigenvalues	$GCCMR$ Re-balance Factors by Material Region
$\alpha_0^1 = 0.479$ (Mat 1)	(1.9893, 0.01071) (eigenvector)
$\alpha_1^1 = -0.0026$ (Mat 2)	(-0.1982, 2.1982) (eigenvector)

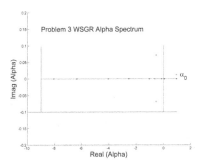

Fig. B.3 Problem #3 $WSGR$ α spectrum at convergence.

From Table B.11 of total inners, we conclude

a. the variable convergence precision reduces the total inners by two-thirds.
b. the $WSGR$ is the more effective, which one would expect from the strong spectral effect of the water.
c. about 15% of the acceleration is coming from the re-balance factors.
d. the best overall improvement($MXOUT5$ scheme) results in 1/4 the computational effort of the standard scheme.

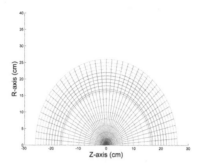

Fig. B.4 Problem #4 material and spatial meshes.

Table B.14 The total number of inner iterations for the schemes.

Scheme	Total Inners
STD	5075
VCP	1536
GCCMR	1287
WSGR	1427 (best re-balance)
HYB	1273
HYBNOF	1315 (re-balance factors accelerate)
MXOUT5	1013

Problem #4

This is a five-material problem with a great deal of spatial inhomogeneity, consisting of concentric spheres of Al (R = 2 cm), U^{235} (R = 16 cm), Fe (R = 17 cm), U^{238} (R = 22 cm), and C (R = 26 cm), all at nominal density. The material and spatial mesh is shown in Fig. B.4.

The same 12-group cross-section set is used. The results for this problem are shown in Table B.14.

At the completion of the first α iteration (for which $k_{eff} = 1.726$), the two re-balance estimates of α are shown in Table B.15.

Table B.15 The two re-balance estimates of α.

Re-balance	α^1
GCCMR	0.620 ($\alpha_0 = 0.826$)
WSGR	0.538 ($\alpha_0 = 0.826$)

Table B.16 The complete eigenvalue spectrum and eigenvectors of re-balance factors of $GCCMR$ scheme.

Eigenvalues	$GCCMR$ re-balance factors by material region
$\alpha_0^1 = 0.620$ (Mat 1)	(1.427, 1.652, 1.142, 0.529, 0.247) (eigenvector)
$\alpha_1^1 = -0.258$ (Mat 2)	(0.119, 0.112, -0.419, -1.50, -2.85) (eigenvector)
$\alpha_2^1 = -0.803$ (Mat 3)	(-0.092, -0.074, 0.480, 1.41, -2.94) (eigenvector)
$\alpha_3^1 = -3.97$ (Mat 4)	(-4.93, 0.012, 0.046, -0.008, 0.001) (eigenvector)
$\alpha_4^1 = -4.82$ (Mat 5)	(0.638, -0.138, 3.64, -0.515, -0.005) (eigenvector)

For the $GCCMR$ at the end of the first α iteration, the complete eigenvalue spectrum and eigenvectors of re-balance factor are shown in Table B.16.

For the $WSGR$ at the complete convergence of the problem, the eigenvalue spectrum for all 12 groups is shown in Fig. B.5. The re-balance factors at the end of the first α iteration are those factors required to convert the k_{eff} solution fluxes to the converged α solution fluxes. By performing a k_{eff} ($\alpha = 0$) calculation, then performing the full α calculation, and then integrating both sets of fluxes over all groups and each coarse mesh zone, it is possible to calculate the exact re-balance factors that are required to convert the k_{eff} solution to the converged α solution. These exact re-balance factors are compared to the approximate re-balance factors at the end of the first α iteration as shown in Fig. B.5.

From Tables B.17 and B.18, we conclude

a. the variable convergence precision reduces the total inners required by 2/3.
b. the $GCCMR$ is more effective, which one might expect from the considerable spatial structure of this problem.

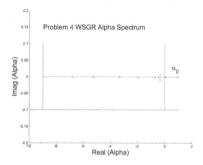

Fig. B.5 Problem #4 $WSGR$ α spectrum at convergence.

Table B.17 The re-balance factors by material zone.

Exact	$GCCMR$
1.581	1.429
1.481	1.653
0.911	1.421
0.634	0.529
0.394	0.247

c. only about 3% of the acceleration is coming from the re-balance factors.

d. the best overall improvement ($MXOUT5$ scheme) results in 1/5 the computational effort of the standard scheme.

Problem #5

This is a simple, homogeneous, three-group, spherical test problem with a radius of 4.0 cm. The problem was run as a 90° segment of the sphere, using 64 mesh cells in the azimuthal direction and 64 mesh cells in the radial direction as shown in Fig. B.6.

The hypothetical three-group macroscopic cross sections, representing two groups and a thermal group are shown in Table B.19.

One of the defects of coarse mesh re-balance is its instability as the

Table B.18 The re-balance factors by group.

Exact	WSGR
2.519	2.265
2.438	2.220
2.272	2.121
1.999	1.936
1.377	1.470
0.723	0.904
0.375	0.549
0.187	0.326
0.108	0.199
3.37 E-4	1.16 E-2
2.39 E-6	1.19 E-4
9.84 E-10	1.56 E-5

re-balance mesh approaches the fine mesh, for some problems. This problem is designed to see if that defect occurs for the α CMR equations. The spatial re-balance mesh is obtained by repeatedly dividing the spatial domain in both the radial and azimuthal directions. Thus, the problem is run with 1 (4096 cells/coarse mesh zone: Mesh A), 4 (1024 cells/coarse mesh zone: Mesh B), 16 (256 cells/coarse mesh zone: Mesh C), and 64 (64 cells/coarse mesh zone: Mesh D) coarse mesh re-balance zones. In addition, the problem is run with a coarse 8 x 8 spatial mesh and 64 (1 cell/coarse mesh zone: Mesh DCM) re-balance zones; thus, fine mesh re-balance. The results for this problem are shown in Table B.20.

At the completion of the first α iteration (for which $k_{eff}=$ 1.948), the two re-balance estimates of α for each mesh are shown in Table B.21.

For the fine mesh re-balance case (Mesh DCM), the first set of outer iterations fails to converge, falling into a two-cycle oscillating mode, which

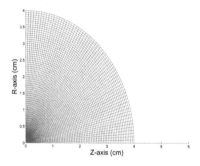

Fig. B.6 Problem #5 fine spatial mesh (64 × 64).

Table B.19 The hypothetical three-group macroscopic cross sections (cm^{-1}), representing two fast groups and a thermal group.

Group	Σ_a	$\nu\Sigma_f$	Σ_t	$\Sigma_{g \to g}$	$\Sigma_{g-1 \to g}$	$\Sigma_{g-2 \to g}$
1	0.25	0.75	0.80	0.25	–	–
2	0.20	0.60	0.70	0.30	0.20	–
3	0.10	0.3	10.	9.9	0.20	0.10

$\chi = (g_1, g_2, g_3)$

$\chi = (0.75, 0.2, 0.05)$

$v = (g_1, g_2, g_3) cm/sh$

$v = (10.0, 0.1, 0.001) cm/sh$

prevents the convergence of λ, except for the $MXOUT5$ scheme in which the outer iteration limit terminates the outers. With the outer iteration limit imposed, both α re-balance schemes, $GCCMR$ and $WSGR$, perform well and give no indication of a stability problem for fine mesh re-balance. The $GCCMR$ gives particularly poor estimates of α for this problem. This might be expected, since the spatial distribution for the k_{eff} problem and the α problem do not differ greatly for homogeneous systems.

For the $GCCMR$ at the complete convergence of the problem, the eigenvalue spectra for the various re-balance meshes are shown in Figs. B.7–B.10.

For this problem, we conclude

Eigenvalue Calculations

Table B.20 The total number of inner iterations for the schemes, notations: (a) fine mesh re-balance: 8 × 8 coarse spatial mesh, (b) first set of outer iterations fails to converge.

Scheme	Total Inners				
	A	B	C	D	DCM^a
STD	1196	729	452	374	b
VCP	218	201	178	137	b
GCCMR	285	244	222	170	b
WSGR	225	119	83	52	b
HYB	225	110	95	61	b
HYBNOF	233	141	97	64	b
MXOUT5	162	111	78	71	114

Table B.21 The two re-balance estimates of α for each mesh, notations: (a) fine mesh re-balance: 8 × 8 coarse spatial mesh.

Re-balance	A	B	C	D	DCM^a
GCCMR	0.00041	0.00041	0.00041	0.00040	0.00040
WSGR	0.425	0.456	0.475	0.476	0.382
α_0	0.485	0.485	0.485		0.491

a. the variable convergence precision reduces the total inners required by 4/5 to nearly 2/3.
b. the $WSGR$ is more effective, which one might expect from the strong spectral effect due to the bottom group.
c. neither the $GCCMR$ or $WSGR$ appears to have stability problems when the re-balance mesh and fine mesh are identical.
d. only a few per cent of the acceleration is coming from the re-balance factors.
e. the best overall improvement ($MXOIJT5$) scheme results in 1/8 to 1/5 the computational effort of the standard scheme.

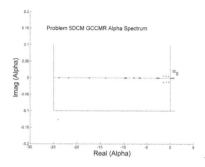

Fig. B.7 Problem #5:5DCM $GCCMR$ α spectrum at convergence.

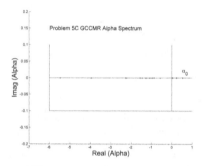

Fig. B.8 Problem #5:5C $GCCMR$ α spectrum at convergence.

B.7 Subcritical Searches

Alpha eigenvalue searches for subcritical systems are notoriously difficult to converge. For highly subcritical systems, a code crash (referred to as a dramatic failure in the code manuals Refs. [B.1–B.3]) is the usual result. In this section, we examine the causes for the code "failure" and describe a remedy for this deficiency.

There are many elegant mathematical papers [B.6–B.10] written on the α eigenvalue spectrum of Eq. (B.3). These can be summarized, for the most interesting case of a finite media, as follows: [B.9] For the continuous velocity variable, $v \in [0, \infty]$, including the limit $v = 0$, there exists a continuum of eigenvalues to the left of α^*, where

$$\alpha^* = -\max_{\vec{r}} \lim_{v \to 0} \left[v \Sigma_t(v) \right], \tag{B.35}$$

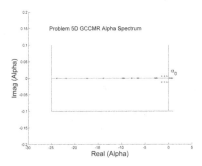

Fig. B.9 Problem #5:5D $GCCMR$ α spectrum at convergence.

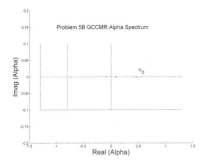

Fig. B.10 Problem #5:5D $GCCMR$ α spectrum at convergence.

and there may exist a discrete spectrum of points and, possibly, curves to the right of α^*, as illustrated in Fig. B.8. The existence of a discrete-eigenvalue spectrum is not guaranteed. If a discrete-eigenvalue spectrum exists, it has been shown [B.18, B.19] there exists a dominant eigenvalue, α_0, with $Im(\alpha_0) = 0$, $\alpha_0 > \alpha^*$, and $\psi_\alpha > 0$.

For velocity space bounded away from zero, $v \in [v_0, \infty]$, $v_0 > 0$, the continuum spectrum in Fig. B.11 becomes a discrete spectrum of points and, possibly, curves.

In the former case in which $v_0 \geq 0$, for sufficiently small bodies, the point spectrum to the right of α^* in Fig. B.11 can disappear. In this situation, the time-dependent flux decay is dominated by the $v \to 0$ limit, by neutrons that are moving very slowly through the medium. There has been considerable mathematical discussion on this "disappearance of the point spectrum into the continuum" and the existence of point eigenvalues

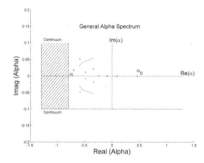

Fig. B.11 General eigenvalue spectrum.

within the continuum. Larsen and Zweifel [B.9] argue that this continuum part of the spectrum is a creature of the mathematics and does not correspond to physical reality; that at velocities $v \to 0$, quantum mechanical effects probably render the transport equation invalid. At these very low velocities, the neutron population density is undoubtedly so low as to make the transport equation inapplicable.

The case of the transport equation in the multi-group approximation has been analyzed by Larsen [B.10] and these esoteric mathematical ambiguities do not exist.

Under simple conditions on the cross sections that are virtually always met in practice (namely, that a neutron or its progeny in any one group can eventually transfer to any other group), he proved the existence of a dominant eigenvalue α_0 and a corresponding positive eigenfunction ψ_α. Thus, the failure of transport codes (which solve the multi-group transport equation) to calculate an α eigenvalue for a subcritical system is not due to the eigenvalue's nonexistence but due to some deficiency in the computational procedure.

For supercritical systems, α_0 (> 0) corresponds to a physically measurable quantity, the exponential growth of the flux at long times after any early transients (with $\alpha_i < 0$) have died out. The possibility of driving a nuclear system sufficiently supercritical to be supercritical in a higher eigen-mode apparently has not been examined.

For subcritical systems ($\alpha_i < 0$), the physical interpretation of the dominant eigenvalue α_0 has been subject to some debate. There are some strong arguments that no valid physical interpretation of the negative dominant eigenvalue (guaranteed by the multi-group transport equation) can be made, no matter how close to critical is the system. For some physical systems, experimenters have actually been unable to measure any pulsed neutron experiment die-away constants. [B.8] However, exponential die-away constants for subcritical $GODIVA$ assemblies [B.17], for water assemblies, [B.8] and for natural uranium systems [B.13] and Rossi-α [B.17] constants for various systems [B.17] have been measured and, in some cases, [B.13] have been found in good agreement with calculations.

For highly absorbing and for subcritical systems, the long-time or asymptotic distribution is dominated by very slowly moving neutrons. In this regime, the multi-group assumption may be a very poor approximation. The neutron population densities may be so low and neutron wave and other quantum effects so large that the results from a multi-group transport calculation have little relation to physical reality. Thus, code users should exercise considerable caution in physical interpretations of the calculated α_0 for far subcritical systems. This is demonstrated in the example problem to follow.

We will assume that the code user, for whatever reason, has a genuine need to calculate an α eigenvalue for a subcritical system. It is then necessary to understand why the present eigenvalue search algorithm fails, in order to devise a remedy.

If, at some point during the search, the α eigenvalue guess becomes sufficiently large and negative, then the effective total cross section,

$$\Sigma_{t,eff} = \Sigma_{t,g} + \frac{\alpha}{v_g}, \tag{B.36}$$

may become negative for some energy groups and mesh cells. Such negative total cross sections obviously have no meaning for the transport equation and the solution algorithms will assuredly fail. Thus, if one constrains the α search procedure to maintain

$$\alpha \geq \alpha^* = -\max_{ij} \min_g [v_g \Sigma_{t,g}], \tag{B.37}$$

which will usually occur in the bottom energy group (g $=$ IGH), then negative effective total cross sections will not occur. Unfortunately, this constraint is insufficient to insure convergence of the α eigenvalue search algorithm. Convergence failure will also occur when an α guess becomes sufficiently small so that the system becomes supercritical in one of the groups. Neglecting leakage, this will occur when

$$\Sigma_{t,g} + \frac{\alpha}{v_g} < \Sigma_{s,g}, \tag{B.38}$$

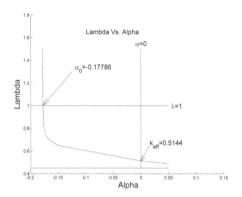

Fig. B.12 $\lambda(\alpha)$ for subcritical uranium sphere.

for some energy group g. When α becomes sufficiently large and negative so that, with leakage, some group becomes supercritical, the inner iterations will fail to converge for that group and, with subsequent additional outer or inner iterations, the flux will diverge until machine overflow occurs, the so-called "dramatic code failure."

To illustrate this situation, consider a homogeneous uranium sphere of radius 8.74 cm, composed of 75% U^{238} and 25% U^{235}, at normal density. For a five-group cross-section set and an $S_2 P_0$ approximation, a plot of λ versus α is shown in Fig. B.12.

For this problem, the system becomes supercritical in the bottom group whenever $\alpha < -0.18$ generations/shake(1 shake $= 10^{-8}s$).

From this curve in Fig. B.12, it is obvious why a conventional α eigenvalue search technique will fail. The linear extrapolation from the first two $\lambda(\alpha)$ points will yield a guess of $\alpha << -0.18 \ gen/sh$, far into the region of iteration divergence ($\alpha < \alpha_{min}$). This is illustrated in Fig. B.13. It should be pointed out that $\alpha^* = -1.018 gen/sh$ for this cross-section set, so that the region of iteration divergence ($\alpha < \alpha_{min}$) is bounded well to the right of the theoretical minimum.

The solution to the failure of a eigenvalue searches for subcritical systems is straightforward. We wish to

1. attempt to determine α_{min} so that any extrapolation will not enter the region of iteration divergence, and
2. restart the α search whenever it extrapolates into this region of iteration divergence.

Eigenvalue Calculations

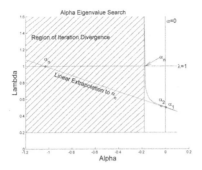

Fig. B.13 Failure of α search procedure.

The modifications to the search procedure are relatively minor:
- Set the inner iteration limit to a moderately large number (say 50).
- Monitor the inner iterations for divergence.

1. If the inner iteration limit is reached in any group, then the current α^n (n = α iteration index) is too far negative. Abort the current outer iteration. Store the current α^n into α_{min} if $\alpha^n >$ current α_{min}.
2. Choose a new α^{n+1} midway between this current α^n (for which the inners diverged) and the last α^{n-1} for which the outers converged

$$\alpha^{n+1} = (\alpha^n + \alpha^{n-1})/2. \tag{B.39}$$

3. Start a new set of outer iterations.
- Constrain the α guesses so that $\alpha^{n+1} > \alpha_{min}$.

1. If the linear or quadratic extrapolation or the re-balance procedure gives an $\alpha^{n+1} < \alpha_{min}$, then choose the next $\alpha^{n+1} = \alpha_{min} + \delta_n$, where δ is some small arbitrary number that changes with α iteration index n (say $\delta = 0.01/n$).

The success of this procedure is illustrated for the subcritical uranium sphere in Fig. B.14, where the λ's for the various α guesses are plotted. In this case, the $GCCMR$ scheme was utilized for the first two α iterations (namely α^2 and α^5), rather than an eigenvalue modifier and a linear extrapolation. Once an α^n for which $\lambda > 1$ is found, the eigenvalue iterations converge very rapidly.

It should be noted that all current discrete-ordinates transport codes cannot calculate α eigenvalues for systems without any fissile material, since the λ eigenvalue is not defined for such systems. To modify these codes whose eigenvalue search procedure is based on roots of the $\lambda(\alpha)$

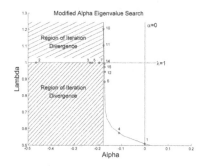

Fig. B.14 Modified search procedure.

curve requires rather major surgery to the code. However, a code whose eigenvalue search is based entirely on a re-balance scheme, which does not require the λ eigenvalue, can be modified with little effort to calculate such eigenvalues.

The α search procedure described above is sufficiently robust to yield the dominant eigenvalue of the multi-group transport equation, $\alpha_0 = -0.178$, for this problem and this multi-group structure. One might ask if this computed α_0 is a physically meaningful quantity; if this α_0 corresponds to an exponential die-away constant for this system. In this case, the answer is probably no.

By changing the cross-section multi-group structure, one finds that the α_0 for this problem is extremely sensitive to changes in the bottom energy group. By changing the lowest group velocity by a factor of 2 ($v_5 = 1.444 cm/sh \rightarrow 2.888 cm/sh$), then $\alpha_0 = -0.178 gen/sh \rightarrow -0.351 gen/sh$, or also changes by a factor of 2, with no change in k_{eff}. Conversely, if one increases the group 1 $\nu\Sigma_f$ by a factor of 2, one noticeably increases k_{eff}, but there is virtually no change in α_0. Thus, for this problem, the calculated α_0 is virtually a function only of the multi-group structure and, in particular, the bottom neutron energy group speed. In this case, it is highly suspect that the calculated α_0 has any meaningful relation to physical reality.

One can continue to increase the $\nu\Sigma_f$ in the cross sections until the point is reached that the α_0 is sensitive to these changes in the cross section. Once this point is reached, one finds that the computed α_0 is now relatively insensitive to changes in the bottom energy group. For example, if the group 3 $\nu\Sigma_f$ is multiplied by 2.5, the system is only slightly subcritical ($k_{eff} = 0.940$). By changing the lowest group velocity by a factor 2 ($v_5 = $

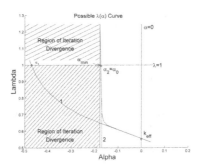

Fig. B.15 Possible $\lambda(\alpha)$ curve.

$1.444 cm/sh \to 2.888 cm/sh$), then $\alpha_0 = -0.087 gen/sh \to -0.090 gen/sh$, or α_0 only changes by 3%.

This behavior of the eigenvalues for this problem leads to the following conjecture. The $\lambda(\alpha)$ curve, as shown in Fig. B.12, may be, perhaps, the superposition of two separate curves, as shown in Fig. B.15. Curve 1 and its eigenvalue α_1 may be associated with some composite properties of the system, while curve 2 and its eigenvalue α_2 ($= \alpha_0$, the dominant eigenvalue, in this case) are associated with the behavior of the most slowly moving neutrons in the bottom energy group. In this situation, the calculated dominant eigenvalue, α_0, has no relation to a physically measurable die-away constant.

As the system is driven towards criticality, curve 1 is moved upward until eventually its associated eigenvalue, α_1, becomes the dominant eigenvalue α_0, a quantity insensitive to the behavior of the neutrons in the bottom group. At this point, the dominant eigenvalue of the multi-group transport may correspond to a physical die-away constant. Whether the curve 1 eigenvalue, α_1, as originally shown in Fig. B.15, corresponds to a "discrete eigenvalue buried in the continuum," one can only guess.

One can summarize the discussion in this section of subcritical α eigenvalues as follows:

- A robust eigenvalue search procedure has been developed to obtain the dominant eigenvalue of the multi-group transport equation, a quantity whose existence is guaranteed under the most general conditions.

The code user must be extremely careful in his interpretation of this dominant eigenvalue as a physically meaningful exponential die-away constant. If his computed eigenvalue is sensitive to the bottom energy group velocity, it is highly suspect. That is to say, if his system is sufficiently subcritical so that the improved search procedure is invoked by the code, the computed dominant eigenvalue most likely has little relation to physical reality. Stated another way, a "dramatic code failure" is nature's way of saying the user is calculating nonsense.

B.8 Implementation of α Re-balance Acceleration

The two α re-balance schemes ($GCCMR$ and $WSGR$) were implemented and tested in the Lagrangian mesh discrete-ordinates code $LaMEDOC$ [B.14]. In this section, we will give the coding details for their implementation in $LaMEDOC$. We will first describe the important variable names. The actual code listing will then be given, broken down into numbered segments. Finally, the purpose of each numbered segment of coding will be described.

The SUBROUTINE AREBAL (ALFAN), the mnemonic for α re-balance, provides the new α guess (ALFAN), using the fluxes stored in common at the completion of the current set of outer iterations.

The important integer variables are

NMAT: number of materials in the problem. The coarse mesh re-balance is performed on the material mesh, so that NMAT is actually the number of coarse mesh regions.

IGM: number of energy group (= IGMD).

LM,KM: number of mesh cells in the two dimensions of the Lagrangian mesh.

The important arrays are

FLUX(NM,KMLM,IGM): the scalar flux and moments NM moments total for each of the KMLM (= KM*LM) Lagrangian mesh cells and for each of IGM energy groups.

FLGS(k,ℓ): the group-summed negative inflows from material zone k to material zone ℓ, with FLGS (k,k) being the total outflow from material zone k. This is the matrix of Eq. (B.14), used in the $GCCMR$.

F(k): the $GCCMR$ re-balance factors, for NMAT material zones.

NL(g): the group-wise net leakage from the system, used in the $WSGR$.

FSS(k): the volume-integrated fission source for the k'th material mesh zone. This is Eq. (B.11), used in the $GCCMR$.

ABSP(k): the volume-integrated absorption for the k'th material mesh zone. This is Eq. (B.10), used in the $GCCMR$.

FV(k): the total neutrons for the k'th material mesh zone. This is

Eq. (B.12), used in the $GCCMR$.
RA(k,ℓ): the FL+AB-FS array on the left-hand side of Eq. (B.9) of the $GCCMR$.
Q(k): the FV*f vector on the right-hand side of Eq. (B.9) of the $GCCMR$.
WSRC(g): the C_g diagonal matrix of Eq. (B.25) for the $WSGR$.
WSRFV(g): the FV_g diagonal matrix of Eq. (B.26) for the $WSGR$.
WSRS(g,g'): the $S_{gg'}$, array of Eq. (B.27) for the $WSCTR$.
WSRFS(g,g'): the $FS_{gg'}$, array of Eq. (B.29) for the $WSGR$.
WSFSS(g): the whole system fission source for each group g. Used in the $WSGR$ to maintain the total fission source normalization of unity.
WSF(g): the re-balance factors for group g of the $WSGR$.
WSRA(g,g'): the NL+C-S-FS array on the left-hand side of Eq. (B.23), for the $WSGR$.
WSQ(g): the FV*f vector on the right-hand side of Eq. (B.23), for the $WSGR$.
The following pages contain the listing of the $AREBAL$ routine.

```
      SUBROUTINE AREBAL(ALFAN)
C
C     PERFORM MATERIAL MESH
C     RE-BALANCE ACCELERATION OF ALFA
C        OPTIMIZE
C        MACRO PARAMC(CLCHFILE)
C        USE PARAMC (CLCHFILE)
C        MACRO CONSTC (CLHFILE)
C        USE CONSTC (CLCHFILE)
C        MACRO FLUXSC (CLCHFILE)
C        USE FLUXSC (CLCHFILE)
C        MACRO XSECC (CLCHFILE)
C        USE XSECC (CLCHFILE)
C        MACRO SETLC (CLCHFILE)
C        USE SETLC (CLCHFILE)
C        MACRO MESHC (CLCHFILE)
C        USE MESHC (CLCHFILE)
C        DIMENSION WSRC(IGMO),WSRS(IGMO,IGMO),WSRFS(IGMO,IGMO)
C       1 ,WSRFV(IGMO),
C       1WSRA(IGMO, IGMO) ,WSO(IGMO),WSF( IGMO),WSFSS(IGMO)
C
        DIMENSION WSRC(12),WSRS(12,12),WSRFS(12,12),WSRFV(12),
```

```
      1 WSRA(12, 12) ,WSO(12),WSF( 12),WSFSS(12)
C
C     the following is done by Wen Ho Lee, 2015/August/30.
C     in Sunnyvale, CA.
      DIMENSION FLUX(5,900,12),FLGS(5,5),F(5),NL(12),FSS(5)
     1,IM(30,30),ANO(30,30),VOL(30,30),MATIX(18,5),
     2FR(18,30,30),ABSP(5),SIGA(12,90),FV(5),RA(5,5)
     3,Q(5),SIGTOT(12,90),SIGNU(12,90),SIGDS(12,12,90)
     4,CHI(12),FISSA(30,30)
C              WHL,2015/Aug/30.
      EQUIVALENCE (WSF(1),WSO(1) )
      INTEGER GP
      INTEGER G
C     DIMENSION VELI(IGMD)
      DIMENSION VELI(12),VEL(12)
      COMMON/CONVG/ EPSL,EPSO,EPSI
C     SCALE FLOWS BY LAST OUTER RE-BALANCE (ORELBAL) FACTOR
C
C
C     beginning of data provided by Wen Ho Lee.
      NMAT = 5
      NM = 5
      igmd = 12
      ala = 1.0
      lm = 30
      km = 30
      eps0 =0.001
      alfa = 0.5
C
C     IX = 1, 2, 3, ...., 90. since IX=MATIX(18,5)=18*5=90
C     maximum.
C     end of data provided by Wen Ho Lee.
C     BLOCK 1 BEGINS
      DO 18 L=1,NMAT
      DO 18 K=1,NMAT
   18 FLGS(K, L)=F(K)*FLGS(K ,L)
C     LAST OUTER REBAL FACTOR = 1/ALA , SCALE NET LEAKGE
      DO 19 G=1,IGM
   19 NL(G)=NL(G)/ALA
```

```
C     ABOVE IS BLOCK 1.
C     BLOCK 2 BEGINS
C     CALCULATE FISSION SOURCE ON MATERIAL
C     MESH.
      CALL CLEAR(0.0, FSS,NMAT)
      DO 20 L=1,LM
      DO 20 K=1,KM
      MAT=IM(K,L)+1
   20 FSS(MAT )=FSS(MAT)+FISSA( K,L)*VOL(K,L)
C     BLOCK 2 ENDS
C     BLOCK 3 BEGINS
C     CALCULATE ABSORPTION (SANS ALFA/VEL) AND FV
C     ON MATERIAL MESH.
      CALL CLEAR(0.0,ABSP,NMAT)
      CALL CLEAR(0.0,FV,NMAT)
      DO 25 G=1,IGM
   25 VELI(G)=1.0 / VEL(G)
      DO 30 L=1,LM
      DO 30 K=1,KM
      KL=(L-1)*KM+K
      MAT=IM(K,L)+1
      DO 28 IPOS=1, 18
      IF(FR(IPOS,K,L).LE.0.0) GO TO 28
      IX=MATIX(IPOS,MAT)
      ATOMS=ANO(K,L)*FR(IPOS,K,L)
      DO 27 G=1,IGM
   27 ABSP(MAT)=ABSP(MAT)+ATOMS*SIGA( G,IX)*FLUX(1,KL,G)
   28 CONTINUE
      DO 30 G=1,IGM
   30 FV(MAT)=FV(MAT )+VELI(G)*FLUX( 1,KL,G)*VOL(K ,L)
C
C     BLOCK 3 ENDS
C     BLOCK 4 BEGINS
      IF(NMAT.EQ.1) THEN
C     ONE MATERIAL, DO NOT ITERATE
      XLA=(FSS( 1)-FLGS(1, 1)-ABSP(1))/FV(1)
      F(1)=1.0
      GO TO 155
      END IF
```

```
C     BLOCK 4 ENDS
C     BLOCK 5 BEGINS
C   BEGIN INVERSE POWER ITERATION FOR ALFA EIGENVALUE.
      XLA=10.0
      T=0.0
      DO 90 K=1,NMAT
   90 T=T+FSS(K)
      CALL CLEAR(T,F,NMAT)
  100 CONTINUE
      DO 104 L=1,NMAT
      DO 102 K=1,NMAT
  102 RA(K,L)=FLGS(K,L)
      Q(L)=FV(L)*F(L)
  104 RA(L,L)=RA(L,L)+ABSP( L)-FSS( L)+XLA*FV(L)
      CALL LSS(NMAT,1,NMAT,RA,Q,DUMY,DET)
      TP=T
      T=0.0
      DO 110 L=1,NMAT
  110 T=T+FSS(L)*F(L)
      TA=1.0 / T
      XLAR=XLA
      XLA=XLAR-TA
      DO 112 L=1,NMAT
  112 F(L)=TA*F(L)
      IF(ABS(1.-XLA/XLAR).GT.EPSO) GO TO 100
C     BLOCK 5 ENDS
C     BLOCK 6 BEGINS
C     DO WHOLE-SYSTEM GROUP-WISE REBALANCE
C     GENERATE WSR MATRICES
  155 CALL CLEAR(0.0,WSRC, IGM)
      CALL CLEAR(0.0, WSRS, IGM*IGM )
      CALL CLEAR(0.0,WSRFV, IGM)
      CALL CLEAR(0.0,WSRFS, IGM* IGM)
C     BLOCK 6 ENDS
C     BLOCK 7 BEGINS
      DO 200 L=1,LM
      DO 200 K=1,KM
      KL=(L-1)*KM+K
      MAT=IM(K,L)+1
```

```
      DO 180 IPOS=1,18
      IF(FR(IPOS,K,L) .LE. 0.0) GO TO 180
      IX=MATIX(IPOS, MAT)
C
C     IX is the 18 (maximum) possible cross-section
C     block I. D. for each material.
C
      ATOMS=ANO(K,L)*FR(IPOS ,K,L)
      DO 160 G=1,IGM
      WSRC(G)=WSRC(G)+ATOMS*SIGTOT(G ,IX)*FLUX(1,KL,G)
      WSRFS(1,G)=WSRFS( 1,G)+ATOMS*SIGNU(G, IX)
     1           *FLUX(1,KL,G)
      DO 160 GP=1,G
      INDXG=(G*(G-1 ))/2+GP
  160 WSRS(G, GP)=WSRS(G,GP)+ATOMS*SIGDS( 1, INDXG,IX)
     1           *FLUX(1,KL ,GP)
  180 CONTINUE
      DO 182 G=1,IGM
  182 WSRFV(G)=WSRFV(G)+FLUX( 1,KL,G)*VOL(K, L)*VELI( G)
  200 CONTINUE
C     BLOCK 7 ENDS
C     BLOCK 8 BEGINS
C     SAVE GROUP-WISE WS FISSION SOURCE FOR NORMALS.
      DO 220 G=1,IGM
  220 WSFSS(G)=WSRFS( 1,G)
C     FILL OUT FISSION MATRIX
      DO 210 G=2,IGM
      DO 210 GP=1,IGM
  210 WSRFS(G, GP)=WSRFS(1 ,GP)
      DO 212 G=1,IGM
      DO 212 GP=1,IGM
  212 WSRFS(G, GP)=CHI (G)*WSRFS(G ,GP)
      XXLA=10.0
      T=0.0
      DO 290 G=1,IGM
  290 T=T+WSFSS(G)
      CALL CLEAR(T,WSF,IGM)
C     BLOCK 8 ENDS
C     BLOCK 9 BEGINS
```

```
C
C      START INVERSE POWER ITERATION.
C
 300   CONTINUE
       DO 320 G=1,IGM
       DO 320 GP=1,IGM
 320   WSRA(G,GP)=-WSRS(G,GP)-WSRFS(G,GP)
       DO 322 G=1,IGM
       WSO(G)=WSF(G)*WSRFV(G)
 322   WSRA(G,G)=WSRA(G,G)+NL(G)+WSRC(G)+XXLA*WSRFV(G)
       CALL LSS(IGM,1,IGMD,WSRA,WSO ,DUMY,DET)
       TP=T
       T=0.0
       DO 310 G=1,IGM
 310   T=T+WSFSS(G)*WSF(G)
       TA=1.0 / T
       XXLAR=XXLA
       XXLA=XXLAR-TA
       DO 312 G=1,IGM
 312   WSF(G)=TA*WSF(G)
       IF(ABS(1.-XXLA/XXLAR).GT.EPSO) GO TO 300
C      BLOCK 9 ENDS
C      BLOCK 10 BEGINS
C
C      EIGENVALUE ITERATION CONVERGED, SCALE FLUX MOMENTS
C      CHOOSE BIGGEST INCREASE IN ALPHA AS BEST NEXT GUESS
C
       IF(ABS(XLA-ALFA) .GE.ABS(XXLA-ALFA )) ALFAN=XLA
       IF(ABS(XXLA-ALFA) .GT.ABS(XLA-ALFA)) ALFAN=XXLA
       DO 130 L=1,LM
       DO 130 K=1,KM
       KL=(L-1)*KM+K
       MAT=IM(K,L)+1
       DO 130 G=1,IGM
       DO 130 N=1,NM
 130   FLUX(N,KL,G)=F(MAT)*WSF(G) *FLUX(N,KL,G)
 150   RETURN
       END
C      BLOCK 10 ENDS
```

```
      subroutine clear (x1,x2,i1)
      return
      end
      subroutine lss (i2,i3,i4,x1,x2,x3,x4)
      return
      end
C
```

The code blocks in the preceding listing perform the following functions:
Block 1: At the completion of the last outer iteration, the outer re-balance factors [leftover in array $F(k)$] contain the factor $1/\lambda$, in order to maintain a fission total of unity. Loop 18 scales the group-summed flow array ($FLGS$) by the factors to make them compatible with the already scaled flux ($FLUX$) and fission source ($FISSA$) array. Loop 19 does the same scaling to the group-wise net leakage (NL) array, using the factor λ (ALA).
Block 2: The volume-integrated fission source on the material mesh is accumulated.
Block 3: The volume-integrated absorption $ABSP$ (without the α/v term) and the total neutrons FV on the material mesh are calculated. Loop 30 cycles over all KM*LM Lagrangian mesh cells, computing the material I.D. (MAT) for each cell. Loop 28 cycles over the 18 possible isotopes for each material. If the isotopic fraction (FR) is nonzero, it computes the cross-section block I.D. (IX) and the total atoms of that isotope (ATOMS) in the cell, based on the total atoms of all isotopes (ANO) in that cell.
Block 4: If there is only one material zone (whole system re-balance), the next α guess (XLA) can be computed explicitly without the inverse power iteration. This equation is equivalent to Eq. (B.18). The re-balance factor is automatically unity, from the normalization.
Block 5: This block is used for multi-material re-balance, using the inverse power iteration to solve the re-balance Eq. (B.9). The α eigenvalue guess (XLA) is set to a large number (10.0 in this case) so the iteration will converge to the eigenvalue of Eq. (B.9) nearest this value, presumably the desired most positive one. The units of α are here assumed in inverse shakes. If velocities are in cm/s, then this value should be changed from 10 to 10^5. Loop 90 computes the system fission total (T) at the start of the inverse power iteration and the initial guess of the re-balance factors set to this value. Loop 100 is the actual inverse power iteration loop.

Following Wachspress [B.16], the power iteration for the eigenvalue equation

$$Mf = \alpha f, \tag{B.40}$$

is given by

$$f^{n+1} = M^{-1}\alpha^n f^n, \tag{B.41}$$

and

$$\alpha^{n+1} = (1, f^n)/(1, f^{n+1}), \tag{B.42}$$

where n is the power iteration index and $(1, f)$ simply represents the inner product of the vector f with the vector of ones. This simple power iteration converges to the α eigenvalue that is largest in magnitude (positive or negative), which is not the one we desire. If we have some estimate of the eigenvalue, α_e, Eq. (B.40) may be written

$$(M - \alpha_e I)f = (\alpha - \alpha_e)f. \tag{B.43}$$

The fractional power iteration is then given by

$$f^{n+1} = (M - \alpha_e I)^{-1}(\alpha - \alpha_e)^n f^n, \tag{B.44}$$

and

$$(\alpha - \alpha_e)^{n+1} = (1, f^n)/(1, f^{n+1}). \tag{B.45}$$

This fractional power iteration will converge to the eigenvalue closest to the guess α_e. If we choose α_e as a large positive number, this will converge to the desired, most positive, α eigenvalue. This iteration is the one coded in Loop 100. In Eqs. (B.44) and (B.45), the normalization is $(1, f) = 1$. In the coding of Loop 100, the normalization is $(FSS, f) = 1$, in order to maintain a fission total of unity.

Loop 104 constructs the matrix

$$RA = FL + AB - FS + \alpha^n FV, \tag{B.46}$$

of Eq. (B.9), where in the above notation,

$$M = FL + AB - FS, \tag{B.47}$$

and the vector on the right-hand side of Eq. (B.9)

$$Q = FV * F. \tag{B.48}$$

The call to subroutine LSS solves the inverse matrix equation

$$(M - \alpha_e I)^{-1}(\alpha - \alpha_e)^n f^n, \tag{B.49}$$

of Eq. (B.44).

Loop 110 calculates a new fission total (T), saving the previous fission total (TP), which is, in fact, 1.0, by the normalization. The next estimate of $(\alpha - \alpha_e)^{n+1}$ in Eq. (B.45) is then given by

$$(\alpha - \alpha_e)^{n+1} = TA = TP/T = 1.0/T. \tag{B.50}$$

The next iteration estimate, α_e, is then set to the previous iteration's eigenvalue

$$\alpha_e = XLAR, \tag{B.51}$$

and the new iteration's eigenvalue of Eq. (B.45) is then

$$\alpha^{n+1} = \alpha_e + TP/T = \alpha_e + 1.0/T. \tag{B.52}$$

Loop 112 then normalizes the re-balance factors f so the fission total T is again unity. The iteration is then terminated when $(1 - \alpha^{n+1}/\alpha^n) < \epsilon_0$.

Block 6: This block initializes the whole-system re-balance ($WSGR$) arrays to zero.

Block 7: This block accumulates the $WSGR$ arrays C (or $WSRC$), FV (or $WSRFV$), S (or $WSRS$), and the first row of array FS (or $WSRFS$), as given in Eqs. (B.24–B.30).

Block 8: Loop 220 stores the group-wise whole system fission source, the first row of the FS array, into the $WSFSS$ array, to be used later in the normalization constraint. Loop 210 fills out the remaining rows of the FS array. Loop 212 then multiplies the FS array by the χ diagonal matrix to obtain the final form of FS, as given in Eqs. (B.29 and B.30). The $WSGR$ α estimate ($XXLA$) is set to a large number (10.0), the fission total (T) again computed, and the initial guess of the re-balance factors equated to this value.

Block 9: Loop 300 is the inverse power iteration for the $WSGR$ equations, virtually identical to the coding of Loop 100, except the $WSGR$ arrays are used.

Block 10: The next guess of the α eigenvalue ($ALFAN$) is chosen between the $GCCMR$ estimate (XLA) and the $WSGR$ estimate ($XXLA$) in order to maximize the change from the previous α eigenvalue estimate ($ALFA$). Loop 130 then applies the $GCCMR$ factors (F) and the $WSGR$ factors (WSF) to the scalar flux and moments. Both re-balance factors are normalized to maintain the fission total of unity.

B.9   Conclusion

The two re-balance schemes are found to accelerate the α eigenvalue calculation by anywhere from a small amount to as much as 50% and more, as compared with the variable convergence precision scheme. Nearly all of the acceleration comes from the improved estimates of α, with very little, in most cases, coming from the re-balance factors themselves.

More importantly, the re-balance scheme makes the iterative solution of the α eigenvalue problem considerably more robust, relieves the code user of much of the burden of providing intelligent input required by the standard search procedure, and permits modifications to the iteration strategy that eliminates many of the unnecessary calculations. By utilizing all the schemes and procedures described in this appendix, we can usually solve the α eigenvalue problem in one-fifth the time required for the present search procedure.

Bibliography

B.1 Hill, TR. (1975). *ONETRAN, A Discrete Ordinates Finite Element Code for Solution of the One-Dimensional Multigroup Transport Equation*, Los Alamos Scientific Laboratory report LA-5990-MS.

B.2 O'Dell, RD., Brinkley, FW. and Marr, DR. (1982). *Users Manual for ONEDANT: A Code Package for One-Dimensional, Diffusion-Accelerated, Neutral-Particle Transport*, Los Alamos National Laboratory report LA-9184-M.

B.3 Lathrop, KD and Brinkley, FW. (1973). *TWOTRAN-I: An Interfaced, Exportable Version of the TWOTRAN Code for Two-Dimensional Transport*, Los Alamos Scientific Laboratory report LA-4848-MS.

B.4 Engle, WW. (1967). *A Users Manual for ANISN, A One-Dimensional Discrete Ordinates Transport Code with Anisotropic Scattering*, Union Carbide Corporation report K-1693.

B.5 Rhoades, WA and Childs, RL. (1982). *An Updated Version of the DOT 4 One- and Two-Dimensional Neutron/Photon Transport Code*, Oak Ridge National Laboratory report ORNL-5851.

B.6 Bell, GI and Glasstone, S. (1970). *Nuclear Reactor Theory*, Van Nostrand and Reinhold Co., New York.

B.7 Milton Wing, G. (1962). *An Introduction to Transport Theory*, John Wiley & Sons Inc., New York.

B.8 Williams, MMR. (1966). *The Slowing Down and Thermalization of Neutrons*, John Wiley & Sons Inc., New York.

B.9 Larsen, EW and Zweifel, PF. (1974). *On the Spectrum of the Linear Transport Operator*, J. Math. Phys., 15(11), 1987.

B.10 Larsen, EW. (1979). *The Spectrum of the Multigroup Neutron Transport for Bounded Spatial Domains*, J. Math. Phys. 20, 8, 1776.

B.11 Duderstadt, JJ and Martin, WR. (1979). *Transport Theory*, John Wiley & Sons Inc., New York.

B.12 Carlson, BG and Lathrop, KD. (1968). *Transport Theory: The Method of Discrete-Ordinates*, in Computing Methods in Reactor Physics, H. Greenspan, C.N., Kelber and D. Okrent, Eds, (Gordon and Breach, New York.)

B.13 Mohan, R, Ahmed, F and Kothari, LS. (1982). *Decay of Fast Neutron Pulses in Uranium Assemblies*, Nucl. Sci. Eng. 81, 532.

B.14 Hill, TR and Paternoster, RR. (1982). *Two-Dimensional Spatial Discretization Methods on a Lagrangian Mesh*, Los Alamos National Laboratory report LA-UR-82-1055.

B.15 Larsen, EW. (1982). *Transport and Reactor Theory Progress Report (July 1 - September 30, 1982) by R. D. ODell and R. E. Alcouf*, Los Alamos National Laboratory.

B.16 Wachspress, EL. (1966). *Iterative Solution of Elliptic Systems*, Prentice-Hall, Englewood Cliffs, New York.

B.17 Keepin, GR. (1965). *Physics of Nuclear Kinetics*, Addison-Wesley Reading, Maine.

B.18 Mingzhu, Y and Guangtian, Z. (1981). *Spectrum of Neutron Transport Operator with Anisotropic Scattering and Fission*, Scientia Sinica 24, 476.

B.19 Marek, I. (1978). *On the Asymptotic Behavior of Solutions of the Homogeneous Transport Equation, in Differential Equations and Numerical Mathematics: Selected Papers Presented to a National Conference*, G. I. Marchuk, Ed. (Novosibirsk, USSR).

Appendix C

Hugoniot Data and JWL EOS

Tables C.1–C.5 list the Hugoniot data.
Hugoniot data fitted by the equation, $u_s = c_0 + s u_p + q u_p^2$.

Notations

u_s the shock velocity (km/s)
u_s particle velocity (km/s)
c_0 material dependent constant (km/s)
p pressure (GPa)
$1 GPa = 10^{-2} Mbar$
$1 Pa = 1 Pascal = 1 N/m^2 = 10^{-5} Bar = 1.0197 \times 10^{-5} atm$

Table C.6 lists the JWL parameters for a number of explosives, along with the C-J detonation parameters used in obtaining this information.

ρ_0^a = undetonated explosive density.
D^b = detonation velocity.
E_0^c = detonation energy.
$^d Data$ = data contributed by J. Jacobson, Group M-4, Los Alamos National Laboratory.

Table C.1 Hugoniot Data.

Material	$\rho_0(g/cm^3)$	$c_0(km/s)$	s	$q(s/km)$	γ_0	Comments
Element						
Antimony	6.700	1.983	1.6527		0.60	
Barium	3.705	0.700	1.600		0.55	Above $P = 115\,(GP_a)$ and $u_s = 2.54\,(km/s)$
Beryllium	1.851	7.998	1.124		1.16	
Bismuth	9.836	1.826	1.473		1.10	
Cadmium	8.639	2.434	1.684		2.27	
Calcium	1.547	3.602	0.948		1.20	
Cesium	1.826	1.048	1.043	0.051	1.62	
Chromium	7.117	5.173	1.473		1.19	
Cobalt	8.820	4.752	1.315		1.97	
Copper	8.930	3.940	1.489		1.99	
Germanium	5.328	1.750	1.750		0.56	Above $P = 300\,(GP_a)$ and $u_s = 4.2\,(km/s)$
Gold	19.240	3.056	1.572		2.97	
Hafnium	12.885	2.954	1.121		0.98	Below $P = 400\,(GP_a)$ and $u_s = 3.86(km/s)$
Hafnium	12.885	2.453	1.353		0.98	Above transition
Indium	7.279	2.419	1.536		1.80	

Table C.2 Hugoniot Data.

Material	$\rho_0(g/cm^3)$	$c_0(km/s)$	s	$q(s/km)$	γ_0	Comments
Element						
Iridium	22.484	3.916	1.457		1.97	
Iron	7.850	3.574	1.920	-0.068	1.69	Above $u_s =$ 3.86 (km/s)
Lead	11.350	2.051	1.460		2.77	
Lithium	0.530	4.645	1.133		0.81	
Magnesium	1.740	4.492	1.263		1.42	
Mercury	13.540	1.490	2.047		1.96	
Molybdenum	10.206	5.124	1.233		1.52	
Nickel	8.874	4.602	1.437		1.93	
Niobium	8.586	4.438	1.207		1.47	
Palladium	11.991	3.948	1.588		2.26	
Platinum	21.419	3.598	1.544		2.40	
Potassium	0.860	1.974	1.179		1.23	
Rhenium	21.021	4.184	1.367		2.44	
Rhodium	12.428	4.807	1.376		1.88	
Rubidium	1.530	1.134	1.272		1.06	
Silver	10.490	3.229	1.596		2.38	
Sodium	0.968	2.629	1.223		1.17	
Strontium	2.628	1.700	1.230		0.41	Above $P =$ 150 (GP_a) and $u_s =$ 3.63 (km/s)
Sulfur	2.020	3.223	0.959			
Tantalum	16.654	3.414	1.201		1.60	
Thallium	11.840	1.862	1.523		2.25	

Table C.3 Hugoniot Data.

Material	$\rho_0(g/cm^3)$	$c_0(km/s)$	s	$q(s/km)$	γ_0	Comments
Element						
Thorium	11.680	2.133	1.263		1.26	
Tin	7.287	2.608	1.486		2.11	
Titanium	4.528	4.877	1.049		1.09	Above transition
Titanium	4.528	5.220	0.767		1.09	Below $P = 175(GP_a)$ and $u_s = 5.74(km/s)$
Tungsten	19.224	4.029	1.237		1.54	
Uranium	18.950	2.487	2.200		1.56	
Vanadium	6.100	5.077	1.201		1.29	
Zinc	7.138	3.005	1.581		1.96	
Zirconium	6.505	3.757	1.018		1.09	Below $P = 260(GP_a)$ and $u_s = 4.63(km/s)$
Zirconium	6.505	3.296	1.271		1.09	Above transition
Alloys						
Brass	8.450	3.726	1.434		2.04	
2024 Aluminum	2.785	5.328	1.338		2.00	
921-T Aluminum	2.833	5.041	1.420		2.10	
Lithium-Magnesium Alloy	1.403	4.247	1.284		1.45	
Magnesium Alloy	1.775	4.516	1.256		1.43	

Table C.4 Hugoniot Data.

Material	$\rho_0(g/cm^3)$	$c_0(km/s)$	s	$q(s/km)$	γ_0	Comments
Stainless Steel	7.896	4.569	1.490		2.17	
U-3 wt%Mo	18.450	2.565	2.20		2.03	
Synthetics						
Adiprene	0.927	2.332	1.536		1.48	
Epoxy resin	1.186	2.730	1.493		1.13	Below $P=$ 240(GP_a) and $u_s=$ 7.0 (km/s)
Epoxy resin	1.186	3.234	1.255		1.13	Above transition
Lucite	1.181	2.260	1.816		0.75	
Neoprene	1.439	2.785	1.419		1.39	
Nylon	1.140	2.570	1.849	0.081	1.07	
Paraffin	0.918	2.908	1.560		1.18	
Phenoxy	1.178	2.266	1.698		0.55	
Plexiglas	1.186	2.598	1.516		0.97	
Polyethylene	0.915	2.901	1.481		1.64	
Polyrubber	1.010	0.852	1.865		1.50	
Polystyrene	1.044	2.746	1.319		1.18	
Polyurethane	1.265	2.486	1.577		1.55	Below $P=$ 220(GP_a) and $u_s=$ 6.5 (km/s)
Silastic						
(RTV-521)	1.372	0.218	2.694	-0.208	1.40	
Teflon	2.153	1.841	1.707		0.59	

Table C.5 Hugoniot Data.

Material	$\rho_0(g/cm^3)$	$c_0(km/s)$	s	$q(s/km)$	γ_0	Comments
Compounds						
Periclase (MgO)	3.585	6.597	1.369		1.32	Above $P = 220\,(GP_a)$ and $u_s = 7.45\,(km/s)$
Quartz	2.204	0.794	1.695		0.90	Stishovite Above $P = 400\,(GP_a)$
Sodium chloride	2.165	3.528	1.343		1.60	Transition ignored
Water	0.998	1.647	1.921	0.096		

Table C.6 JWL equation of state parameters.

HE	Comp B	Comp C-4	Cyclotol 77/23	H-6	HMX	LX-04-1
$\rho_0^a(Mg/m^3)$	1.717	1.601	1.754	1.760	1.891	1.865
$D^b(km/s)$	7.98	8.19	8.25	7.47	9.11	8.47
$E_0^c(GPa)$	8.50	9.00	9.20	10.3	10.5	9.5
$A(GPa)$	524.2	609.8	603.4	758.1	778.3	849.8
$B(GPa)$	7.678	12.95	9.924	8.513	7.071	15.277
R_1	4.20	4.50	4.30	4.90	4.20	4.65
R_2	1.10	1.40	1.10	1.10	1.00	1.30
ω	0.34	0.25	0.35	0.20	0.30	0.35
HE	LX-07	LX-09	LX-10	LX-11	LX-13	LX-14-0
$\rho_0^a(Mg/m^3)$	1.865	1.838	1.875	1.875	1.540	1.835
$D^b(km/s)$	8.64	8.84	8.82	8.32	7.35	8.80
$E_0^c(GPa)$	10.0	10.5	10.4	9.0	6.6	10.2
$A(GPa)$	871.0	868.4	880.2	779.1	2714.0	826.1
$B(GPa)$	13.896	18.711	17.437	10.668	17.930	17.240
R_1	4.60	4.60	4.60	4.50	7.00	4.55
R_2	1.15	1.25	1.20	1.15	1.60	1.32
ω	0.30	0.25	0.30	0.30	0.35	0.36

Table C.7 JWL equation of state parameters.

HE	LX-17-0	Nitro-methane	Octol	PBX-9010	PBX-9011	PBX-9404-3
$\rho_0^a(Mg/m^3)$	1.900	1.128	1.821	1.787	1.777	1.840
$D^b(km/s)$	7.60	6.28	8.48	8.39	8.50	8.80
$E_0^c(GPa)$	6.90	5.10	9.60	9.00	8.90	10.12
$A(GPa)$	446.0	209.2	748.6	581.4	634.7	852.4
$B(GPa)$	13.399	5.689	13.380	6.801	7.998	20.493
R_1	3.85	4.40	4.50	4.10	4.20	4.60
R_2	1.30	1.20	1.20	1.00	1.00	1.35
ω	0.46	0.30	0.38	0.35	0.30	0.25

HE	PBX-9407	PBX-9501^d	PBX-9502^d	Pentolite	PETN	Tetryl
$\rho_0^a(Mg/m^3)$	1.600	1.840	1.895	1.670	1.770	1.730
$D^b(km/s)$	7.91	8.80	7.62	7.47	8.30	7.91
$E_0^c(GPa)$	8.60	10.12	7.07	8.0	10.10	8.20
$A(GPa)$	573.19	852.4	460.3	491.1	617.0	586.8
$B(GPa)$	14.639	18.02	9.544	9.061	16.926	10.671
R_1	4.60	4.60	4.00	4.40	4.40	4.40
R_2	1.40	1.30	1.70	1.10	1.20	1.20
ω	0.32	0.38	0.48	0.30	0.25	0.28

HE	TNT	Detasheet C^d
$\rho_0^a(Mg/m^3)$	1.630	1.48
$D^b(km/s)$	6.93	7.00
$E_0^c(GPa)$	7.00	3.69
$A(GPa)$	371.2	349.0
$B(GPa)$	3.231	4.524
R_1	4.15	4.10
R_2	0.95	1.20
ω	0.30	0.30

Appendix D

Supplementary Materials

The supplementary materials for this book is available from the website of World Scientific Publishing. The download process is described in Section A.1 below. The supplementary materials are computer programs written in Fortran for the calculation of high explosive burn time and burn distance, shear modulus and yield strength for many materials, as well as the MATLAB plotting programs for many perforators.

A.1 Downloading the supplementary materials

Begin the download process of obtaining the supplementary materials by entering the World Scientific website using the URL https://www.worldscientific.com/worldscibooks/10.1142/10966#t=suppl. When you enter the site, you will be asked to login. If you have not previously registered on the World Scientific site, a simple (free) registration process will be required. Once you have successfully logged in and have reached the page for *Computational Solid Mechanics for Oil Well Perforator Design,* select the "Supplementary" tab. This tab contains a direct link to the supplementary materials. Clicking on the link will allow you to download the zipped file containing the computer programs. If, at this point, entry of a specific access token is requested, you may use the following access token:

https://www.worldscientific.com/r/10966-supp

If your computer automatically places downloaded files in a folder named "Downloads", you will want to relocate the zipped file to a convenient permanent directory before unzipping. The zipped file is named "Computer Programs.zip" and after unzipping, it will produce an unzipped folder of the same name.

Index

A

Ahmed, F, 357
Ahrens, T, 171
angular fluxes, 310
appendix, 241, 309, 359
artificial viscosity, 36

B

Bell, GI, 357
Bird, RB, 129
Blewett, PJ, 36
Boris, JP, 129
Brinkley, FW, 356
Browne, BG, 307
Burton, D, 171

C

C-J detonation, 146, 359
Carlson, BG, 357
Chapman–Jouguet, 145
Clark, RA, 129
Clifton, R, 172
coarse mesh re-balance, 314
Cochran, SG, 171
cohesive energy, 171
complex eigenvalues, 316
compression, 9, 39
continuous Eulerian, 38
Cook, WH, 171
critical size search, 310

D

decay rate of stress, 168
diamond difference, 312
dilitation, 11
directed kinetic energy, 248, 305
discrete-ordinates, 312, 313
dominant eigenvalue, 311, 316
down-scatter, 320
Duderstadt, JJ, 357
Duvall, G, 171

E

eigenvalue modifier, 314
eigenvalue spectrum, 311
eigenvector, 315
energy, 9, 39
Engle, WW, 357
EOSGY, 142, 143
equation of state (*see also* EOS), 137, 216
equivalent plastic strain, 117, 139
Eringen, AC, 36
EULE2D-Fig, 183
Eulerian, 4

F

fast supercritical assemblies, 324
first invariant, 53
fission spectrum, 310
force equilibrium, 56
fracture, 60

G

Glasstone, S, 357
GODIVA, 341
Grüneisen, 133
Guangtian, Z, 357
Guinan, MW, 171
Guinea, F, 172

H

Hageman, LJ, 129
Harlow, FH, 128
HE programmed burn, 145
HEDET2, 147
HEDET3, 150
Herrmann, W, 36
Hill, TR, 356, 357
Hodge,PG, 36
hollow liner, 2
Hookes, 29
Horning, H, 171
Hugoniot, 5, 133, 134
Hugoniot relations, 151
hydrocrbon, 1
hydrostatic pressure, 53

I

ICCG, 232
inner iterations, 322
intermolecular potential, 168
inverse power iteration, 317

J

Jacobson, J, 171
jerk, 212
Johnson and Cook, 142
Johnson, GR, 171
Johnson, WE, 128
JWL Equation of State, 145

K

Keepin, GR, 357
Kershaw, DS, 240
Kolsky, HG, 128

Kothari, LS, 357
Kronecker delta, 9, 39

L

Lagrangian, 8
Lagrangian stress gradient, 164
Larsen, EW, 357
Lathrop, KD, 356–357
leakage operator, 311
Lee, WH, 6, 36, 206, 240
Lee, E, 171
Legendre polynomial, 310
Lindemann law, 140
line of sight, 147
Lund, CM, 171

M

Mader, CL, 128
Mandell, D, 171
Marek, I, 357
Marr, DR, 356
Marsh, S, 171
Martin, WR, 357
mass, 9, 39
MATLAB, 174
matrix eigenvalue equation, 315
McLemore, RL, 6
McQueen, R, 171
melt, 59
Micro-mechanical effects, 159
Milton Wing, G, 357
Mingzhu, Y, 357
Mohan, R, 357
momentum, 9, 39
Monte Carlo, 209
Monte Carlo Method, 233
Morse potential, 168
most effective iteration, 323

O

O'Dell, RD, 356
objective stress, 31
Osborne Model, 135
outer and inner iterations, 313

P

particle velocity, 134
Paternoster, RR, 357
PBX-W-113, 203
perforators, 3, 174
PIC, 5, 38
Pomraning, GC, 240
Prager, W, 36
Prandtl–Reuss, 52, 55
Prandtl–Reuss plastic flow, 220
probability, 234
prompt fission, 310

R

radiation diffusion, 215
radiation hydrodynamic, 209
radiation transport, 215
random number, 238
re-balance acceleration, 314
re-balance schemes, 323
reactive armor, 198
Reynold's transport theorem, 66
REZONE, 243
rigid body, 31
rigid body rotation, 55
root-finding, 322
Rose, JH, 172

S

scattering function, 310
Schulz, WD, 240
second invariant, 54
second-order PIC, 82, 119
shake, 212
shaped charges, 1
shear modulus, 136
shock wave velocity, 134
shock-change, 158
Smith, JR, 172
sound speed, 35
spall, 59
specific internal energy, 44, 48
specific kinetic energy, 44
specific total energy, 44
Steinberg, DJ, 171

Steinberg–Guinan Model, 137
Stewart, WE, 129
stress deviator, 9, 39
stress power, 45
stress tensor, 9, 39
Swegle, JW, 128

T

tantalum EFP, 200
theoretical spall strength, 168
thermal reactor systems, 324
Thompson, SL, 129
Tillotson, J, 136, 171
time absorption eigenvalue, 310
total cross section, 310
transport equation, 310
Tresca, 55
tri-diagonal, 316
truncation error analysis, 98
two-detonators, 149

U

up-scatter, 319

V

von Mises criterion, 53
von Mises flow stress, 142

W

Wachspress, EL, 357
Wallick, KB, 307
Wallick, ML, 240
Walters, WP, 6
wellbore, 1
Wilkins, ML, 128
Williams, MMR, 357
Wilson, HL, 240

Y

yield criterion, 53
yield strength, 136

Z

Zweifel, PF, 357

CPSIA information can be obtained
at www.ICGtesting.com
Printed in the USA
LVHW081606160719
624279LV00002B/32/P